pay 23 & 24
Xin chap 8 set of laws
8.1 & 8.5
45 & 46 & 47 of Fellinger pub.

8/56
$7.20

Dynamics of
MACHINERY

Dynamics of MACHINERY

A. R. HOLOWENKO

Associate Professor of Mechanical Engineering
Purdue University

JOHN WILEY & SONS, INC., NEW YORK
CHAPMAN & HALL, LTD., LONDON

Copyright, 1955
BY
John Wiley & Sons, Inc.

All Rights Reserved

This book or any part thereof must not be reproduced in any form without the written permission of the publisher.

Library of Congress Catalog Card Number: 54-10969

PRINTED IN THE UNITED STATES OF AMERICA

PREFACE

An aspect of design which is receiving more and more attention, necessitated by the higher speeds sought for in machines, is the dynamic effects resulting from the higher speeds. Even at relatively low speeds, balancing of engines and critical speeds are dynamic problems confronting the designer. More and more often, engineering curricula are being changed to include a required course in dynamics of machinery, although some schools include the material as part of the course content of machine design. At Purdue University the machine design sequence of courses is: Mechanism, a three-credit course in the fifth semester; Dynamics of Machinery, a three-credit course in the sixth semester; and Machine Design, a five-credit course in the seventh semester. This book, *Dynamics of Machinery*, has been written principally for use in the sixth semester, with the topics extended in coverage so that the book will be suitable for use in the senior year, if the course is a technical elective, as it is in some schools.

If machine design is considered generalized into the four areas of kinematics, force analysis, material selection, and proportioning of parts, this book is concerned with only the first two items. A departure from the usual textbook presentation has been made. An attempt has been made to give all the steps in a process, with all the step-by-step free body diagrams as they might be made in a lecture, and as they should be made by the student in his analysis. Free body diagrams have not been restricted to forces; they are a necessary part of all topics treated in the book. It is hoped that the spirit of making and using free body diagrams will be contagious to the student, since the habit of making them is such a vital part of any engineering analysis.

A variety of problems have been selected for illustration, although the slider-crank mechanism has been used extensively for velocity, acceleration, force, and dynamic analyses. It is hoped that the student will recognize principles rather than specific problems and will recognize that procedures are applicable to other types of problems not discussed. For instance, although the material on flywheels has been expressed in terms of application to an internal combustion engine and a punch press, the method of attack is applicable to any type of engine where an energy reservoir is needed or where speed control within a cycle is desired.

Balancing of engines, with emphasis on means of balancing, together with balancing of rotating masses, are subjects which stem from the presence of accelerations, which involves dynamics. Critical speeds or whirling speeds and torsional oscillations of shafts are just two topics in the field of vibration analysis, which is usually a technical elective course. The student is given an introduction to the field in the chapter on critical speeds. Gyroscopes are included in the book because of the dynamic effect from gyroscopic action.

No claim is made to any new material in the book, the basic principles having been formulated many years ago. The philosophy of approach to a problem has been the major concern of this book. Consequently, more than one approach to a problem has been used. Both analytical and graphical developments have been presented, not for duplication but for a better understanding of the relations developed. It is hoped that the student, even if not assigned two methods of development, will spend time on the alternate method. Since instructors have individual preferences for methods of presentation of material, it is hoped that greater flexibility will result.

A considerable number of problems have been included to provide for selection and variety, and an attempt has been made to classify the problems in a relative order of difficulty.

Calculus, both differential and integral, has been used. No attempt has been made to avoid mathematics.

I have the very good fortune to have been at one time an assistant to Professor J. A. Dent at the University of Pittsburgh, and I have drawn heavily on the benefits of that association.

The response of industry to my requests for material and illustrations has been most gratifying. Their cooperation and their generosity are very gratefully acknowledged. Particular thanks are due Mr. C. A. Rasmussen of the Cadillac Division, General Motors Corporation, for his kindness in furnishing me with a very complete description of balancing the Cadillac engine.

I wish also to thank my colleagues in the Machine Design Department at Purdue University for their valuable discussions and contributions, Professor A. S. Hall, Jr., for the many problems which he suggested, and my wife, Virginia, for her patience with me during the preparation of the book.

A. R. HOLOWENKO

West Lafayette, Indiana
February, 1955

CONTENTS

1. Introduction 1
2. Velocities and Accelerations 3
3. Relative Velocities 8
4. Application of the Relative Velocity Equation 17
5. Special Methods of Velocity Solution 44
6. Relative Accelerations 58
7. Application of the Relative Acceleration Equation for Two Points on a Rigid Link 71
8. Acceleration Equation for Two Coincident Points 96
9. Special Methods of Acceleration Solution 118
10. Equivalent Mechanisms 127
11. Review of Static Forces and Graphic Statics 135
12. Static Forces in Machines 150
13. Inertia Forces 188
14. Dynamic Analysis 219
15. Analytic Determination of Accelerations in a Slider-Crank Mechanism 238
16. Flywheel Analysis 243
17. Balancing Rotating Masses 259
18. Balancing Machines 282
19. Balancing Masses Reciprocating in a Plane 294
20. Balancing Masses Reciprocating in Several Planes 341
21. Vibrations in Shafts 383
22. Gyroscopes 442
Index 457

CHAPTER 1

Introduction

Dynamics of machinery is defined as the study of motions and forces in machines. A curriculum in machine design includes such subjects as kinematics, statics, kinetics, strength of materials, and dynamics of machinery as prerequisites to a course in machine design. Such fundamental, basic subjects are some of the working tools of the designer engineer. Dynamics of machinery is not a new, different field to be treated independently of the other phases of engineering, but must be considered as a development of the basic concepts.

The concepts of this book, like those in books on statics, kinematics, strength of materials, and other related subjects, are based on Newton's laws. Since the laws are so important, it is advisable to give them here.

I. Every particle remains in a state of rest or moves with a constant velocity in a straight line unless an unbalanced force acts on it.

II. The acceleration of a particle is directly proportional to the resultant force acting on it and inversely proportional to its mass, and the sense of the acceleration is the same as that of the resultant force.

III. To every action there is an equal and opposite reaction.

The relations may be expressed by the following derived equations for a body in plane motion:

$$\Sigma F_x = MA_x \tag{1}$$

$$\Sigma F_y = MA_y \tag{2}$$

$$\Sigma T = I\alpha \tag{3}$$

where the resultant force acting on a body in a given direction is equal to the mass of the body times the acceleration of the center of gravity of the body, with the acceleration being in the same direction as the resultant force, and where the moment of the forces about the center of gravity is equal to the mass moment of inertia of the body about the center of gravity times the angular acceleration of the body, with the direction of angular acceleration corresponding to the direction of the moment about the center of gravity.

If the body is moving at constant velocity, or is at rest, which is a special case of constant velocity, $A_x = 0$, $A_y = 0$, and $\alpha = 0$. For such a case, which is called equilibrium, the following equations are derived:

$$\Sigma F_x = 0 \qquad (4)$$

$$\Sigma F_y = 0 \qquad (5)$$

$$\Sigma T = 0 \qquad (6)$$

Dynamics of machinery continues the application of the above basic equations. Analysis of engines cannot be made, however, without consideration of velocities since the rate of change of velocity defines acceleration, which, in turn, is proportional to the dynamic forces. It will be shown that even though one link of a mechanism is rotating at a constant speed, other links in the same mechanism can experience accelerations. Also, it will be shown that a dynamic analysis of an engine can be considered as an analysis of equilibrium.

The advent of higher-speed engines has brought with it a host of new problems as a result of the dynamic effects. The effect of the dynamic loads on bearing loads and on stresses in the various links in a given machine is, often, too great to be neglected, as was done for slow-speed engines.

Other aspects of dynamics to be included in this book are balancing of rotating masses, balancing of reciprocating engines, critical speeds, and gyroscopic action. This book will not take up the problem of design of machine members, but is restricted to what may be called the two initial considerations in design: kinematics and force analysis. The other two aspects, material selection and proportioning of parts, is left to textbooks on machine design.

CHAPTER 2

Velocities and Accelerations

Since dynamic forces are a function of accelerations, and since accelerations are a function of velocities, it is advisable to present a brief discussion of the fundamental concepts and definitions of velocities and accelerations, both linear and angular.

2.1 Linear velocity and acceleration

Velocity is defined as the rate of change of displacement, or, if the displacement is considered to be a vector quantity, velocity is defined as the vector rate of change of the displacement. If a point is moving in a plane, the motion may be considered to be the result of motion in any two directions. Analysis of the motion in a given direction will eliminate the necessity of picturing a vector operation, and, for that reason, linear motion will be assumed for the initial presentation of the concepts.

For a point moving at a constant rate or constant velocity, the rate of change of displacement is constant. Figure 2.1 illustrates a possible curve for such a case. The change of displacement (Δs) per unit of time is the same for any time element, or $\frac{\Delta s}{\Delta t}$ is a constant quantity. $\frac{\Delta s}{\Delta t}$ is, by definition, velocity; but it is also the slope of the curve in Fig. 2.1.

If the rate of change of displacement is not a constant quantity, the change in displacement, Δs, will be different for any given time element, Δt, as shown in Fig. 2.2. To apply the definition of velocity for this case, we must take the limit of $\frac{\Delta s}{\Delta t}$ as Δt approaches 0, which is expressed by $\lim_{\Delta t \to 0} \frac{\Delta s}{\Delta t} = \frac{ds}{dt}$, the mathematical expression for the slope of the displacement-time curve.

In a similar fashion, acceleration is defined as the instantaneous rate of change of velocity, which can be expressed by $\dfrac{dv}{dt}$, which can be interpreted, also, as the slope of the velocity-time curve.

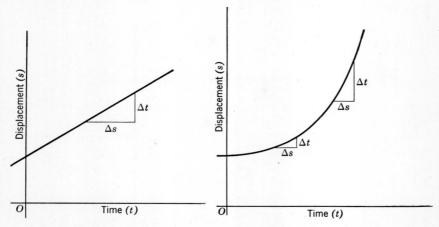

Fig. 2.1. Constant velocity.   Fig. 2.2. Variable velocity.

Since $\dfrac{ds}{dt} = v$, and $\dfrac{dv}{dt} = a$, a third form may be derived for the acceleration:

$$a = \frac{dv}{dt} = \frac{d\left(\dfrac{ds}{dt}\right)}{dt} = \frac{d^2s}{dt^2}$$

2.2 Angular velocity and angular acceleration

Angular velocity* is defined as the rate of change of angle, expressed by $\omega = \dfrac{d\theta}{dt}$, and angular acceleration is defined as the rate of change of angular velocity, expressed by $\alpha = \dfrac{d\omega}{dt}$. Or angular velocity is the slope of the angular displacement-time curve, and angular acceleration is the slope of the angular velocity-time curve.

For the concepts of angular velocity and angular acceleration, we need not be concerned with a center of rotation for a line, inasmuch

---

* Angular velocity and angular speed are used interchangeably in this book for *plane* motion. Strictly speaking, angular velocity is a vector quantity, whereas angular speed is a scalar quantity.

as changes of angles of lines are involved. For instance, in Fig. 2.3, line A–B moves to a position A'–B' with a motion that need not be prescribed. The change of angle, $\Delta\theta$, is as shown, regardless of what reference line is taken. Angles are always measured between lines, and for this reason we cannot express the angular velocity or the angular acceleration of a point.

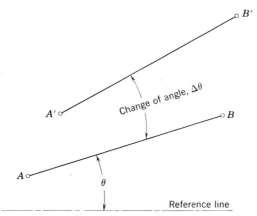

Fig. 2.3. Rate of change of angle defines angular velocity.

The relation of angular velocity and angular acceleration may be expressed by

$$\alpha = \frac{d\omega}{dt} = \frac{d\left(\frac{d\theta}{dt}\right)}{dt} = \frac{d^2\theta}{dt^2}$$

Velocities and accelerations, both linear and angular, are vector quantities and may be operated upon as vectors, in exactly the same fashion as force vectors are handled.

PROBLEMS

(Unless specified otherwise, all figures referred to in problems are problem figures.)

2.1. The equation for the angular velocity, in radians per second, of a link moving about a fixed point is given as a function of time, in seconds, by the following:

$$\omega = 3t^2 - 20t + 6$$

(a) What is the expression for the angular acceleration?
(b) Sketch the angular velocity-time curve and determine from the curve whether the angular velocity is increasing or decreasing at $t = 2$ sec.

2.2. The equation for the angular displacement, in radians, of a link moving about a fixed point is given by

$$\theta = (2t + 3)^{1/2}$$

Velocities and Accelerations

(a) Sketch the curve of θ as a function of time and determine from the curve whether the link is increasing in speed or decreasing in speed at $t = 3$ sec.

(b) Determine the equation for the angular speed as a function of time.

(c) Sketch the curve of angular speed as a function of time.

(d) Determine from the angular speed curve whether the link is increasing or decreasing in angular speed at $t = 3$ sec.

(e) Determine the angular acceleration at $t = 3$ sec.

2.3 (Fig. 2.3). If link A–B, which is 6 in. long, moves to a new position A'–B' in 2 sec, and A–A' is 3 in., what is the average angular velocity of the link?

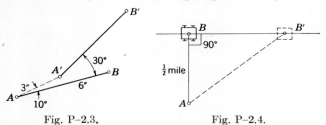

Fig. P–2.3.　　　　　　Fig. P–2.4.

2.4 (Fig. 2.4). A car, B, moves at a constant velocity of 60 mph from B to B', a straight path. An observer stationed at A watches the car. Determine an expression for the angular speed in radians per second of the line from the observer to the car as a function of time. What is the angular speed of the line from the observer to the car after the car has traveled 1 mile? 2 miles?

Plot the curve of angular speed of the line from the observer to the car as a function of time and determine from the curve whether the angular speed is increasing or decreasing.

2.5 (Fig. 2.5). An observer at A watches a car B moving along a straight path B–B'. The line joining the observer and the car is rotating at a constant angular speed of $\frac{1}{80}$ rad/sec. Determine the velocity of the car at point B. Determine also the velocity of the car 5 sec after passing point B.

Determine the acceleration of the car for the two positions.

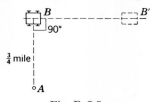

Fig. P–2.5.

2.6. The expression for simple harmonic motion can be given by

$$x = R \cos \omega t$$

Sketch the displacement as a function of time, for constant values of R and ω. By examination of the curve, sketch a probable velocity curve $\left(\dfrac{dx}{dt}\right)$ as a function of time. Verify your curve by determining the equation for the velocity and plotting it.

Similarly, from the velocity curve, sketch a probable acceleration curve $\left(\dfrac{d^2x}{dt^2}\right)$ as a function of time, and verify your curve by determining the equation for the acceleration and plotting it.

2.7. The simplified equations for the x and y displacement components of a projectile, neglecting air resistance, is given by:

$$x = V_x t$$
$$y = -gt^2 + V_y t$$

where V_x, V_y, and g are constants.

Determine an expression for the total velocity of the projectile as a function of time. Determine also the total acceleration of the projectile as a function of time.

2.8. An approximate equation for the displacement of the slider of a slider-crank mechanism is given by

$$x = R\cos\theta + L - \frac{1}{2}\frac{R^2}{L}\sin^2\theta$$

where R and L are constant quantities for a given slider crank. Sketch the curve for the vicinity $\theta = 90°$ to $\theta = 150°$, and determine from the curve whether the slider is slowing down or speeding up for $\theta = 120°$.

CHAPTER 3

Relative Velocities

Several methods are available for determining velocities in kinematic systems: centro method (or instant center method), phorograph method, relative velocity method, and analytical method. Each one has its place, but in this book only the last two will be considered. The relative velocity method has the one important advantage that the techniques used for velocity solutions can be extended to acceleration solutions.

3.1 Relative velocity of two distinct points

Let us define the relative velocity between two moving points, A and B, as the velocity one point would have if its motion is considered with respect to the second point, where the second point is considered as stationary. Figure 3.1a shows two points A and B moving in a plane where A has an absolute velocity, V_A, and B has an absolute velocity, V_B. If the same motion is pictured as being imparted to each point, no change in relative velocity is made. Thus let the velocity $-V_A$ be given to each point, as shown in Fig. 3.1b. Point A will have zero velocity, whereas point B will have a motion equal to the vector sum of $-V_A$ and V_B, which is defined as the relative velocity of B with respect to A, written as V_{BA}. The vector sum of V_B and $-V_A$ is shown in Fig. 3.1c, and is labeled V_{BA}. Further investigation of the vector triangle thus formed reveals that $V_B = V_A \looparrowright V_{BA}$, the relative velocity equation.

3.2 Relative velocity of two points on a rigid link

The relative velocity equation applies to any two points,* whether the two points are moving independently or whether they are on a rigid link. The following analysis is presented to determine the relation of velocities of two points on a rigid link.

---

* The interpretation of the relative velocity equation, where a point moves with respect to a body, is slightly different. It is discussed in Chapter 4.

Relative Velocity of Two Points on a Rigid Link

Link rotating about a fixed center. Consider a rigid link rotating about a fixed center, O, as shown in Fig. 3.2a. The distance between O and B is R, and the line O–B makes an angle of θ with

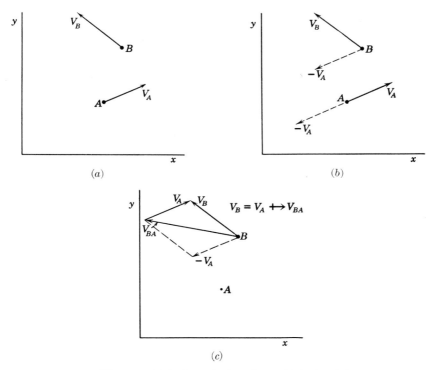

Fig. 3.1. Relative velocity of two distinct points.

respect to the x-axis. The displacement of point B in the x-direction is

$$x = R \cos \theta$$

and in the y-direction is

$$y = R \sin \theta$$

Differentiation of each equation with respect to time gives the following, with R kept constant:

$$\frac{dx}{dt} = R(-\sin \theta)\frac{d\theta}{dt}$$

$$\frac{dy}{dt} = R \cos \theta \frac{d\theta}{dt}$$

But $\frac{dx}{dt} = V_B{}^x$, the velocity of point B in the x-direction, $\frac{dy}{dt} = V_B{}^y$, the velocity of point B in the y-direction, and $\frac{d\theta}{dt} = \omega$, the angular speed of line O–B. Therefore,

$$V_B{}^x = -R\omega \sin \theta$$
$$V_B{}^y = R\omega \cos \theta$$

The total velocity of point B is found by adding vectorially the two rectangular components, with the result seen in Fig. 3.2b:

$$V_B = R\omega \sin \theta \nrightarrow R\omega \cos \theta$$
$$= [(R\omega \sin \theta)^2 + (R\omega \cos \theta)^2]^{1/2}$$
$$= R\omega$$

since $(\sin^2 \theta + \cos^2 \theta)^{1/2} = 1$.

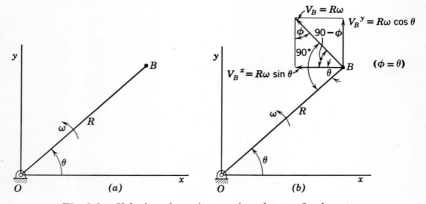

Fig. 3.2. Velocity of a point moving about a fixed center.

Referring to Fig. 3.2b, one may show that the velocity of point B is perpendicular to the line O–B by showing that $\phi = \theta$ from the equation below:

$$\tan \phi = \frac{R\omega \sin \theta}{R\omega \cos \theta} = \tan \theta$$

Note the relation of the direction of $R\omega$ and ω.

Relation of velocities of two points on a rigid link. The relative velocity equation for two points on a rigid link can be developed by extension of the procedure of the preceding analysis.

Relative Velocity of Two Points on a Rigid Link

Consider a line A–B, as shown in Fig. 3.3a, having a motion of combined translation and rotation in the plane of the paper. To locate point B, let the coordinates of point A be X_A and Y_A, let R be

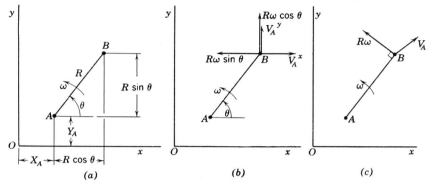

Fig. 3.3. Relation of velocities of two points on a rigid link.

the distance between A and B, and let θ define the angle which the line makes with the x-axis. The coordinates of point B are

$$x_B = X_A + R \cos \theta$$
$$y_B = Y_A + R \sin \theta$$

Differentiate each of the equations above with respect to time, remembering that R is the only constant quantity:

$$\frac{dx_B}{dt} = \frac{dX_A}{dt} - R \sin \theta \frac{d\theta}{dt}$$

$$\frac{dy_B}{dt} = \frac{dY_A}{dt} + R \cos \theta \frac{d\theta}{dt}$$

The interpretation of each quantity is as follows:

$\dfrac{dx_B}{dt}$ is the velocity of point B in the x-direction, represented by $V_B{}^x$.

$\dfrac{dX_A}{dt}$ is the velocity of point A in the x-direction, represented by $V_A{}^x$.

$\dfrac{dy_B}{dt}$ is the velocity of point B in the y-direction, represented by $V_B{}^y$.

$\dfrac{dY_A}{dt}$ is the velocity of point A in the y-direction, represented by $V_A{}^y$.

$\dfrac{d\theta}{dt}$ is the angular velocity of the line A–B, ω.

Thus
$$V_B{}^x = V_A{}^x - R\omega \sin\theta$$
$$V_B{}^y = V_A{}^y + R\omega \cos\theta$$

These vectors are shown in position in Fig. 3.3b. The vectors on the right side of the two equations must be added vectorially to give the total velocity of point B. The order of vector addition is immaterial; let us group the quantities to suit our convenience:

$$V_B = (V_A{}^x \nrightarrow V_A{}^y) \nrightarrow (R\omega \sin\theta \nrightarrow R\omega \cos\theta)$$

But $(V_A{}^x \nrightarrow V_A{}^y)$ is the total velocity of point A, V_A; and $(R\omega \sin\theta \nrightarrow R\omega \cos\theta) = R\omega$, as shown on page 10.

Thus the above equation may be expressed by

$$V_B = V_A \nrightarrow R\omega$$

where $R\omega$ is a velocity vector perpendicular to line A–B and directed in the same sense as the sense of the angular velocity.

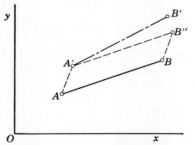

Fig. 3.4. Motion considered a combination of translation and rotation.

Let us now interpret the significance of $R\omega$. Consider point A as being fixed ($V_A = 0$), and think of the line A–B as if it were rotating about A as a fixed point (Fig. 3.3c). The velocity of point B would then be $R\omega$, and would be the velocity relative to A, which is defined by V_{BA}. Consequently, the relative velocity of point B with respect to A, V_{BA}, is $R\omega$.

For two points on a rigid link, therefore, either of the two expressions may be used:
$$V_B = V_A \nrightarrow R\omega$$
$$V_B = V_A \nrightarrow V_{BA}$$

The full significance of the above relative velocity equation may be had by reference to Fig. 3.4, where a line A–B moves to a new position

A'–B'. The motion can be considered as made up of two parts: translation from A–B to A'–B'' and rotation about A' to A'–B'. Actually, the two motions occur simultaneously in the general case. Note, also, that it is not necessary to determine the actual point about which the link is rotating, the concept which is used in centro (or instant center) analysis.

Graphical determination of the relative velocity equation.
An alternate graphical proof of the relative velocity equation for two points on a rigid link is presented here.

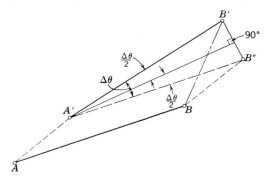

Fig. 3.5. Relative velocity equation determined graphically by consideration of translation and rotation.

Consider a line A–B in Fig. 3.5 moving to a new position A'–B'. The motion is composed of two parts, translation to A'–B'' and rotation to A'–B'. The vector equation for the displacement of point B is

$$\mathbf{BB'} = \mathbf{BB''} \nrightarrow \mathbf{B''B'}$$

But
$$\mathbf{BB''} = \mathbf{AA'}$$

Therefore
$$\mathbf{BB'} = \mathbf{AA'} \nrightarrow \mathbf{B''B'}$$

Divide through by Δt, the time interval:

$$\frac{\mathbf{BB'}}{\Delta t} = \frac{\mathbf{AA'}}{\Delta t} \nrightarrow \frac{\mathbf{B''B'}}{\Delta t}$$

where the distances divided by the time interval define velocities.

To obtain instantaneous velocities, we must take the limit of each side of the equation as the time interval approaches zero:

$$\lim_{\Delta t \to 0} \frac{\mathbf{BB'}}{\Delta t} = \lim_{\Delta t \to 0} \frac{\mathbf{AA'}}{\Delta t} \nrightarrow \lim_{\Delta t \to 0} \frac{\mathbf{B''B'}}{\Delta t}$$

But $\lim_{\Delta t \to 0} \dfrac{BB'}{\Delta t}$ is the velocity of point B, V_B, and $\lim_{\Delta t \to 0} \dfrac{AA'}{\Delta t}$ is the velocity of point A, V_A. However,

$$\mathbf{B''B'} = 2\mathbf{AB} \sin \frac{\Delta \theta}{2}$$

Divide through by Δt, and take the limit as the time interval approaches zero:

$$\lim_{\Delta t \to 0} \frac{\mathbf{B''B'}}{\Delta t} = \lim_{\Delta t \to 0} \frac{2\mathbf{AB} \sin \dfrac{\Delta \theta}{2}}{\Delta t}$$

For small angles, in the limit

$$\sin \frac{\Delta \theta}{2} = \frac{\Delta \theta}{2}$$

Therefore,

$$\lim_{\Delta t \to 0} \frac{2\mathbf{AB} \sin \dfrac{\Delta \theta}{2}}{\Delta t} = \lim_{\Delta t \to 0} \frac{2\mathbf{AB} \dfrac{\Delta \theta}{2}}{\Delta t} = \lim_{\Delta t \to 0} \mathbf{AB} \frac{\Delta \theta}{\Delta t} = \mathbf{AB} \frac{d\theta}{dt} = \mathbf{AB}\omega$$

Thus, $\qquad V_B = V_A + \!\!\!\!+ \mathbf{AB}\omega$

where $\mathbf{AB}\omega$, or V_{BA}, is at right angles to the line A–B since the angle A'–B''–B' approaches 90 degrees as Δt approaches zero, and in the limit is 90 degrees.

PROBLEMS

3.1. Car A is traveling due north at a rate of 88 ft/sec, while car B is traveling due east at 44 ft/sec. What is the relative velocity of car A with respect to car B, and what is the direction of the relative velocity?

What is the relative velocity of car B with respect to car A, and what is the direction of the relative velocity?

3.2. If a ship is traveling due north with a velocity of 8 knots, and its relative velocity with respect to a second ship is 16 knots in a direction north 60° west, what is the direction and magnitude of velocity of the second ship?

3.3. Show that the relation of angular speeds of two links A and B moving in a plane may be expressed as

$$\omega_a = \omega_b + \omega_{ab}$$

where ω_a is the angular speed of link A, ω_b is the angular speed of link B, and ω_{ab} is the relative angular speed of link A with respect to link B.

3.4. A sailor on watch at the top of a 42.2-ft mast on a ship lets a wrench drop from his hands. The ship is moving at 20 ft/sec. The captain, 6 ft tall, is standing 30 ft behind the mast. Will the captain be hit on the head? If not, where will the wrench land with respect to the captain? Air resistance is negligible, and the sea is calm.

3.5. A ship leaves port A, and at the same time a ship leaves port B 20 miles due north from port A. The ship leaving port A sails north 60° east at 15 knots while the ship leaving port B sails due east at 20 knots. What is the shortest distance the two ships will be apart, and when will it occur?

3.6 (Fig. 3.6). A bomber is flying in a path at 300 mph while a jet plane is traveling at 600 mph in a path at right angles to that of the bomber at the same elevation.

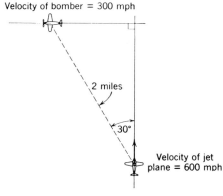

Fig. P–3.6.

When the two planes are 2 miles apart, the pilot of the jet plane decides to shoot at the bomber. If the muzzle velocity of the bullets leaving the machine gun is 2500 ft/sec (relative to the gun), in what direction must the machine gun be aimed to hit the bomber? How long after firing will it take for the bomber to be hit? How far will the bomber have traveled? Neglect gravity.

(1) Solve the problem graphically.

(2) Solve the problem analytically, using components of velocity, and check the graphical solution.

3.7 (Fig. 3.7). Link A-B, part of a four-link mechanism, has been analyzed and it has been found that the velocity of A is 30 ft/sec as shown. It is also

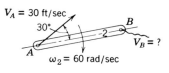

Fig. P–3.7.

known that the angular velocity of the link for the instant being considered is 60 rad/sec clockwise. If link A–B is 4 in., what is the total velocity of point B, in magnitude and direction. Solve, using $V_A = V_B \nrightarrow V_{AB}$; and solve using $V_B = V_A \nrightarrow V_{BA}$.

3.8. (Fig. 3.8) A link A-B is 8 in. long. The components of velocity of points A and B are shown. What is the angular velocity of the link, in magnitude and direction? (Note that the components of velocity along the rod must be equal.)

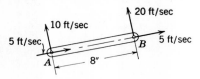

Fig. P–3.8.

3.9 (Fig. 3.9). Velocity of point A of link 2 is known in direction and magnitude. The relative velocity of point B with respect to point A is known in direction and magnitude. Show how the velocity of point B and the velocity of a third point C may be determined.

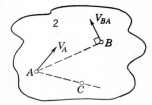

Fig. P–3.9.

CHAPTER 4

Application of the Relative Velocity Equation

Let us consider application of the relative velocity equation to various types of mechanisms. If we were to try to analyze each available mechanism as a separate problem for illustration of the basic principles involved, the task would be excessively difficult. The logical approach is to consider some of the basic mechanisms, or components of a mechanism, from which all mechanisms are derived. Some of them might be classified as follows:

(1) Slider-crank mechanism.
(2) Four-link mechanism.
(3) Shaper mechanism.
(4) "Floating link."
(5) Cams.
(6) Gears.
(7) Any combination of the above.

The velocity analysis of the basic types will be considered in this chapter and the next, and several mechanisms composed of the basic components will be analyzed.

4.1 Slider-crank mechanism

The simplest mechanism to consider is the in-line slider-crank mechanism shown in Fig. 4.1a. We shall assume for this problem, as we shall in all the problems discussed hereafter, that all the dimensions of the mechanism are known and that the linkage is drawn to scale in the position for which the analysis is to be made, inasmuch as all vector diagrams are based on the position of the links. Let us assume, also, that an initial condition is specified, which, for this case, is the angular velocity of link 2. Link 2 is assumed to be rotating clockwise with a known angular velocity, ω_2 radians per second.

The velocity of point A, since point A is rotating about a fixed point, is given by $V_A = \mathbf{O_2A}\omega_2$, which can be calculated.

18 Application of the Relative Velocity Equation

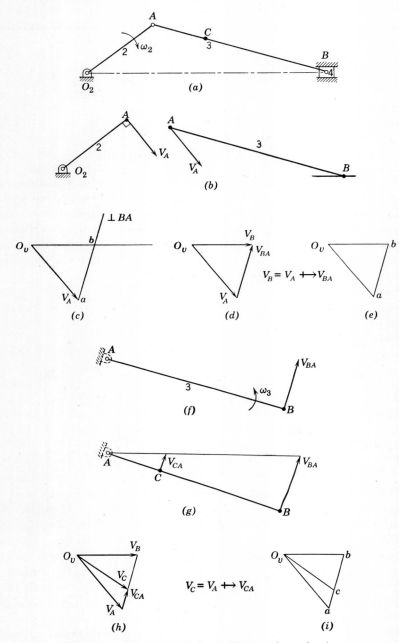

Fig. 4.1. Velocity analysis of slider-crank mechanism.

Links 2 and 3 are shown isolated in Fig. 4.1b, with the known information specified. The velocity of point B may be determined from

$$V_B = V_A \nrightarrow V_{BA}$$

or
$$V_B = V_A \nrightarrow \mathbf{BA}\omega_3$$

provided there are no more than two unknowns in the vector equation. Examination shows that there are two unknowns* in the equation, magnitude of V_B and magnitude of V_{BA}. The known quantities are direction of V_B (point B is moving in a straight line), direction and magnitude of V_A, and direction of V_{BA} (perpendicular to the line between B and A, as seen in Chapter 3, for the case of two points on a rigid link). A vector diagram is started at the pole of the velocity diagram, O_v, in Fig. 4.1c. V_A is drawn in position to a scale. The vector equation states that a vector, perpendicular to the line between points B and A, must be added to V_A, the resultant of which is a vector in the direction of the motion of point B. Point b in Fig. 4.1c satisfies the interpretation of the vector equation.

Figure 4.1c is redrawn as shown in Fig. 4.1d to show the necessary direction of each vector. The usual figure drawn is as shown in Fig. 4.1e, where the arrows are omitted and the velocity of each point of the mechanism is represented by the corresponding letter, as a and b.

Figure 4.1d (or Fig. 4.1e) can be used to find the angular velocity of link 3. Link 3 is isolated in Fig. 4.1f, where V_{BA} is used inasmuch as point A is considered stationary. For the direction of V_{BA}, link 3 is rotating counterclockwise for the position shown, with a magnitude of

$$\omega_3 = V_{BA}/\mathbf{BA}$$

where ω_3 must be expressed as radians per unit time. If V_{BA} is expressed as feet per second, $\mathbf{BA}$ must be expressed as feet, and ω_3 as radians per second.

It is now required to find the velocity of point C on link 3, in Fig. 4.1a. Again we may apply the relative velocity equation for the two points C and A:

$$V_C = V_A \nrightarrow V_{CA}$$

The direction and magnitude of V_C is unknown, and the magnitude of V_{CA} is unknown. We must obtain additional information before

---

* A general procedure to be followed in the solution of any vector equation is to determine the number of *unknowns*. A vector equation, to be soluble, can have no more than *two* unknowns.

we can proceed. The additional information is available if we consider the relation of V_{CA} and V_{BA}:

$$V_{CA} = \mathbf{CA}\omega_3$$

and
$$V_{BA} = \mathbf{BA}\omega_3$$

Dividing one equation by the other, we obtain

$$\frac{V_{CA}}{V_{BA}} = \frac{\mathbf{CA}\omega_3}{\mathbf{BA}\omega_3} = \frac{\mathbf{CA}}{\mathbf{BA}}$$

Note that the angular velocity of line C–A is the same as that of line B–A, which is the angular velocity of link 3.

In the above equation, V_{CA} may be calculated or determined graphically, as shown in Fig. 4.1g, by the principle of similar triangles. The solution giving V_C is shown in Fig. 4.1h.

Since relative velocities are proportional to distances, it can be shown that the relative velocity of any other point on line A–B of link 3 will be located on line a–b of the velocity polygon of Fig. 4.1i; or the velocity of any point on line A–B of link 3 will start at the pole, O_v, of Fig. 4.1i and end on the line a–b. Thus the vector a-b of the velocity polygon corresponds to the "image" of line A–B of link 3. It is for this reason that the relative velocity method is sometimes called the "velocity image" method, where relative velocity vectors become the "images" of the corresponding links of the original mechanism.

Consider, next, the slider mechanism shown in Fig. 4.2a. Link 3 is pictured expanded to A–B–D, primarily for illustration in determining the velocity of such a point as D. The simplest procedure is to solve the two vector equations:

$$V_D = V_B \leftrightarrow V_{DB}$$

$$V_D = V_A \leftrightarrow V_{DA}$$

where there are four unknowns: direction and magnitude of V_D, magnitude of V_{DB}, and magnitude of V_{DA}. Two vector equations permit the determination of four unknowns, the solution of which is shown in Fig. 4.2b. It is to be noted that a–b–d in Fig. 4.2b is the image of A–B–D, link 3.

4.2 Four-link mechanism

A four-link mechanism is shown in Fig. 4.3a. A complete velocity polygon is desired. Again we assume that the mechanism is drawn to scale in the position for which the velocity analysis is to be made. It is also assumed that the angular velocity link 2, ω_2, is known. The steps in the solution are as follows:

(1) $$V_A = O_2A\omega_2$$

perpendicular to line O_2–A, and as shown in Fig. 4.3b.

(2) $$V_B = V_A \mathbin{+\mkern-8mu+} V_{BA}$$

where V_B is known to be perpendicular to O_4–B, but is unknown in magnitude; where V_A has been found completely; where V_{BA} is known to be perpendicular to line A–B, since A and B are two points on a rigid link, but is unknown in magnitude. Figure 4.3c shows the construction to determine V_B and V_{BA}, and to satisfy the vector equation. Figure 4.3d shows the polygon without the arrows, where, again, the

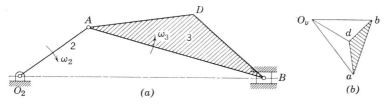

Fig. 4.2. Velocity solution gives images of corresponding links.

distance from the velocity pole represents the absolute velocity of the corresponding point in the mechanism.

The angular velocity of link 3 is clockwise, as seen from Fig. 4.3e, and is given by

$$\omega_3 = V_{BA}/\mathbf{BA}$$

The angular velocity of link 4 is counterclockwise, and is given by

$$\omega_4 = V_B/\mathbf{O}_4\mathbf{B}$$

Let it now be required to find the velocity of a point C on link 3 of the four-link mechanism already discussed and now redrawn in Fig. 4.3f.

Relate the velocity of points C and A by

$$V_C = V_A \mathbin{+\mkern-8mu+} V_{CA}$$

where there are three unknowns: magnitude and direction of V_C and magnitude of V_{CA}. Since A, B, and C are three points on a rigid link, the following may be written:

$$\frac{V_{CA}}{V_{BA}} = \frac{\mathbf{CA}}{\mathbf{BA}}$$

22   Application of the Relative Velocity Equation

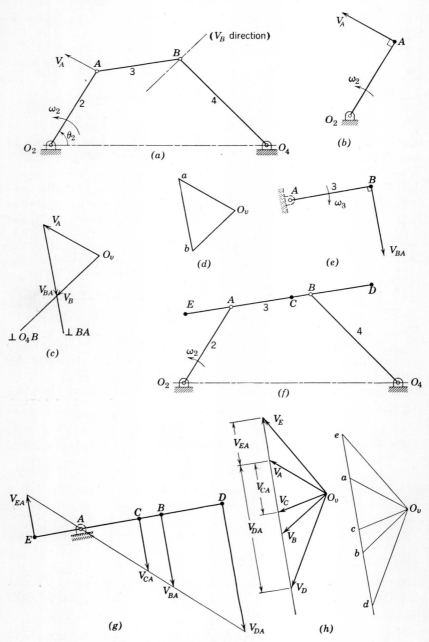

Fig. 4.3.   Velocity solution of four-link mechanism.

Use of similar triangles, as shown in Fig. 4.3g, gives V_{CA}. V_C is shown in position in Fig. 4.3h.

The above procedure can be used to find the velocity of any point, as D and E, for instance, in Fig. 4.3f, as illustrated in Figs. 4.3g and 4.3h.

4.3 Powell engine

A mechanism which combines the slider-crank mechanism and the four-link mechanism is shown in Fig. 4.4a. The angular velocity of link 2, ω_2, is assumed known. The velocity of point E on link 5 is desired. The steps in the solution are:

(a) $V_A = \mathbf{O_2A}\omega_2$, as shown in Fig. 4.4b.
(b) $V_B = V_A \nrightarrow V_{BA}$, the solution for which is shown in Fig. 4.4c.
(c) $\dfrac{V_C}{\mathbf{O_4C}} = \dfrac{V_B}{\mathbf{O_4B}}$, as illustrated in Fig. 4.4d.
(d) $V_D = V_C \nrightarrow V_{DC}$, the solution for which is shown in Fig. 4.4e.
(e) $V_E = V_C \nrightarrow V_{EC}$, where $(V_{EC}/V_{DC}) = (\mathbf{EC/DC})$, and where V_{EC} is found in Fig. 4.4f.

Figure 4.4g shows the combined figures, and Fig. 4.4h shows the combined figures with the arrows omitted.

It is left to the student to show that links 3, 4, and 5 are rotating clockwise.

4.4 Joy locomotive valve gear

The Joy locomotive valve gear shown schematically in Fig. 4.5a is taken as an exercise for the student. Piston B is the main piston; piston H controls the flow of steam to the main cylinder. Assuming that the angular velocity of link 2 is known, show that Fig. 4.5b is the velocity polygon. Write all equations in the order of solution, isolating links, and show how the angular velocity of links 3, 7, and 9 are determined in magnitude and direction.

4.5 Relative velocity of two coincident points

So far the relative velocity of two distinct points and the relative velocity of two points on a rigid link have been discussed. The problem now is to determine the relative velocity equation for two coincident points where one point is moving with respect to another moving link.

To determine the magnitude of the total velocity of a point moving with respect to a moving body, consider a point B which is moving

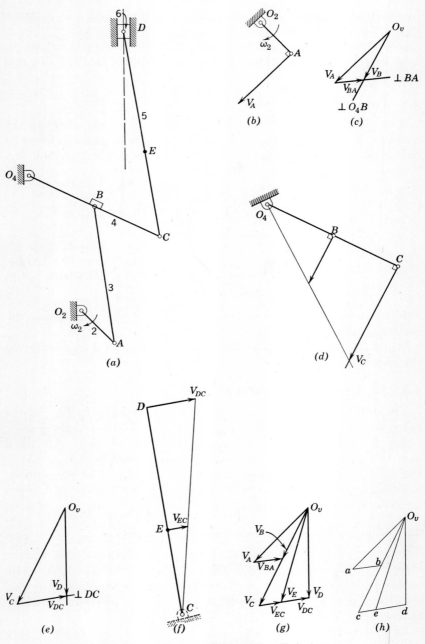

Fig. 4.4. Velocity analysis of Powell engine.

Relative Velocity of Two Coincident Points

with respect to body M at the same time that body M is moving in a plane, as shown in Fig. 4.6a. A coordinate system of axes, X and Y, are fixed and will be used to determine the absolute position of any point in the plane of X and Y. A second system of axes, c and d, *are fixed on body M* and move in the same way that body M moves. The angle θ gives the angular position of the c-axis with the X-axis.

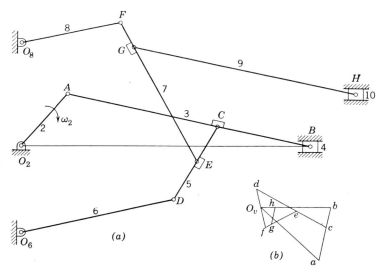

Fig. 4.5. Velocity solution of Joy valve gear.

Inspection of Fig. 4.6a shows that the X and Y displacements of point B may be expressed by the following, where A (any reference point on body M) is a fixed point on M:

$$X_B = X_A + c \cos \theta - d \sin \theta$$
$$Y_B = Y_A + c \sin \theta + d \cos \theta$$

Differentiate the above, remembering that $\dfrac{d\theta}{dt}$ is defined as ω, the angular speed of body M, and that c and d are variables:

$$\frac{dX_B}{dt} = V_B{}^x = \frac{dX_A}{dt} - c\omega \sin \theta + \frac{dc}{dt} \cos \theta - d\omega \cos \theta - \frac{dd}{dt} \sin \theta$$

$$\frac{dY_B}{dt} = V_B{}^y = \frac{dY_A}{dt} + c\omega \cos \theta + \frac{dc}{dt} \sin \theta - d\omega \sin \theta + \frac{dd}{dt} \cos \theta$$

Application of the Relative Velocity Equation

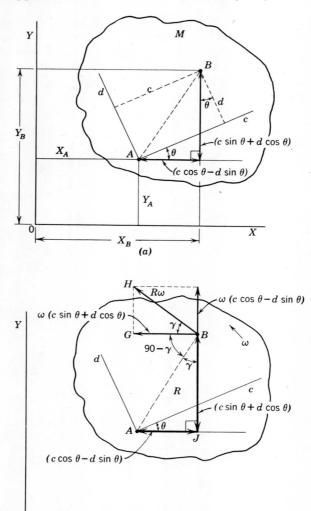

Fig. 4.6. Relative velocity of two coincident points, point B and a coincident point on link M.

Relative Velocity of Two Coincident Points

Set $\dfrac{dc}{dt} = u_c$, the time rate of change of displacement along the c-axis,

as if body M were stationary, which is defined as a relative velocity.

$\dfrac{dd}{dt} = u_d$, the time rate of change of displacement along the d-axis,

as if body M were stationary, which is defined as a relative velocity.

$\dfrac{dX_A}{dt} = V_A{}^x$, the velocity of point A in the x-direction.

$\dfrac{dY_A}{dt} = V_A{}^y$, the velocity of point A in the y-direction.

Therefore, the components of velocity of point B may be expressed by the following, with the above substitutions, and with the terms rearranged:

$$V_B{}^x = V_A{}^x - \omega(c \sin\theta + d \cos\theta) + u_c \cos\theta - u_d \sin\theta$$
$$V_B{}^y = V_A{}^y + \omega(c \cos\theta - d \sin\theta) + u_c \sin\theta + u_d \cos\theta$$

The order of vector addition is immaterial. The vectors will be added in a particular fashion to give a simplified final result:

(a) $V_A{}^x \nrightarrow V_A{}^y = V_A$, the absolute velocity of point A.
(b) $\omega(c \sin\theta + d \cos\theta) \nrightarrow \omega(c \cos\theta - d \sin\theta) = \omega(c^2 + d^2)^{1/2}$,

inasmuch as the components are perpendicular to each other, as shown in Fig. 4.6b. Since the instantaneous distance from point B to point A is given by $(c^2 + d^2)^{1/2}$, which may be called R, $\omega(c^2 + d^2)^{1/2} = R\omega$. We may see that the direction of $R\omega$ is perpendicular to the line joining A and B by the following equations. First, it can be shown that angle ABJ and angle HBG are equal by:

$$\tan \angle ABJ = \frac{(c \cos\theta - d \sin\theta)}{(c \sin\theta + d \cos\theta)}$$

$$\tan \angle HBG = \frac{\omega(c \cos\theta - d \sin\theta)}{\omega(c \sin\theta + d \cos\theta)}$$

If the above angles are called γ, angle GBA is equal to $(90° - \gamma)$ since angle GBJ is equal to $90°$ and angle ABJ is equal to γ. Thus angle $HBA = \gamma + (90° - \gamma) = 90°$. Therefore, $R\omega$ is perpendicular to the line joining point A and B. Also, if the angular speed of body M is taken positive, that is, the angle θ is increasing with time, or the angular speed is counterclockwise, the component $R\omega$ is directed to correspond with the sense of ω, as shown.

(c) $(u_c \cos \theta \leftrightarrowtail u_c \sin \theta) = [(u_c \cos \theta)^2 + (u_c \sin \theta)^2]^{1/2} = u_c.$

The above is the relative velocity of point B with respect to body M, as though M were stationary, along the c-axis.

(d) $(u_d \sin \theta \leftrightarrowtail u_d \cos \theta) = [(u_d \sin \theta)^2 + (u_d \cos \theta)^2]^{1/2} = u_d.$

The above is the relative velocity, along the d-axis, of point B with respect to body M, as though body M were stationary.

Summing up, the velocity of point B may be expressed by:

$$V_B = V_A \leftrightarrowtail R\omega \leftrightarrowtail u_c \leftrightarrowtail u_d \tag{1}$$

But $V_A \leftrightarrowtail R\omega = V_B{}^m$, the velocity of a point on body M coincident with the point B, since A and a point on body M coincident with B are two points on a rigid link. Also, $u_c \leftrightarrowtail u_d = u$, the relative velocity of B with respect to body M. Therefore, Eq. 1 may be written:

$$V_B = V_B{}^m \leftrightarrowtail u$$

In conclusion, the interpretation of the above equation is that the velocity of a point which is moving with respect to a moving body is determined by adding vectorially the velocity of the *coincident* point on the moving body and the relative velocity of the point with respect to the *body*, as though the body were stationary.

The equation may be written in the form already used by calling the moving point B point B_3 and the coincident point on the body M B_4, with the equation then being in this form:

$$V_{B_3} = V_{B_4} \leftrightarrowtail V_{B_3 B_4}$$

It is to be noted that even though the relative velocity of two coincident points are considered in the equation, the actual interpretation, as seen in the development of the equation, is that the relative velocity, $V_{B_3 B_4}$, is to be considered by analyzing the *path of motion of point B_3 relative to link 4* (the link on which the point B_3 is moving), that is, by thinking of *link 4* as stationary. It is to be noted, also, that the relative velocity interpretation as used for two points on a rigid link is not the same as for two coincident points. In the former case, the relative velocity is reckoned with respect to a *point*, whereas in the latter case the relative velocity is reckoned with respect to a *body*.

One important application of the relative velocity equation, where a point is considered moving with respect to a moving body, is the determination of the relative velocity of a particle of steam with respect to a steam turbine blade and the determination of the exit velocity of the particle of steam from the turbine blade, V_{SB} and V_S' in Fig. 4.7.

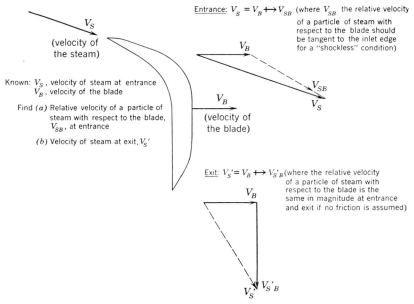

Fig. 4.7. Application of the relative velocity equation to a turbine blade.

4.6 Shaper mechanism

A schematic arrangement of the shaper mechanism is shown in Fig. 4.8a. Link 2 rotates about a fixed center, O_2. A block, link 3, is pinned at the end of link 2, which block can rotate with respect to link 2. The block, link 3, is slotted so that link 3 may slide along link 4. The mechanism, as shown or in a modified form, is used to give a slow cutting stroke and a fast return. Let us analyze the mechanism for velocities in the position shown, assuming that the angular velocity of link 2 is given. The steps in the solution are as follows.

(a) Consider the isolated links 2, 3, and 4 shown in Fig. 4.8b. Three specific points are to be considered: point A_2, a point at the end of the crank; point A_3, a point on link 3; and point A_4, a point on link 4 coincident with points A_2 and A_3. Points A_2 and A_3 have the

Application of the Relative Velocity Equation

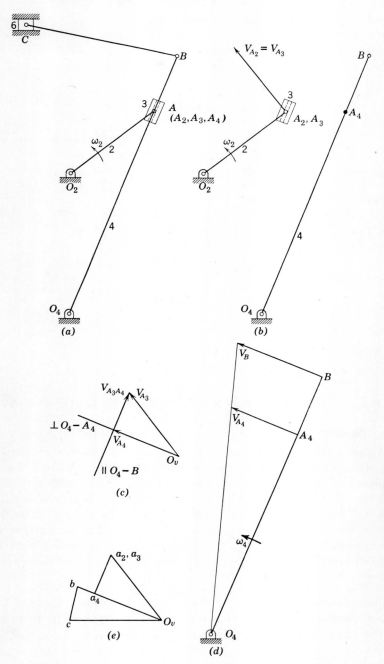

Fig. 4.8. Velocity analysis of the shaper mechanism. Two coincident points are considered in the solution.

Floating Link Mechanism

same velocity, which can be expressed by

$$V_{A_2} = V_{A_3} = O_2A\omega_2$$

(b) As shown in the preceding section, the relation of the velocities of two coincident points can be expressed by

$$V_{A_3} = V_{A_4} \nrightarrow V_{A_3A_4}$$

where point A_3 is pictured as a point moving with respect to a body which, in this case, is link 4. The relative motion of point A_3 *with respect to link 4* must be along the line O_4–B. That is, link 4 is pictured stationary, and the path traced on link 4 by point A_3 must be along the line O_4–B.

The velocity of point A_4 is known in direction (perpendicular to the line O_4–A_4) but unknown in magnitude. The magnitude of $V_{A_3A_4}$ is unknown. With only two unknowns in the equation, the vector equation can be solved, as shown in Fig. 4.8c.

It is desirable to point out that other relations, such as

$$V_{A_4} = V_{A_3} \nrightarrow V_{A_4A_3} \tag{1}$$

$$V_{A_4} = V_{A_2} \nrightarrow V_{A_4A_2} \tag{2}$$

among others, could have been used for a solution with a proper interpretation of the equations. In Eq. 1 above, point A_4 is pictured as moving with respect to link 3, where $V_{A_4A_3}$ is along the slot of *link 3*. In Eq. 2 above, point A_4 is pictured moving with respect to *link 2*. The complete path of motion of A_4 on link 2 is not a simple one, although *at the instant shown* the path of motion of A_4 on link 2 is tangent to line O_4–B at A_4. Even though the velocity analyses with all the interpretations given yield the same final results, it is extremely important that the thinking is proper for simplification in the acceleration analysis of two coincident points in Chapter 8.

(c) The velocity of point B is proportional to the velocity of point A_4, as shown in Fig. 4.8d.

(d) The velocity of point C may be found by application of the relative velocity equation for points C and D. The final velocity polygon is shown in Fig. 4.8e. Note that the angular velocity of link 3 is the same as that for link 4.

4.7 Floating link mechanism

The analysis of mechanisms using a "floating link" will be deferred to the next chapter. A special method will be given for such cases.

4.8 Special positions

The reader should realize that the relative velocity method can be applied to a mechanism in a particular position of the mechanism. If the links change position, a new velocity diagram should be drawn. Thus, to obtain a complete velocity analysis, it is necessary to determine the velocity polygons for various positions of the mechanism

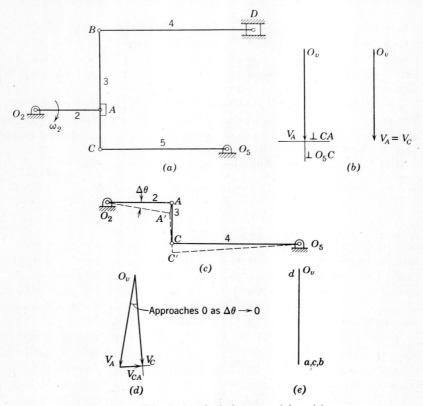

Fig. 4.9. Velocity analysis for a special position.

and then to plot or tabulate the results to give a comprehensive picture.

During such an analysis, a so-called special position may arise. Figure 4.9a is such a case. It is desired to find the velocity of point D for a given angular velocity of link 2. The steps in the solution are:

(a) $V_A = O_2 A \omega_2$.
(b) $V_C = V_A \nrightarrow V_{CA}$.

There are two unknowns in the equation: magnitude of V_C and magnitude of V_{CA}. The solution is shown in Fig. 4.9b, where $V_A = V_C$ and $V_{CA} = 0$. If we have difficulty in seeing the vector diagram, we may picture rotating link 2 a small angle, $\Delta\theta$, as shown in Fig. 4.9c. The velocity diagram for Fig. 4.9c is shown in Fig. 4.9d, and we may visualize V_{CA} approaching zero as $\Delta\theta$ approaches zero, and V_C becoming coincident with V_A. The angular speed of link 3 is zero.

(c) Since $V_{CA} = 0$, $V_{BA} = 0$, or $V_B = V_A$.
(d) $V_D = V_B + V_{DB}$.

The final velocity polygon is shown in Fig. 4.9e. $V_D = 0$.

PROBLEMS

4.1 (Fig. 4.1). A single-cylinder diesel engine has a crank 3 in. long and a connecting rod length of $11\frac{1}{4}$ in. If the crank rotates counterclockwise at 2500 rpm, determine graphically the velocity of the piston when the crank makes an angle of 60 degrees with the line through the crank bearing and piston pin.

Determine, also, the velocity of the point C, which is located on the connecting rod one third the distance to the piston pin from the crank pin.

What is the angular speed of the connecting rod, in magnitude and direction?

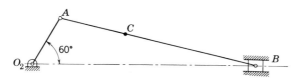

Fig. P-4.1. $O_2A = 3''$; $AB = 11\frac{1}{4}''$; $AC = 3\frac{3}{4}''$.

4.2 (Fig. 4.2a, b). The figures show the Walschaert valve gear as used on steam locomotives. This problem, however, concerns itself only with the motion of the main driving piston and connecting rod on one side of the locomotive.

Determine the velocity of the piston and the angular speed of the connecting rod in Fig. 4.2a for three positions:

(1) The position shown.

(2) The position of the piston when the line from the crank pin to the center of the wheel makes an angle of 30 degrees (measured clockwise) with the centerline of the wheels.

(3) The position of the piston when the line from the crank pin to the center of the wheel makes an angle of 120 degrees with the centerline of the wheels.

Note that the engine is an offset type of slider crank mechanism.

The locomotive is moving at 60 mph to the right. The driving wheels are 80 in. in diameter.

What is the velocity of the piston, for the three positions, relative to the frame of the locomotive? What is the velocity of the main crank pin relative to the frame of the locomotive for the three positions?

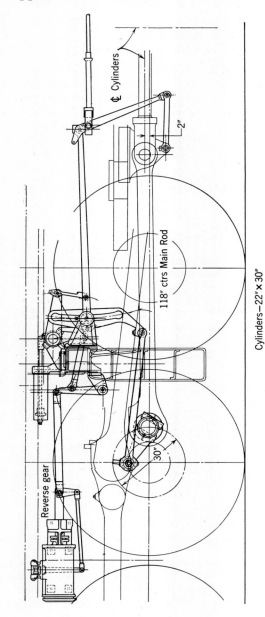

Fig. P-4.2a. Walschaert Valve Gear as applied to New York, New Haven & Hartford 4-6-4 Locomotives. Built by The Baldwin Locomotive Works. Reproduced with permission from *Locomotive Cyclopedia*, Simmons-Boardman Corp.

Problems

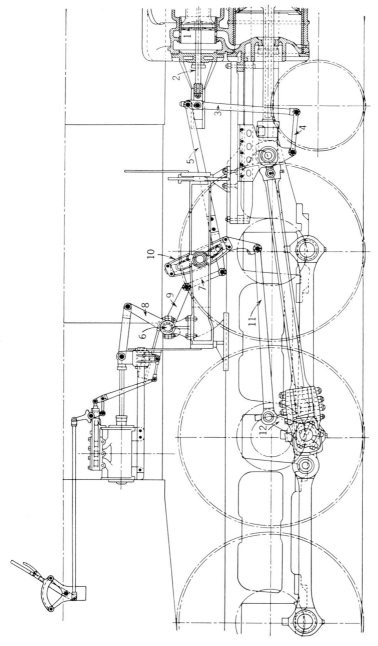

Fig. P-4.2b. Elevation showing general arrangement of a Walschaert Valve Gear for a 4-6-2 (Pacific) Type Locomotive. (1) Valve. (2) Valve stem. (3) Combination lever. (4) Crosshead link. (5) Radius rod. (6) Reverse shaft. (7) Lifting link. (8) Reverse shaft link. (9) Reverse shaft arm. (10) Reverse link. (11) Eccentric rod. (12) Eccentric crank. Reproduced with permission from *Locomotive Cyclopedia*, Simmons-Boardman Corp.

4.3 (Fig. 4.3). The velocity of A is 10 ft/sec, downwards. Find the velocity of points B, C, and D, and the angular velocity of links 3 and 5.

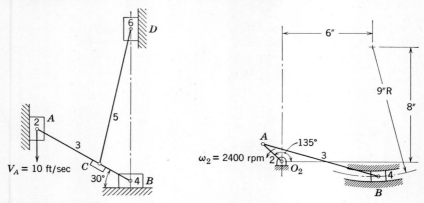

Fig. P–4.3. $AB = 6''$; $AC = 4''$; $CD = 8''$.

Fig. P–4.4. $O_2A = 1\tfrac{3}{4}''$; $AB = 8''$.

4.4 (Fig. 4.4). Find the angular velocity of link 3, the velocity of point B of the slider, and the angular velocity of link 4.

4.5 (Fig. 4.5). Find the velocity of points B, C, and D for the two-cylinder V-engine, using an articulated rod and master rod. Determine also the angular speed of links 3 and 5.

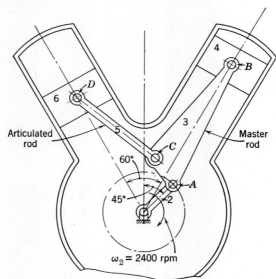

Fig. P–4.5. $O_2A = 2\tfrac{1}{4}''$; $AB = 7\tfrac{1}{2}''$; $BC = 6\tfrac{3}{4}''$; $AC = 1\tfrac{3}{4}''$; $CD = 5\tfrac{1}{2}''$.

4.6. Refer to Fig. 4.3a of the text. Link 2 is 4 in. long, link 3 is 6 in. long, and link 4 is 8 in. long. The distance between O_2 and O_4 is 14 in. θ_2 is 45 degrees.

If link 2 is rotating at 1800 rpm counterclockwise, find the velocity of point B, the velocity of the midpoint of link 3, and the angular velocity of links 3 and 4.

Without redrawing the velocity polygon, determine the above for the same magnitude of angular velocity for link 2, except in a clockwise direction.

4.7 (Fig. 4.7). Determine, for the crossed-link mechanism, the velocity of points B and C and the angular velocity of links 3, 4, and 5.

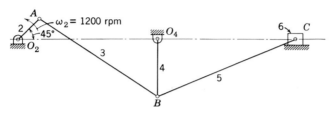

Fig. P–4.7. $O_2A = 2''$; $AB = 9\frac{1}{2}''$; $O_4B = 4''$; $BC = 10''$.

4.8 (Fig. 4.8). Draw the velocity polygon for the crossed-link mechanism and determine the velocity of point D and the angular velocity of link 5. Link 2 has an angular velocity of 1500 rpm counterclockwise and an angular acceleration of 10 rad/sec² clockwise.

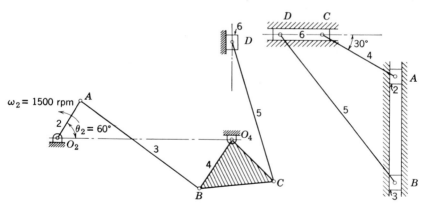

Fig. P–4.8. $O_2A = 3''$; $AB = 10''$; $O_4B = 4''$;   Fig. P–4.9. $AC = 4''$;
$O_4C = 4''$; $BC = 5''$; $CD = 10''$. $DC = 2''$; $BD = 9''$.

4.9 (Fig. 4.9). The relative velocity of point A with respect to point B, V_{AB}, is known: 10 ft/sec downwards. Determine the velocity of points A, B, and C. Determine, also, the angular velocity of links 4 and 5.

4.10 (Fig. 4.10). The slider, link 3, is pinned to the crank at A and can move with respect to link 4. For the position shown, determine the angular velocity

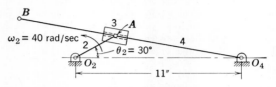

Fig. P–4.10. $O_2A = 3''$; $O_4B = 15''$.

of links 3 and 4 and determine the velocity of point B. What is the relative angular velocity of link 3 with respect to link 4?

4.11 (Fig. 4.11). A schematic sketch of a pump is shown in the figure, the arrangement being basically a Scotch yoke. The frame, on which the pump is

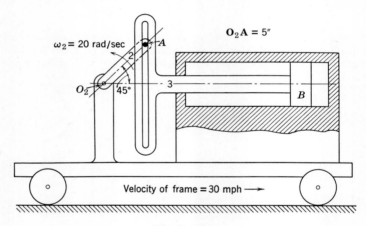

Fig. P–4.11.

mounted, is moving to the right at a constant velocity of 30 mph. The crank, link 2, is rotating counterclockwise at 20 rad/sec. Determine the velocity of the piston, B, and the velocity of the piston with respect to the moving frame.

4.12 (Fig. 4.12). Steam leaving a fixed nozzle at 2000 ft/sec is directed at a steam turbine blade, which is moving at 500 ft/sec. For what angle must the inlet edge of the blade be designed for "shockless" entrance, that is, to have the relative velocity of the steam tangent to the blade?

Assuming that the magnitude of the relative velocity of the steam with respect to the blade is unchanged as the steam moves with respect to the blade, determine the proper angle of the exit edge so that the absolute velocity of the steam will be tangent to the inlet edge of the stationary blade shown.

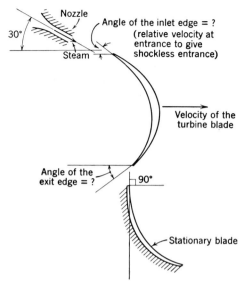

Fig. P–4.12.

4.13 (Fig. 4.13). The velocity of point A is 40 ft/sec, with counterclockwise rotation of the crank, link 2.

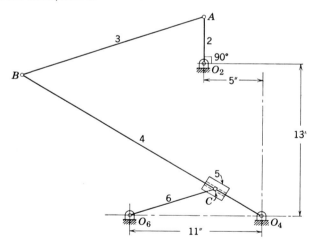

Fig. P–4.13. $O_2A = 4''$; $AB = 16''$; $O_4B = 23\frac{3}{16}''$; $O_6C = 7\frac{1}{2}''$.

Draw the velocity polygon for the position shown. Determine the angular velocity of links 4, 5, and 6 and the velocity of point B.

4.14 (Fig. 4.14). Assume a velocity for point A. Can the velocity polygon be drawn? Explain.

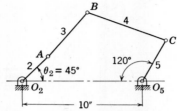

Fig. P–4.14. $O_2A = 3''$; $AB = 5''$; $BC = 7''$; $O_5C = 4''$.

4.15 (Fig. 4.15). Link 2 is rotating at the instant shown at 20 rad/sec counterclockwise. Draw the velocity polygon and determine the angular velocity of links 3, 4, 5, and 6, and the relative angular velocity of link 5 with respect to link 6. What is the relative angular velocity of link 6 with respect to link 5?

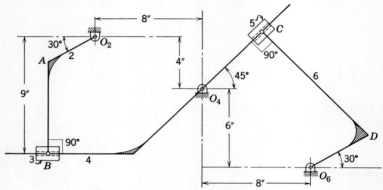

Fig. P–4.15. $O_2A = 4''$; $O_6D = 5''$.

4.16 (Fig. 4.16). Link 4 is rotating at 40 rad/sec counterclockwise for the position shown. Determine the angular velocity of link 5 and the velocity of point C.

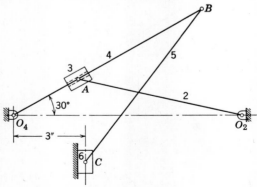

Fig. P–4.16. $O_2A = 7''$; $O_4B = 9''$; $BC = 8''$.

4.17 (Fig. 4.17). Link 5 has an angular velocity of 20 rad/sec clockwise. Determine the velocity of point A, and the relative angular velocity of link 3 with respect to link 4.

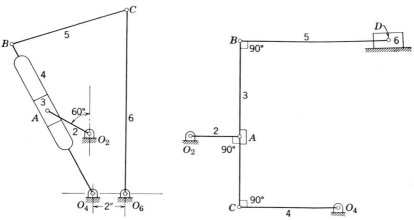

Fig. P-4.17. $O_6C = 12''$; $BC = 7\frac{1}{2}''$; $O_4B = 11''$; $O_4A = 6''$; $O_2A = 3''$. Angle O_4–O_6–$C = 90°$.

Fig. P-4.18. $O_2A = 3''$; $AC = 4\frac{1}{2}''$; $CO_4 = 6''$; $AB = 6''$; $BD = 9''$.

4.18 (Fig. 4.18). In the mechanism shown, link 2 rotates at 30 rad/sec clockwise. Draw the complete velocity polygon and determine the angular velocity of all the links.

4.19 (Fig. 4.19). Link 2 is rotating at a constant velocity of 20 revolutions per second counterclockwise. In the position shown, A–B and D–F are horizontal, and O_2–A, O_4–D, and O_6–F are vertical. Draw the velocity polygon and determine the velocity of point F and the angular velocity of each link.

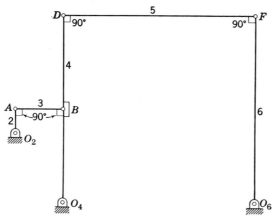

Fig. P-4.19. $O_2A = 2''$; $AB = 4''$; $O_4C = 16''$; $O_6D = 16''$; $O_4B = 8''$; $CD = 16''$; $CE = 12''$.

4.20 (Fig. 4.20). Link 3 has an angular velocity of 45 rad/sec clockwise. Draw the velocity polygon and determine the angular velocity of links 2 and 4.

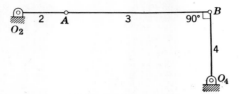

Fig. P–4.20. $O_2A = 4''$; $AB = 12''$; $O_4B = 6''$.

4.21 (Fig. 4.21). Internal gear E of the epicyclic gear train is fixed. Arm A turns counterclockwise at 60 rad/sec. Draw the complete velocity polygon. What is the angular velocity of gear B? gear C? gear D?

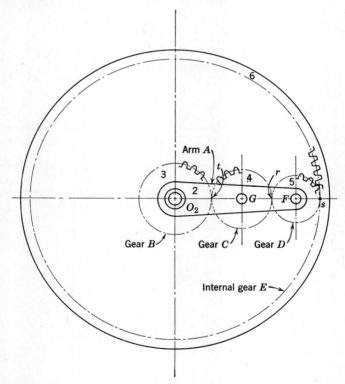

Fig. P–4.21. Diameter of gear $B = 4.50''$. Diameter of gear $C = 3.75''$. Diameter of gear $E = 18.00''$.

What is the relative velocity of point r, the point of contact of gears C and D, with respect to a coincident point on the arm?

Problems

(*Note.* The velocity of the point of contact of gears D and E, point s, is zero. *Hint.* Find the velocity of the center of gear D first, and with the information of the velocity of point s, find the velocity of point r, the point of contact of gears C and D.)

4.22. The information and requirements are the same as for problem 4.21, except that gear E is rotating at 25 rad/sec clockwise. (Note that the velocity of point s is not zero for this case.)

4.23. (Fig. 4.23). Part of the shaper quick-return mechanism is shown in two different special positions in (a) and (b). If link 2 is rotating counterclockwise at

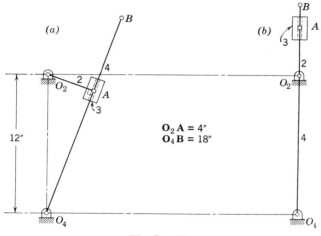

Fig. P-4.23.

40 rad/sec, determine the relative velocity of sliding of point A on link 3 with respect to link 4, the velocity of point B, and the angular velocity of link 4 for the two positions shown.

CHAPTER 5

Special Methods of Velocity Solution

Sometimes it is necessary to analyze a mechanism for which the relative velocity equation as treated in the previous chapter does not permit a direct solution from link to link. There are various methods of handling such cases, such as the combination of the centro (or instant center) method with the relative velocity method; the centro method alone; or the velocity polygon may be started at a point in the mechanism from which the velocity analysis can be made from link to link, with the scale being determined as a last step. The last procedure is based on the fact that the velocity polygons are always similar for a given mechanism regardless of the initial velocity condition. However, the various techniques used for velocity analysis are not directly applicable to acceleration analysis. Therefore, let us examine a direct approach by use of so-called auxiliary points* which permit a direct solution not only for velocities but also for accelerations. The procedure will be illustrated by several examples.

5.1 Watt "walking beam" mechanism

Figure 5.1a shows the Watt "walking beam" mechanism, which cannot be handled directly for velocities by the methods of Chapter 4 alone if the motion of link 2 is specified. Link 4 may be considered to be a "floating link," a link which has no fixed center of rotation. The velocity of point D is desired for a known angular velocity of the crank, ω_2. The steps in the solution are as follows.

(a)
$$V_A = O_2A\omega_2 \tag{1}$$

(b)
$$V_B = V_A \leftrightarrow V_{BA} \tag{2}$$

---

* See "Auxiliary Points Aid Acceleration Analysis," by A. S. Hall and E. S. Ault, *Machine Design*, November, 1943.

In Eq. 2 there are three unknowns, magnitude and direction of V_B, and magnitude of V_{BA}. Further information cannot be obtained immediately to effect a solution, nor can we start at any other point, as C or D, to obtain a direct solution.

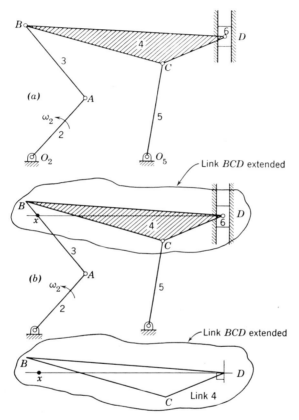

Fig. 5.1a, b. Velocity analysis, using an auxiliary point.

(c) Let us picture link 4 isolated and *extended*, as shown in Fig. 5.1b, *with a point x on link 4*, as indicated. The reason for the selection of point x will be apparent shortly. Let us write the equation for velocity of two points, points x and B, on the rigid link 4:

$$V_x = V_B \leftrightarrow V_{xB} \tag{3}$$

where the only known information is that V_{xB} is perpendicular to the line between x and B.

46 Special Methods of Velocity Solution

(d) Substitute Eq. 2 into Eq. 3:
$$V_x = (V_A \leftrightarrow V_{BA}) \leftrightarrow V_{xB} \qquad (4)$$

(e) The above may be written as
$$V_x = V_A \leftrightarrow \text{(vector perpendicular to the line } B\text{-}A) \qquad (5)$$

since V_{BA} is perpendicular to the line B–A, and V_{xB} is perpendicular to the line x–B, and the lines B–A and x–B are coincident. Note that line B–A is a part of link 3, whereas line x–B is a part of link 4. So far, there are three unknowns in Eq. 5: direction and magnitude of V_x, and magnitude of the vector perpendicular to the line B–A.

(f) Consider next the two points x and D, both points on link 4:
$$V_x = V_D \leftrightarrow V_{xD} \qquad (6)$$

(g) The above equation may be written as
$$V_x = \text{(vector in the direction of motion of point } D) \qquad (7)$$

since the line between points x and D is perpendicular to the motion of point D, so that V_{xD} and V_D are in the same direction.

(h) The results, so far, are:
$$V_x = V_A \leftrightarrow \text{(vector perpendicular to } A\text{-}B)$$
$$V_x = \text{(vector in the direction of motion of point } D)$$

The two equations may be solved simultaneously to give V_x, as shown in Fig. 5.1c. Note that point x is *not* a part of link 3, and, for that

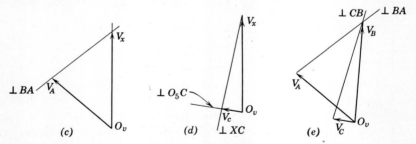

Fig. 5.1. (c) Velocity of auxiliary point determined. (d) Velocity of point C determined. (e) Velocity of point B determined.

reason, it is *not* correct to say that the velocity of point B with respect to point A is proportional to the velocity of point x with respect to point A. The proportionality of relative velocities is true when the points considered are on the same rigid link.

(i) The velocity of points B or D cannot be found at this point. It is necessary to go to point C, a point on link 4:

$$V_C = V_x \mathbin{+\!\!\!+} V_{Cx} \tag{8}$$

The solution is shown in Fig. 5.1d.

(j) The velocity of point B may be found from the simultaneous solution of

$$V_B = V_C \mathbin{+\!\!\!+} V_{BC}$$
$$V_B = V_A \mathbin{+\!\!\!+} V_{BA} \tag{9}$$

as shown in Fig. 5.1e.

(k) The velocity of point D may be found from

$$V_D = V_C \mathbin{+\!\!\!+} V_{DC} \tag{10}$$

The complete velocity polygon is shown in Fig. 5.1f. A check of the solution is that V_{DB} is perpendicular to the line D–B, which fact was not used in the solution.

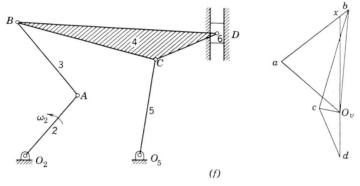

Fig. 5.1f. Complete velocity polygon.

The significance of the selection of point x should be apparent now. The point had to be selected in such a way that the relative velocity of the point with respect to another point on the same link corresponded to the direction of motion of the reference point.

Alternate auxiliary point. It is possible to solve the problem by using an auxiliary point y on link 4, shown in Fig. 5.1g. The equations for solution, given without discussion, are:

$$V_A = \mathbf{O_2 A} \omega_2 \tag{1}$$

$$V_B = V_A \mathbin{+\!\!\!+} V_{BA} \tag{2}$$

$$V_y = V_B \mathbin{+\!\!\!+} V_{yB} \tag{3}$$

$$V_y = (V_A \nrightarrow V_{BA}) \nrightarrow V_{yB}, \text{ or}$$

$$V_y = V_A \nrightarrow \text{(vector perpendicular to the line } B\text{--}A) \quad (4)$$

$$V_y = V_C \nrightarrow V_{yC}, \text{ or}$$

$$V_y = \text{(vector perpendicular to } O_5\text{--}C) \quad (5)$$

By using Eqs. 4 and 5 we find V_y. Note that points B, C, D, and y

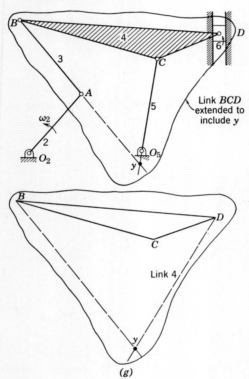
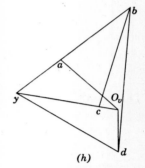

Fig. 5.1g, h. Use of another auxiliary point.

are on link 4, whereas points A, B, and y are not on the same link. Also, points O_5, C, and y are not points on the same rigid link.

$$V_D = V_y \nrightarrow V_{Dy} \quad (6)$$

$$\left.\begin{array}{l} V_B = V_D \nrightarrow V_{BD} \\ V_B = V_A \nrightarrow V_{BA} \end{array}\right\} \quad (7)$$

$$V_C = V_B \nrightarrow V_{CB} \quad (8)$$

The complete velocity polygon is shown in Fig. 5.1h.

5.2 Modified shaper mechanism

Figure 5.2a shows another mechanism that cannot be solved for velocities by the direct link-to-link method for a prescribed motion of link 2. Link 4 can be considered to be a "floating" link. Let us

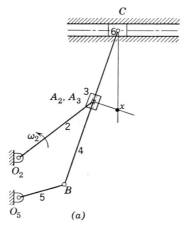

Fig. 5.2a. Velocity analysis of the modified shaper mechanism, using an auxiliary point.

determine the complete velocity polygon for a known angular velocity of link 2. The steps in the solution are as follows.

(a) Consider the point A_2, a point on the crank:

$$V_{A_2} = O_2A_2\omega_2 \tag{1}$$

(b) Consider the point A_3, a point on the slider, which has the same velocity as point A_2:

$$V_{A_2} = V_{A_3} \tag{2}$$

(c) Write the equation for point A_3 and a coincident point A_4 on link 4:

$$V_{A_4} = V_{A_3} \nrightarrow V_{A_4A_3} \tag{3}$$

where the unknowns are three: magnitude and direction of V_{A_4}, and magnitude of the relative velocity of point A_4 with respect to *link* 3. (The direction of relative velocity of point A_4 with respect to link 3 is known to be parallel to the slot of link 3.)

(d) Let us introduce an auxiliary point, point x, whose location is unknown, but will be found from the proper interpretation of the equations to be written:

$$V_x = V_{A_4} \nrightarrow V_{xA_4} \tag{4}$$

(e) Substitute Eq. 3 into Eq. 4:

$$V_x = (V_{A_3} \leftrightarrow V_{A_4A_3}) \leftrightarrow V_{xA_4} \tag{5}$$

If point x is selected so that $V_{A_4A_3}$ and V_{xA_4} are the same, in direction, Eq. 5 could be simplified. It is known that $V_{A_4A_3}$ is parallel to the slot of link 3; thus point x should be located so that V_{xA_4} is in the same direction. This can be done if, first, points x and A_4 are considered to be on the same link, link 4, and if, second, the line between points

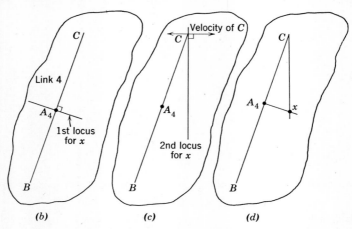

Fig. 5.2 b, c, d. Two loci used to locate auxiliary point.

x and A_4 is perpendicular to $V_{A_4A_3}$. The first locus for point x is thus set, as shown in Fig. 5.2b. Equation 5 may be written

$$V_x = V_{A_3} \leftrightarrow \text{(vector parallel to the slot of link 3)} \tag{6}$$

(f) A second locus for point x may be found from consideration of point x and another point on link 4. Point B or point C can be taken. Let us take point C:

$$V_x = V_C \leftrightarrow V_{xC} \tag{7}$$

If point x is selected so that V_C and V_{xC} have the same direction, a second locus for x can be found. It is necessary that the line x–C should be perpendicular to the motion of point C so that V_C and V_{xC} have the same directions. Figure 5.2c shows the second locus, and Fig. 5.2d shows point x determined. Equation 7 may be written

$$V_x = \text{(vector directed along the motion of point } C\text{)} \tag{8}$$

(g) Simultaneous solution of Eqs. 6 and 8 gives V_x, as shown in Fig. 5.2e.

Stephenson Mechanism

The remainder of the equations for solution are:

$$V_B = V_x \leftrightarrow V_{Bx} \qquad (9)$$

$$V_C = V_B \leftrightarrow V_{CB} \qquad (10)$$

Figure 5.2f shows the complete velocity polygon, with point a_4 determined by proportionality on b–c. A check of the work is that the line a_2–x intersects line b–c at a_4.

A solution, using point y in Fig. 5.2g, is left up to the student.

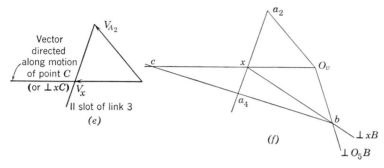

Fig. 5.2. (e) Velocity of auxiliary point determined. (f) Complete velocity polygon.

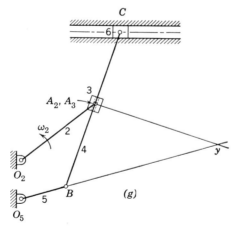

Fig. 5.2g. An alternate auxiliary point.

5.3 Stephenson mechanism

Part of the Stephenson valve gear is shown in Fig. 5.3a; Fig. 5.3b and Fig. 5.3c show the actual arrangement as used on locomotives. The angular velocity of link 2 is specified. It is left up to the student

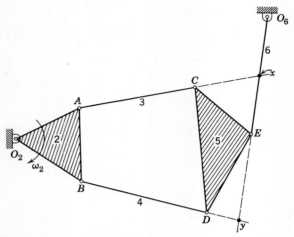

Fig. 5.3a. Schematic arrangement of Stephenson mechanism.

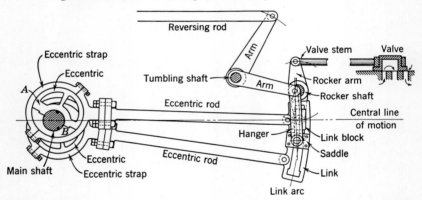

Fig. 5.3b. Stephenson Valve Gear with link block connected directly to rocker arm. Reproduced with permission from *Locomotive Cyclopedia*, Simmons Boardman Publishing Corp.

to write the necessary equations for the velocity polygon solution, using either point x on link 5, or point y on link 5.

5.4 Wanzer needle-bar mechanism

A schematic sketch of the Wanzer needle-bar mechanism is shown in Fig. 5.4. The slider, link 3, moves relative to the slot s of link 2, whereas the slider, link 5, moves relative to the slot t of link 2. Link 2 rotates about a fixed center, O_2. Assuming that the angular velocity of link 2 is specified, draw the velocity polygon, using the auxiliary point x. The solution is left up to the student.

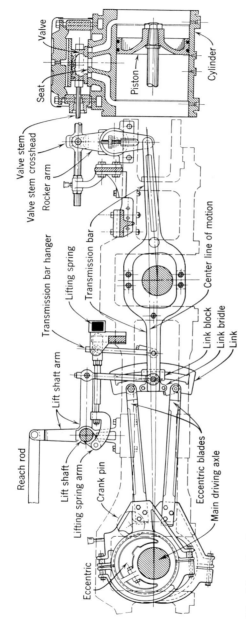

Fig. 5.3c. Stephenson Valve Gear with transmission bar connecting link block to rocker arm. Reproduced with permission from *Locomotive Cyclopedia*, Simmons-Boardman Publishing Corp.

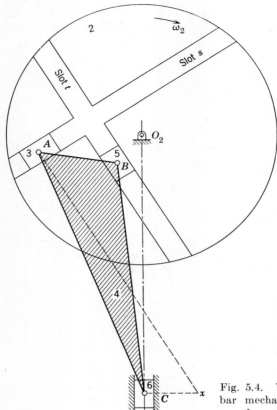

Fig. 5.4. Wanzer needle bar mechanism requiring use of an auxiliary point.

PROBLEMS

5.1 (Fig. 5.1). For the given velocity of point D, find the angular velocity of links 2, 3, and 4, and the velocity of point B.

Draw the complete velocity polygon, and make use of an auxiliary point.

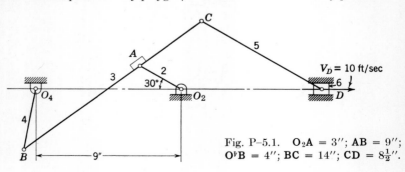

Fig. P-5.1. $O_2A = 3''$; $AB = 9''$; $O_4B = 4''$; $BC = 14''$; $CD = 8\frac{1}{2}''$.

5.2 (Fig. 5.2). For the given velocity of point E, draw the complete velocity polygon and determine the angular velocity of links 3 and 4.

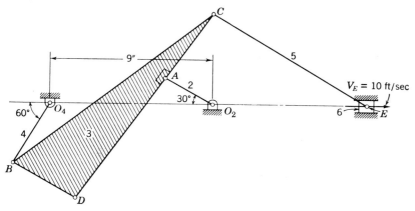

Fig. P–5.2. $O_2A = 3''$; $BC = 14''$; $O_4B = 4''$; $AC = 4\frac{1}{2}''$; $DC = 13''$; $BD = 4''$
$CE = 10''$.

5.3 (Fig. 5.3). Link 2 is rotating at 1200 rpm clockwise. Draw the complete velocity polygon. Determine the velocity of points C, D, and E and the angular velocity of links 3, 4, and 5.

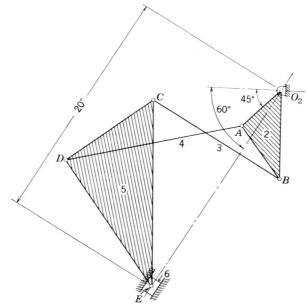

Fig. P–5.3. $O_2A = 4\frac{1}{2}''$; $O_2B = 7\frac{1}{2}''$; $AB = 5\frac{1}{2}''$; $CB = 12\frac{1}{2}''$; $AD = 15''$;
$DE = 13''$; $CE = 16''$.

5.4 (Fig. 5.4). Determine, for the two-cylinder V-engine, the velocity of piston B and the angular velocity of links 2 and 3.

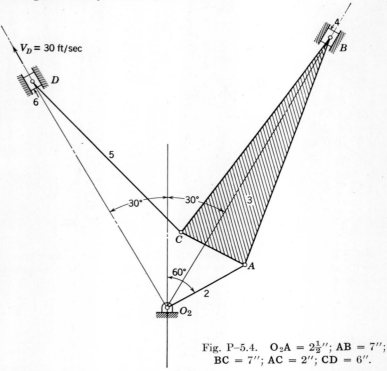

Fig. P–5.4. $O_2A = 2\frac{1}{2}''$; $AB = 7''$; $BC = 7''$; $AC = 2''$; $CD = 6''$.

5.5 (Fig. 5.5). Part of a Walschaert valve gear is shown simplified for purposes of this problem. If the velocity of point A is assumed to be 8 ft/sec, what is the velocity of points C and D, and the angular velocity of links 2, 3, and 4?

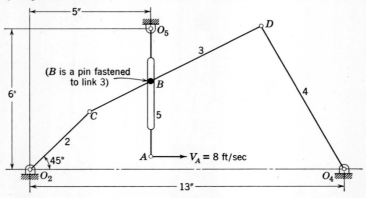

Fig. P–5.5. $O_2C = 3\frac{1}{2}''$; $CD = 8''$; $CB = 3''$; $O_4D = 7''$; $O_5A = 8''$.

5.6 (Fig. 5.6). Link 2 is rotating, for the instant shown, at 10 rad/sec counterclockwise. Find the angular velocity of link 3 and the velocity of point C. Compare the angular velocity of links 2 and 3.

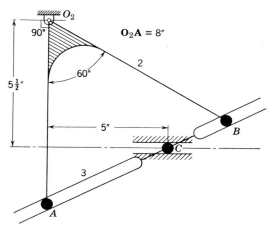

Fig. P–5.6.

CHAPTER 6

Relative Accelerations

In acceleration analysis, the three situations may be encountered that were discussed in velocity analysis: (1) the relation of acceleration of two distinct and separate points, (2) the relation of acceleration of two points on a rigid link, and (3) the relation of acceleration of a point to a body, where the point moves with respect to the body. The form of the equation for the first case is comparable to that for the second case, and inasmuch as application would be made to problems outside the scope of this book, case 1 will not be discussed. This chapter is concerned primarily with the development and interpretation of the relative acceleration equation for two points on a rigid link; Chapter 8 will be devoted to the development and interpretation of the acceleration equation for a point moving with respect to a body.

For a complete picture of analysis and interpretation, a link rotating about a fixed point and a link moving in a plane will be discussed, both analytically and graphically.

6.1 Acceleration of a point on a link rotating about a fixed center with a constant radius. Analytic analysis

A link, as shown in Fig. 6.1a, is rotating about a fixed center, O_2, with an angular velocity ω radians per second, counterclockwise, and is accelerating counterclockwise with an angular acceleration α. The distance between O_2 and B is given by R. Line O_2–B makes an angle of θ with the x-axis. The total acceleration of point B is desired.

As shown in Chapter 3, page 10, the velocity of point B in the x- and y-directions is given by

$$V_B{}^x = -R\omega \sin \theta$$

$$V_B{}^y = R\omega \cos \theta$$

A Point on a Link Rotating about a Fixed Center

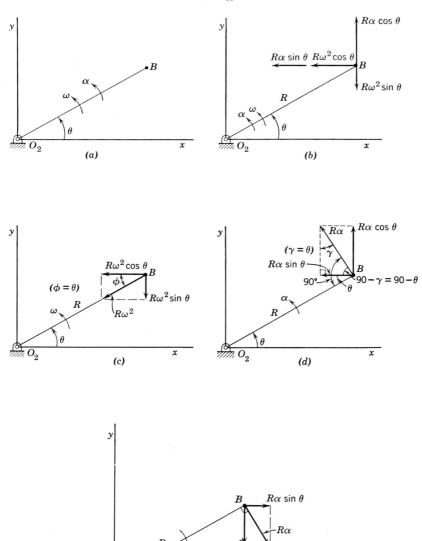

Fig. 6.1. Acceleration of a point on a link rotating about a fixed center.

Relative Accelerations

Differentiation of the above equations with respect to time, remembering that R is constant, gives:

$$\frac{dV_B^x}{dt} = -R\left[\omega(\cos\theta)\frac{d\theta}{dt} + (\sin\theta)\frac{d\omega}{dt}\right] \quad (1)$$

$$\frac{dV_B^y}{dt} = R\left[\omega(-\sin\theta)\frac{d\theta}{dt} + (\cos\theta)\frac{d\omega}{dt}\right] \quad (2)$$

Set

$\dfrac{dV_B^x}{dt} = A_B^x$, the acceleration of point B in the x-direction

$\dfrac{dV_B^y}{dt} = A_B^y$, the acceleration of point B in the y-direction

$\dfrac{d\omega}{dt} = \alpha$, the rate of change of the angular velocity, which is called angular acceleration

Rewrite Eqs. 1 and 2:

$$A_B^x = -R\omega^2\cos\theta - R\alpha\sin\theta \quad (3)$$

$$A_B^y = -R\omega^2\sin\theta + R\alpha\cos\theta \quad (4)$$

Figure 6.1b shows the vectors in position, the plus and minus signs being taken care of by the vector directions. The order of vector addition is immaterial in obtaining the total acceleration of point B. Let us express the total acceleration of point B as

$$A_B = (R\omega^2\cos\theta \mathrel{+\mkern-10mu+} R\omega^2\sin\theta) \mathrel{+\mkern-10mu+} (R\alpha\sin\theta \mathrel{+\mkern-10mu+} R\alpha\cos\theta) \quad (5)$$

The two rectangular components in the first parentheses, shown in Fig. 6.1c, give a resultant equal to $R\omega^2$, which can be shown to be directed from point B to the center of rotation of the link. The two rectangular components in the second parentheses, shown in Fig. 6.1d, give a resultant equal to $R\alpha$, which can be shown to be perpendicular to the line B–O_2, and to correspond in direction to the sense of the direction of the angular acceleration of the link. Figure 6.1e shows the effect of reversing the direction of angular acceleration. Note that $R\omega^2$ is a vector which is a function numerically of the angular velocity, but is independent of the direction of rotation of the link.

Two Points on a Rigid Link. Analytic Analysis

The total acceleration of point B can be expressed, therefore, by

$$A_B = R\omega^2 \nrightarrow R\alpha \tag{6}*$$

where $R\omega^2$ is called the normal or radial acceleration component and $R\alpha$ is called the tangential acceleration component. Note that the angular speed must be expressed as radians per unit time, as radians per second; and the angular acceleration must be expressed as radians per unit time per unit time, as radians per second per second.

Since the components of Eq. 6 are at right angles to each other, A_B might be expressed as

$$A_B = [(R\omega^2)^2 + (R\alpha)^2]^{1/2} \tag{7}$$

However, the form of Eq. 7 is not one which readily lends itself to solution of problems, and will not be used in this book.

6.2 Relative acceleration of two points on a rigid link. Analytic analysis

It has been pointed out that a complete velocity analysis of a mechanism could be made by use of the relative velocity equation. Is there a comparable method for acceleration analysis? Let us investigate the relation of accelerations of two points on a rigid link. Consider a line A–B, in Fig. 6.2a, which is part of a rigid link moving in a plane with any arbitrary motion. A coordinate system of axes will be used to define the location of point B:

$$X_B = X_A + R \cos \theta$$
$$Y_B = Y_A + R \sin \theta$$

As shown in Chapter 3, page 12, the velocity of point B in the x- and y-directions is:

$$V_B{}^x = V_A{}^x - R\omega \sin \theta$$
$$V_B{}^y = V_A{}^y + R\omega \cos \theta$$

---

* While the equation has been developed for a point moving at a constant distance from the center of rotation, the equation is still applicable if R is variable, *provided that R is the radius of curvature at the instant being considered.* A proof of this will not be given, but the reader is referred to *Introduction to the Calculus*, by W. F. Osgood, The Macmillan Co., 1931, page 261, for a discussion of the osculating circle.

Relative Accelerations

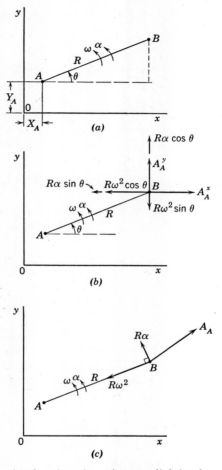

Fig. 6.2. Acceleration of a point on a link in plane motion.

$$A_B = A_A \nrightarrow R\omega^2 \nrightarrow R\alpha$$
$$A_B = A_A \nrightarrow \mathbf{BA}\omega^2 \nrightarrow \mathbf{BA}\alpha$$
$$A_B = A_A \nrightarrow A_{BA}$$
$$A_B = A_A \nrightarrow \frac{V_{BA}^2}{BA} \nrightarrow \mathbf{BA}\alpha$$
$$A_B = A_A \nrightarrow A_{BA}{}^n \nrightarrow A_{BA}{}^t$$

[Identical expressions for the acceleration of point B, where B and A are two points on a rigid link.]

Two Points on a Rigid Link. Analytic Analysis

Differentiation of the above equations with respect to time, remembering that the only constant quantity is R, gives

$$\frac{dV_B{}^x}{dt} = \frac{dV_A{}^x}{dt} - R\left[\omega(\cos\theta)\frac{d\theta}{dt} - (\sin\theta)\frac{d\omega}{dt}\right]$$

$$\frac{dV_B{}^y}{dt} = \frac{dV_A{}^y}{dt} + R\left[\omega(-\sin\theta)\frac{d\theta}{dt} + (\cos\theta)\frac{d\omega}{dt}\right]$$

But $\dfrac{dV_B{}^x}{dt}$ is defined as the acceleration of point B in the x-direction, $A_B{}^x$; $\dfrac{dV_A{}^x}{dt}$ is the acceleration of point A in the x-direction, $A_A{}^x$; similarly, $\dfrac{dV_B{}^y}{dt}$ is $A_B{}^y$; and $\dfrac{dV_A{}^y}{dt}$ is $A_A{}^y$. The remainder of the terms are discussed in the previous section. Rewriting the equations gives

$$A_B{}^x = A_A{}^x - R\omega^2 \cos\theta - R\alpha \sin\theta$$

$$A_B{}^y = A_A{}^y - R\omega^2 \sin\theta + R\alpha \cos\theta$$

The total acceleration of point B, A_B, is determined by the vector addition of the two rectangular components:

$$A_B = A_B{}^x \mathbin{+\mkern-8mu+} A_B{}^y$$

Figure 6.2b shows each vector in position. The order of vector addition is immaterial. Thus, consider adding the vectors as follows:

$$A_B = (A_A{}^x \mathbin{+\mkern-8mu+} A_A{}^y) \mathbin{+\mkern-8mu+} (R\omega^2 \cos\theta \mathbin{+\mkern-8mu+} R\omega^2 \sin\theta)$$
$$\mathbin{+\mkern-8mu+} (R\alpha \cos\theta \mathbin{+\mkern-8mu+} R\alpha \sin\theta)$$

The first quantity is the total acceleration of point A; the second quantity, as shown in the previous section, is equal to $R\omega^2$, a vector directed from B to A; the third quantity, as shown in the previous section, is equal to $R\alpha$, and is perpendicular to the line B–A, the direction of the vector corresponding to the sense of the angular acceleration. The acceleration of point B, therefore, may be expressed by

$$A_B = A_A \mathbin{+\mkern-8mu+} R\omega^2 \mathbin{+\mkern-8mu+} R\alpha$$

However, if point B is considered as rotating about point A as if point A were a fixed point, the acceleration of point B will be equal to $R\omega^2 \mathbin{+\mkern-8mu+} R\alpha$, as shown in the previous section. Therefore, the equation may be written as

$$A_B = A_A \mathbin{+\mkern-8mu+} A_{BA}$$

where A_{BA} is defined as the relative acceleration of point B with respect to point A, or A_{BA} is defined as the acceleration point B would have if point A is considered stationary.

Figure 6.2c shows the vectors in position.

We may express the equation differently by noting that $\omega = \dfrac{V_{BA}}{\mathbf{BA}}$, and that $R\omega^2 = \mathbf{BA}\left(\dfrac{V_{BA}}{\mathbf{BA}}\right)^2 = \dfrac{V_{BA}{}^2}{\mathbf{BA}}$:

$$A_B = A_A \nrightarrow \frac{V_{BA}{}^2}{\mathbf{BA}} \nrightarrow \mathbf{BA}\alpha$$

6.3 Acceleration of a point rotating about a fixed center with a constant radius. Graphical analysis

A graphical development of the equation for the acceleration of a point rotating about a fixed center at a constant radius is presented for a further appreciation of the quantities involved. The next section will be devoted to the determination of the relative acceleration equation for two points on a rigid link moving in a plane.

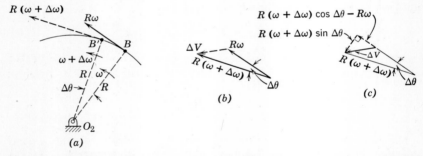

Fig. 6.3. Graphical analysis for the acceleration of a point on a link rotating about a fixed center.

Figure 6.3a shows a point B moving along the circular arc, with constant radius R, to a new position B'. The initial velocity of the point is $R\omega$, and the velocity of the point after a change of angle of $\Delta\theta$ of the radial line is $R(\omega + \Delta\omega)$, where $\Delta\omega$ is the change of angular velocity of the radial line. The change of velocity, as seen in Fig. 6.3b, is the vector difference of the final and initial velocity, which change of velocity is labeled ΔV. We may consider any components of ΔV for the determination of an analytical expression. The components selected here are the two shown in Fig. 6.3c, where one component,

Two Point on a Rigid Link. Graphical Analysis

$[R(\omega + \Delta\omega) \cos \Delta\theta - R\omega]$, is directed along the "$R\omega$" vector, and the other component, $R(\omega + \Delta\omega) \sin \Delta\theta$, is perpendicular to the $R\omega$ vector.

Thus the component of change of velocity in the tangential direction, that is, along the $R\omega$ vector, designated by ΔV^t, is

$$\Delta V^t = R(\omega + \Delta\omega) \cos \Delta\theta - R\omega$$

and the component of change of velocity in the normal or radial direction, that is, perpendicular to the $R\omega$ vector, designated by ΔV^n, is

$$\Delta V^n = R(\omega + \Delta\omega) \sin \Delta\theta$$

Divide through each expression by Δt, and take the limit as Δt approaches zero:

$$\lim_{\Delta t \to 0} \frac{\Delta V^t}{\Delta t} = \lim_{\Delta t \to 0} \frac{R(\omega + \Delta\omega) \cos \Delta\theta}{\Delta t} - \lim_{\Delta t \to 0} \frac{R\omega}{\Delta t}$$

$$\lim_{\Delta t \to 0} \frac{\Delta V^n}{\Delta t} = \lim_{\Delta t \to 0} \frac{R(\omega + \Delta\omega) \sin \Delta\theta}{\Delta t}$$

Recognizing that $\lim_{\Delta t \to 0} \frac{\Delta V^t}{\Delta t} = A^t$, the acceleration in the tangential direction, that $\lim_{\Delta t \to 0} \frac{\Delta V^n}{\Delta t} = A^n$, the acceleration in the normal direction, that in the limit $\cos \Delta\theta = 1$, that in the limit $\sin \Delta\theta = \Delta\theta$, that in the limit $\frac{\Delta\omega}{\Delta t} = \frac{d\omega}{dt} = \alpha$, the angular acceleration, and that differentials of higher order approach zero, we may write

$$A^t = R\alpha$$

$$A^n = R\omega^2$$

Inspection of Fig. 6.3c shows that, as Δt approaches zero, the tangential component, $R\alpha$, coincides with $R\omega$ in direction. It is in the same sense if the angular velocity of the radial line is increasing and is in the opposite sense to $R\omega$ if the angular velocity of the radial line is decreasing. Also, the normal component, $R\omega^2$, is always directed towards the center of rotation regardless of the direction of rotation of the radial line.

6.4 Relative acceleration of two points on a rigid link. Graphical analysis

Figure 6.4a shows a rigid link, represented by A–B, in a given position where the link is rotating counterclockwise with an angular

Relative Accelerations

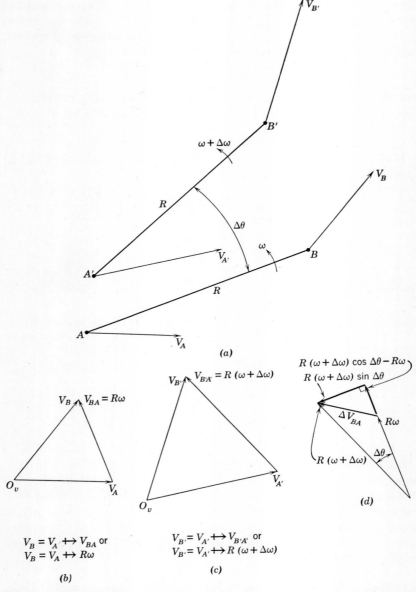

Fig. 6.4. Graphical analysis for the acceleration of a point on a link in plane motion.

Two Point on a Rigid Link. Graphical Analysis

velocity ω. After a period of time, Δt, line A–B moves to a position A'–B', with a change of angle of $\Delta\theta$, and in the new position the line has an angular velocity equal to $(\omega + \Delta\omega)$. Figure 6.4b shows the velocity vector polygon for the equation

$$V_B = V_A \mathbin{+\!\!+} R\omega \qquad (1)$$

and Fig. 6.4c shows the velocity vector polygon for the equation

$$V_{B'} = V_{A'} \mathbin{+\!\!+} R(\omega + \Delta\omega) \qquad (2)$$

Subtract Eq. 1 from Eq. 2:

$$(V_{B'} \to V_B) = (V_{A'} \to V_A) \mathbin{+\!\!+} (R(\omega + \Delta\omega) \to R\omega) \qquad (3)$$

In the above, $V_{B'} \to V_B = \Delta V_B$, the change of velocity of point B; $V_{A'} \to V_A = \Delta V_A$, the change of velocity of point A; $R(\omega + \Delta\omega) \to R\omega = \Delta V_{BA}$, the change of relative velocity.

Substituting the above into Eq. 3, we obtain

$$\Delta V_B = \Delta V_A \mathbin{+\!\!+} \Delta V_{BA}$$

Dividing through by Δt, and taking the limit as Δt approaches zero, we obtain

$$A_B = A_A \mathbin{+\!\!+} A_{BA}$$

The question now is what is the magnitude of A_{BA}. This magnitude may be determined from re-examination of $\lim\limits_{\Delta t \to 0} \dfrac{\Delta V_{BA}}{\Delta t}$, which can be expressed by:

$$\lim_{\Delta t \to 0} \frac{\Delta V_{BA}}{\Delta t} = \lim_{\Delta t \to 0} \frac{R(\omega + \Delta\omega)}{\Delta t} \to \lim_{\Delta t \to 0} \frac{R\omega}{\Delta t}$$

Comparison of Fig. 6.3c and Fig. 6.4d shows that the change of velocity of a point on a link rotating about a fixed center is exactly the same as the change of relative velocity of two points of a link moving in a plane. Therefore, using the results of the preceding section, we may write the following equation:

$$A_{BA} = \lim_{\Delta t \to 0} \frac{\Delta V_{BA}}{\Delta t} = R\omega^2 \mathbin{+\!\!+} R\alpha$$

Summing up, we may express the relation of acceleration of two points on a rigid link by

$$A_B = A_A \mathbin{+\!\!+} A_{BA}$$

or by

$$A_B = A_A \mathbin{+\!\!+} R\omega^2 \mathbin{+\!\!+} R\alpha$$

or by

$$A_B = A_A \nrightarrow A_{BA}{}^n \nrightarrow A_{BA}{}^t$$

where $R\omega^2$, the normal or radial component, is directed from B to A; and $R\alpha$, the tangential component, is in the direction of the relative velocity and is directed in the same sense as the relative velocity if the relative velocity is increasing and is directed in the opposite sense to the relative velocity if the relative velocity is decreasing.

The relative acceleration of point B with respect to point A, A_{BA}, can now be expressed as the acceleration point B would have if point A is considered stationary, as seen from the interpretation of the derived equation.

The interpretation of the relative acceleration equation, applied to various mechanisms, will be considered in detail in Chapter 7.

PROBLEMS

6.1. A link 5 in. long is rotating at 400 rpm clockwise, with one of the ends fixed, and 5 sec later is rotating at 1800 rpm, with constant angular acceleration. What is the acceleration of the midpoint of the link at the time when the link is rotating at 1400 rpm?

6.2 (Fig. 6.2). If the total acceleration of point A is as shown, what are the angular velocity and the angular acceleration of the link for the position shown? What is the direction of angular acceleration? Can the direction of angular velocity be determined?

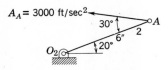

Fig. P–6.2.

6.3 (Fig. 6.3). Five different positions of acceleration vectors are shown, one of which is not possible, for a link rotating about a fixed center. Which one is incorrect? Why?

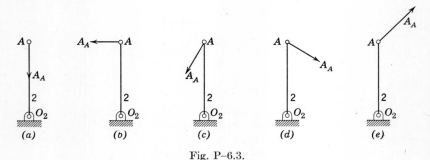

Fig. P–6.3.

Specify the conditions of motion necessary to obtain the acceleration vectors for the other four cases.

6.4 (Fig. 6.4). A link is rotating clockwise at 30 rad/sec and increasing in speed at the rate of 1200 rad/sec². The motion of point B is restricted, as shown. If the acceleration of point A is 800 ft/sec², what is the total acceleration of point B?

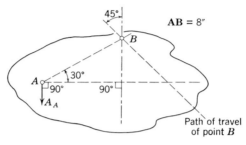

Fig. P–6.4.

6.5 (Fig. 6.5). If the normal acceleration of point B with respect to point A is 400 ft/sec² and if the tangential acceleration of point B with respect to point A is 800 ft/sec², what are the angular speed and the angular acceleration of the link?

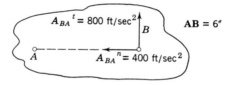

Fig. P–6.5. $AB = 6''$.

6.6 (Fig. 6.6). If the total acceleration of points A and B are known and are as shown, what are the angular speed and angular acceleration of the link? Consider in two ways: the relation of the acceleration of point B with respect to point A, and the relation of the acceleration of point A with respect to point B.

Determine also the acceleration of point C.

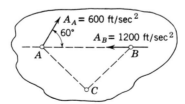

Fig. P–6.6. $AB = 5''$; $AC = 4''$; $BC = 3''$.

6.7 (Fig. 6.7). Nine different combinations of absolute accelerations of two points on a rigid link are shown. Some are possible values and directions; the rest are impossible values and directions. Which are impossible? Determine the angular acceleration for each possible case.

70　　　　　　　　　　Relative Accelerations

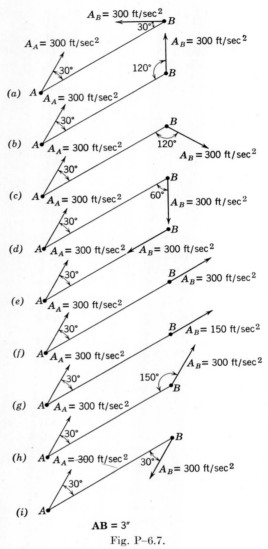

Fig. P-6.7.

6.8. It was stated in problem 3.8 that the components of velocity along the line joining two points on a rigid link must be equal. This is true since the two points cannot separate if the body is rigid. Can the same reasoning be applied to acceleration analysis to state that the components of accelerations along the line joining two points on a rigid link must be equal? What factor is present in acceleration analysis that is not present in velocity analysis to preclude such reasoning? What special type of motion would permit the components of acceleration along the line joining two points on a rigid link to be equal?

CHAPTER 7

Application of the Relative Acceleration Equation for Two Points on a Rigid Link

The relative acceleration equation derived for the relation of acceleration of two points on a rigid link in the preceding chapter will now be applied to various mechanisms, such as the slider-crank mechanism, four-link mechanism, and mechanisms using a combination of the preceding two. The types of problems that can be handled are restricted, at this point, because of the limitation set up in the derivation of the equation that the relation of accelerations of two points on a rigid link be considered.

7.1 Slider-crank mechanism

The slider-crank mechanism analyzed for velocities in Chapter 4 will be taken as an illustrative problem for acceleration analysis. The mechanism is shown in Fig. 7.1a, where link 2 is assumed to be rotating clockwise with a constant angular velocity, and where the mechanism is drawn to scale in the position for which the analysis is to be made. Figure 7.1b shows the velocity polygon.

For the first step, isolate link 2, as shown in Fig. 7.1c. The acceleration of point A, since point A is rotating about a fixed center, is given by:

$$A_A = R\omega_2^2 \mathrel{+\mkern-10mu+} R\alpha_2$$

Since R and ω_2 are known, the normal acceleration, $R\omega_2^2$, may be calculated. The direction of $R\omega_2^2$ is along the line A–O_2 and is directed *from A to O_2*. Since α_2 is zero because ω_2 is constant, $R\alpha_2 = 0$. A_A is laid off to an arbitrary acceleration scale in Fig. 7.1d.

Consider, next, link 3 as isolated. The acceleration of point A may be related to the acceleration of point B, or the acceleration of point B may be related to that of point A, with the same final results.

Application of the Relative Acceleration Equation

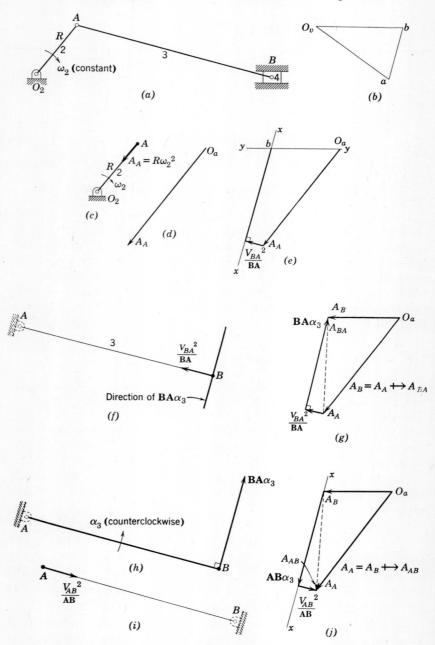

Fig. 7.1. Acceleration analysis of a slider-crank mechanism.

Slider-Crank Mechanism

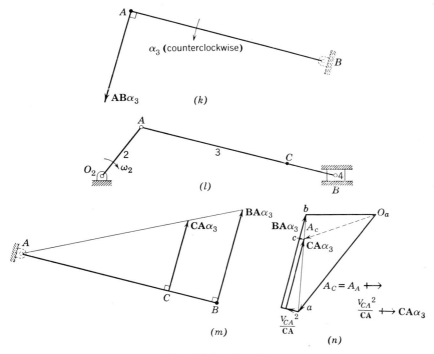

Fig. 7.1 (continued).

Let us consider the latter, and express the relation of accelerations by any of these equations:

$$A_B = A_A \mapsto A_{BA}$$

or
$$A_B = A_A \mapsto A_{BA}{}^n \mapsto A_{BA}{}^t$$

or
$$A_B = A_A \mapsto \mathbf{BA}\omega_3{}^2 \mapsto \mathbf{BA}\alpha_3$$

or
$$A_B = A_A \mapsto \frac{V_{BA}{}^2}{\mathbf{BA}} \mapsto \mathbf{BA}\alpha_3$$

All the forms are synonymous, with the same interpretation for each. Let us use the last form, since the relative velocity, V_{BA}, can be obtained directly from the velocity polygon and will eliminate the necessity of calculating the angular velocity of link 3. The important step now is the interpretation of each term in the equation:

(a) A_B is known in direction, inasmuch as point B is moving with pure translation, and it can have an acceleration only in the direction of motion. The magnitude of A_B is not known.

(b) $V_{BA}{}^2/\mathbf{BA}$ can be determined completely, both in magnitude and in direction. V_{BA} can be determined from the velocity polygon, $\mathbf{BA}$ is known, and the normal component of acceleration is directed *from B to A* since the acceleration of B with respect to A is being determined.

(c) $\mathbf{BA}\alpha_3$ is known to be perpendicular to the line from B to A, but its magnitude is unknown.

Therefore, there are two unknowns: magnitude of A_B and magnitude of $\mathbf{BA}\alpha_3$, which can be found from the solution of a vector polygon. A vector polygon may be drawn with the vectors added in any fashion as long as the vector equation is satisfied. However, all vectors representing total accelerations will be drawn from a common point, called the pole of the acceleration polygon, for simplification later on in determining accelerations of points on a link.

Start the acceleration diagram for link 3 at O_a, as shown in Fig. 7.1e, by drawing A_A to scale and in the proper direction. As the equation states, add $V_{BA}{}^2/\mathbf{BA}$ to A_A. (Note the direction of the normal component of acceleration. Figure 7.1f shows point B as if it were rotating about point A, as if A were a fixed point. The normal component of acceleration of point B with respect to point A is directed *from* point B *to* point A.)

The next component, $\mathbf{BA}\alpha_3$, is known in direction and may be indicated to lie along line x-x in Fig. 7.1e.

This is as far as we may go in working the right side of the equation. However, we know that the acceleration of point B must start at O_a and end somewhere on the line x-x. We also know that the resultant of the three components must be a vector whose direction is known, that is, in the direction of motion of the piston. The resultant direction is shown by line y-y in Fig. 7.1e. The only point which will satisfy the vector equation in all respects is given by the intersection of x-x and y-y, that is, point b.

Figure 7.1e is redrawn in Fig. 7.1g, with each component labeled. Either figure will give all the information necessary for a complete picture of accelerations in the mechanism. A_B may be scaled off directly. The magnitude and direction of the angular acceleration of link 3 may be determined quickly. If the tangential acceleration of point B with respect to point A is placed on link 3, isolated as shown in Fig. 7.1h, we shall note that the direction of the angular acceleration is counterclockwise. Since the product of $\mathbf{BA}$ and α_3, that is $\mathbf{BA}\alpha_3$, can be scaled off from the acceleration diagram, and since $\mathbf{BA}$ is known, α_3 can be determined from

$$\alpha_3 = \frac{(\mathbf{BA}\alpha_3)}{\mathbf{BA}}$$

Slider-Crank Mechanism

If the units used are feet, seconds, the angular acceleration would be expressed as radians per second per second, written as rad/sec². If the units used are inches, seconds, the angular acceleration would be expressed as in./sec² (inches per second per second).

Attention is called to the fact that, if A_A is expressed as a function of A_B,

$$A_A = A_B \nrightarrow A_{AB}$$

or

$$A_A = A_B \nrightarrow \frac{V_{AB}^2}{AB} \nrightarrow \mathbf{AB}\alpha_3$$

the same results would be obtained for magnitudes and directions with proper interpretation of the equation. Note that the acceleration of point A with respect to point B means that point B is now considered stationary, as shown in Fig. 7.1i, where $V_{AB}^2/\mathbf{AB}$ is directed from A to B. The acceleration polygon of Fig. 7.1j is obtained in these steps:

(a) Draw the direction of A_B, which is known, from O_a in Fig. 7.1j.
(b) Draw A_A.
(c) Lay off $V_{AB}^2/\mathbf{AB}$ so that it will be the last of three components to be added to give A_A.
(d) The tangential component is perpendicular to the line A–B and may be drawn from the beginning of the normal component vector, as represented by line x–x in Fig. 7.1j.

Note that the polygon found in Fig. 7.1g is obtained. The only vectors which are reversed are those of the relative acceleration components. However, putting in the tangential component on link 3 isolated in Fig. 7.1k shows that the angular acceleration of link 3 is counterclockwise, as determined previously, with the same magnitude.

It will be shown now that the acceleration of any point on link 3 may be found in the acceleration polygon. Let us assume that the acceleration of point C, three quarters of the distance from A to B, as shown in Fig. 7.1l, is desired. Using the following,

$$A_C = A_A \nrightarrow \frac{V_{CA}^2}{CA} \nrightarrow \mathbf{CA}\alpha_3$$

we cannot proceed directly since there are three unknowns: direction and magnitude of A_C and magnitude of $\mathbf{CA}\alpha_3$. There are two methods of approach:

(a) Calculate α_3, and then in turn calculate $\mathbf{CA}\alpha_3$.

76 Application of the Relative Acceleration Equation

(b) Recognize that $\dfrac{CA\alpha_3}{BA\alpha_3} = \dfrac{CA}{BA}$, a proportionality which may be obtained from similar triangles, as shown in Fig. 7.1m.

Also, the normal acceleration of point C with respect to point A is proportional to the normal acceleration of point B with respect to point A. Thus, in Fig. 7.1n, the acceleration of point C will start at the pole and end on the vector a–b so that the relative acceleration of point C with respect to point A is proportional to the relative acceleration of point B with respect to point A. It can be shown that the acceleration of any point on line A–B will be directed from the acceleration pole to a corresponding point on the relative acceleration vector in the acceleration polygon, as a–b. For this reason, we have a line in the acceleration polygon which is the "image" of the corresponding line of a link. The "image" is, however, nothing more than the relative acceleration of one point on the link with respect to another point on the same link.

7.2 Four-link mechanism

The four-link mechanism of Fig. 7.2a is selected for further illustration of the principles in determining accelerations. It is assumed that the mechanism is drawn to scale in the position for which the analysis is to be made, that link 2 is rotating with an angular velocity of ω_2 rad/sec counterclockwise at the instant and is decreasing in speed with an angular acceleration of α_2 rad/sec^2. Or, the angular acceleration is clockwise. The velocity polygon is shown in Fig. 7.2b.

The acceleration of point A is:

$$A_A = O_2A\omega_2^2 \nrightarrow O_2A\alpha_2$$

These vectors are drawn in Fig. 7.2c.

To determine the acceleration of point B, express the relation between B and A by

$$A_B = A_A \nrightarrow \frac{V_{BA}^2}{BA} \nrightarrow BA\alpha_3$$

Each quantity is interpreted in these steps:

(a) A_B: neither direction nor magnitude is known.
(b) A_A: this quantity is known completely.
(c) $\dfrac{V_{BA}^2}{BA}$: this quantity is known completely. Its magnitude can be calculated since V_{BA} can be determined from the velocity diagram

and since **BA** is known. It is directed from B to A, since the relative acceleration of B with respect to A is specified.

(d) **BA**α_3: direction only is known. The tangential acceleration of B with respect to A is perpendicular to the line between B and A.

Therefore, there are three unknowns in the vector equation above: direction and magnitude of A_B; magnitude of **BA**α_3. It is necessary to obtain another condition in order to be able to solve the equation. To do this, isolate link 4, as shown in Fig. 7.2d. Since B is rotating about a fixed point, O_4, the acceleration of B may be expressed by

$$A_B = \mathbf{BO}_4 \omega_4{}^2 \looparrowright \mathbf{BO}_4 \alpha_4$$

or by the following, since $\omega_4 = V_B/\mathbf{BO}_4$:

$$A_B = \frac{V_B{}^2}{\mathbf{BO}_4} \looparrowright \mathbf{BO}_4 \alpha_4$$

Therefore, the following equations may be written by substituting the value of A_B from the equation above into the equation for $A_B = A_A \looparrowright A_{BA}$:

$$\frac{V_B{}^2}{\mathbf{BO}_4} \looparrowright \mathbf{BO}_4 \alpha_4 = A_A \looparrowright \frac{V_{BA}{}^2}{\mathbf{BA}} \looparrowright \mathbf{BA} \alpha_3$$

By further investigation of the quantities in the equation above we obtain the following:

(e) $V_B{}^2/\mathbf{BO}_4$: this quantity is known completely, since V_B can be found from the velocity diagram and since $\mathbf{BO}_4$ is known. The vector is directed from B to O_4, as shown in Fig. 7.2d.

(f) $\mathbf{BO}_4 \alpha_4$: direction is known in that it must be perpendicular to link 4. The magnitude is not known. Thus an equation is derived wherein there are only two unknowns, magnitude of the tangential acceleration of B with respect to A and the magnitude of the tangential acceleration of B with respect to O_4.

The solution of the vector equation is shown in Fig. 7.2e. The procedure of construction is

(a) Draw A_A from the pole, O_a.
(b) Draw $V_{BA}{}^2/\mathbf{BA}$.
(c) Draw x–x perpendicular to the line B–A. A_B must start from O_a and end somewhere along x–x.
(d) Draw $V_B{}^2/\mathbf{BO}_4$ from the pole, O_a.
(e) Draw a line y–y perpendicular to line B–O_4. A_B must start at O_a and end somewhere along y–y.

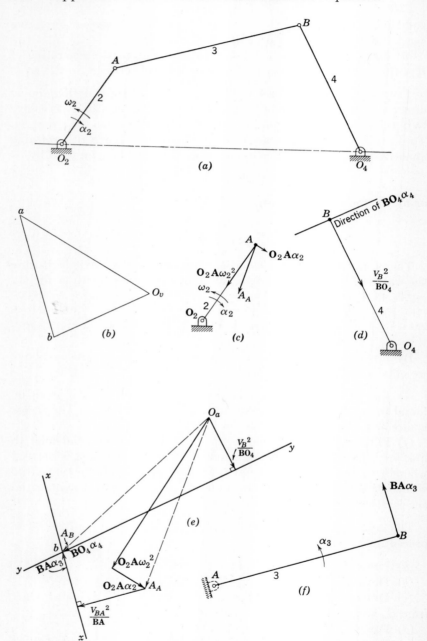

Fig. 7.2. Acceleration analysis of a four-link mechanism.

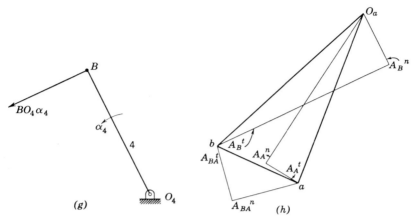

Fig. 7.2 (continued).

The only point which will satisfy all the conditions is point b. A_B is therefore determined by the line from O_a to b, as shown.

The angular accelerations of links 3 and 4 are now easily determined both in direction and in magnitude. The magnitudes are determined by scaling off $\mathbf{BA}\alpha_3$ and $\mathbf{BO}_4\alpha_4$, and calculating the angular accelerations by

$$\alpha_3 = \frac{(\mathbf{BA}\alpha_3)}{\mathbf{BA}}$$

$$\alpha_4 = \frac{(\mathbf{BO}_4\alpha_4)}{\mathbf{BO}_4}$$

The directions of angular acceleration of link 3 is counterclockwise as shown by isolating link 3 in Fig. 7.2f and indicating the proper direction of the tangential component of acceleration of B with respect to A.

The direction of angular acceleration of link 4 is counterclockwise as shown by isolating link 4 in Fig. 7.2g and indicating the tangential component of acceleration of B with respect to O_4.

Figure 7.2h shows the final acceleration diagram in simplified form. It is to be noted that a–b represents the image of line A–B, and the acceleration of any point on line A–B will start from the pole, O_a, and end at the corresponding point on a–b.

7.3 Powell engine

The mechanism selected using a slider-crank and four-link combination is the Powell engine of Fig. 7.3a. Link 2 is assumed to be rotating

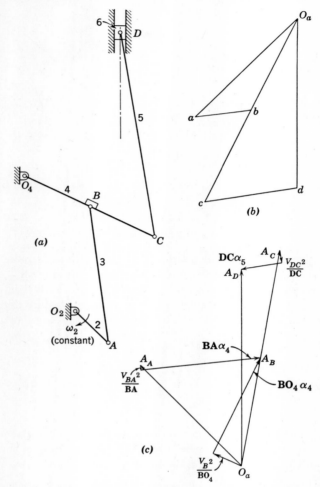

Fig. 7.3. Velocity and acceleration solution of the Powell engine.

at a constant angular velocity, ω_2, clockwise. The velocity polygon is shown in Fig. 7.3b. The acceleration polygon is shown in Fig. 7.3c. Show that the following equations will give the solution:

$$A_A = \mathbf{O_2A}\omega_2{}^2 \tag{1}$$

$$\left.\begin{array}{l} A_B = A_A \leftrightarrow \dfrac{V_{BA}{}^2}{\mathbf{BA}} \leftrightarrow \mathbf{BA}\alpha_3 \\[2ex] A_B = \dfrac{V_B{}^2}{\mathbf{BO_4}} \leftrightarrow \mathbf{BO_4}\alpha_4 \end{array}\right\} \tag{2}$$

Jaw Crusher Mechanism

$$\frac{A_C}{A_B} = \frac{CO_4}{BO_4} \tag{3}$$

$$A_D = A_C \nrightarrow \frac{V_{DC}{}^2}{DC} \nrightarrow DC\alpha_5 \tag{4}$$

Making use of isolated links, show that the angular acceleration of link 3 is clockwise, the angular acceleration of link 4 is counterclockwise, and that the angular acceleration of link 5 is counterclockwise.

7.4 Jaw crusher mechanism

Another mechanism to be analyzed is the jaw crusher mechanism. Figure 7.4a is a drawing of the mechanism, and Fig. 7.4b shows a schematic arrangement, for which the analysis will be made. The mechanism was drawn originally to a scale of 1 in. = 2 feet. Link 2 is rotating at a constant angular velocity of 500 rpm counterclockwise. The velocity diagram shown in Fig. 7.4c was drawn originally to a scale of 1 in. = 20 ft/sec. The scale for the acceleration diagram was originally 1 in. = 600 ft/sec². It is desired to obtain the acceleration of G_6, the center of gravity of link 6. The solution is as follows:

(a) $\quad A_A = O_2 A \omega_2{}^2 = (0.75)\left[\frac{500(2\pi)}{60}\right]^2 = 2060 \text{ ft/sec}^2.$

(b) $\quad A_B = A_A \nrightarrow \frac{V_{BA}{}^2}{BA} \nrightarrow BA\alpha_3$

$$A_B = \frac{V_B{}^2}{BO_4} \nrightarrow BO_4 \alpha_4$$

where $\quad \dfrac{V_{BA}{}^2}{BA} = \dfrac{(21.5)^2}{3.5} = 132 \text{ ft/sec}^2$

and $\quad \dfrac{V_B{}^2}{BO_4} = \dfrac{(42.0)^2}{2.0} = 882 \text{ ft/sec}^2.$

(c) There are several ways by which the acceleration of point C may be found. Two methods will be described:

(1) Express the relation of $A_C = A_A \nrightarrow \dfrac{V_{CA}{}^2}{CA} \nrightarrow CA\alpha_3$, where A_A has been found, $V_{CA}{}^2/CA$ can be computed, and $CA\alpha_3$ can be found in magnitude from $\dfrac{CA\alpha_3}{BA\alpha_3} = \dfrac{CA}{BA}.$

82   Application of the Relative Acceleration Equation

(2) Recognize that the relative acceleration of point C with respect to point A is proportional to the acceleration of point B with respect to point A, and also that the relative acceleration of point C with respect to point B is proportional to the acceleration of point A with respect to point B. Therefore, the figure c–a–b in

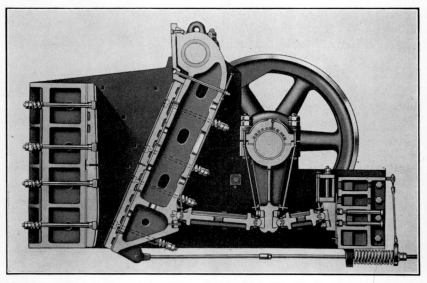

Fig. 7.4a. Jaw crusher mechanism. Courtesy of Allis-Chalmers A-C Industrial Press, Milwaukee, Wisconsin.

Fig. 7.4d in the acceleration polygon is similar to the figure C–A–B in the original mechanism. The similarity of figures was used in the solution.

(d) $$A_D = A_C \nrightarrow \frac{V_{DC}^2}{DC} \nrightarrow DC\alpha_5$$

$$A_D = \frac{V_D^2}{DO_6} \nrightarrow DO_6\alpha_6$$

where $\dfrac{V_{DC}^2}{DC} = \dfrac{(50.0)^2}{2.0} = 1225 \text{ ft/sec}^2$

and $\dfrac{V_D^2}{DO_6} = \dfrac{(36.0)^2}{2.5} = 518 \text{ ft/sec}^2.$

(e) The acceleration of point E and G_6 is found from similar triangles, the similar triangles being shown in Fig. 7.4e. Note that in

Special Positions 83

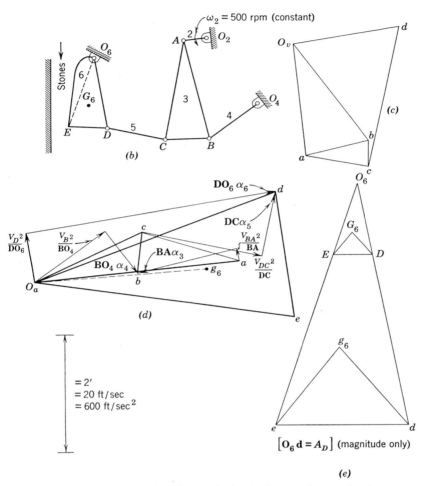

Fig. 7.4b, c, d, e. Acceleration analysis of a jaw crusher mechanism.

the complete acceleration polygon of Fig. 7.4d the true directions of the accelerations are given, the construction in Fig. 7.4e being used to determine magnitudes only.

7.5 Special positions

One mechanism discussed under special positions in velocity analysis will be presented for acceleration analysis. The mechanism is shown in Fig. 7.5a, where link 2 is assumed to be rotating at a constant angular velocity of ω_2 rad/sec clockwise. Figure 7.5b shows the velocity solution, and Fig. 7.5c shows the acceleration solution. It is left up to the student to isolate the links, write the necessary equations for solution, and verify the results. Show that the angular acceleration of link 4 and of link 5 is each zero for the position shown.

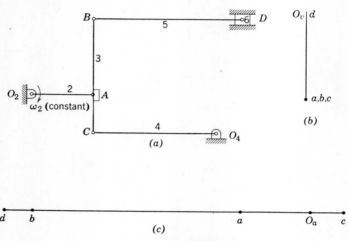

Fig. 7.5. Acceleration solution for a special position.

7.6 Completely graphical solution

The analysis presented so far allows one to determine completely the acceleration polygons for mechanisms where the relative acceleration equation can be applied. The only calculations necessary are for the normal components, the tangential component being determined by the vector solution of the equations. The purpose of this section is to present a method whereby the normal component of acceleration may be determined graphically, eliminating the numerical calculation.

The expression for the normal component of acceleration of B with respect to A, two points on a rigid link, is

$$A_{BA}{}^n = \frac{V_{BA}{}^2}{\mathbf{BA}}$$

which may be rewritten as $\dfrac{A_{BA}{}^n}{V_{BA}} = \dfrac{V_{BA}}{\mathbf{BA}}.$

The equation above represents proportionality of quantities, which may be obtained by similar triangles. Figure 7.6a shows a link BA, with V_{BA} drawn at B perpendicular to the line BA, and the angle ACD made 90°. The intercept BD is the *magnitude* of $A_{BA}{}^n$ (the direction must be determined separately), if the quantities have been drawn to the proper scales.

A certain relation of scales must be used if the construction is to give the proper relations of quantities, as may be seen from the following analysis:

Figure 7.6b shows a construction similar to Fig. 7.6a, except that x, y, and z are indicated as the true distances of the lengths shown.

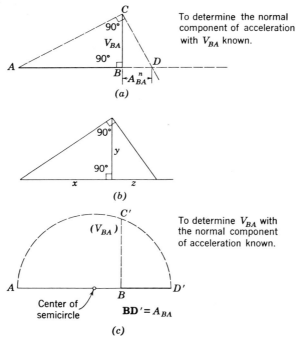

Fig. 7.6. Normal component of acceleration determined graphically.

It is known from geometry that

$$\frac{x}{y} = \frac{y}{z} \qquad (1)$$

The scales are represented by the following symbols:

k_s = space scale (1 in. = k_s ft, that is, if 1 in. = 3 in., for instance, 1 in. = $\frac{3}{12}$ ft, or 1 in. = $\frac{1}{4}$ ft; or $k_s = \frac{1}{4}$).
k_v = velocity scale (1 in. = k_v ft/sec, that is, if 1 in. = 200 ft/sec, for instance, k_v = 200).
k_a = acceleration scale (1 in. = k_a ft/sec², that is, if 1 in. = 4000 ft/sec², for instance, k_a = 4000).

If the distance x is proportional to the length of line B–A according to the space scale k_s, the true length of line B–A is xk_s.

Also, if the distance y is proportional to V_{BA} according to the velocity scale k_v, the actual magnitude of V_{BA} is yk_v.

Finally, if the distance z is considered to be proportional to $A_{BA}{}^n$ according to the acceleration scale k_a, the actual magnitude of $A_{BA}{}^n$ is zk_a.

Thus
$$x = \frac{\mathbf{BA}}{k_s}$$

$$y = \frac{V_{BA}}{k_v}$$

$$z = \frac{A_{BA}{}^n}{k_a}$$

Substitution of the equations above into Eq. 1 gives

$$\frac{\mathbf{BA}/k_s}{V_{BA}/k_v} = \frac{V_{BA}/k_v}{A_{BA}{}^n/k_a}$$

Clearing terms, we obtain

$$A_{BA}{}^n = \left(\frac{V_{BA}{}^2}{\mathbf{BA}}\right)\left(\frac{k_s k_a}{k_v{}^2}\right)$$

Thus $k_s k_a / k_v{}^2$ must be equal to 1 to avoid introduction of an additional scale factor.

Or, finally, the relation of the scales must be such as to satisfy $k_s k_a = k_v{}^2$ for the method of similar triangles to give the correct quantity for the normal acceleration. It is to be noticed that any two scales may be chosen arbitrarily, but the third must be determined from the equation.

There are two types of problems encountered:

(a) to determine the normal component of acceleration graphically, knowing the relative velocity. The construction for this type is shown already in Fig. 7.6a.

(b) to determine the relative velocity when the normal component of acceleration is known. The construction for this is shown in Fig. 7.6c, where $A_{BA}{}^n$ is set off to the proper scale along the link, as shown by BD'. A semicircle is drawn with AD' as diameter. The intercept, BC'', on the line perpendicular to AD' drawn through B, is the relative velocity, V_{BA}. Note that the angle $AC''D'$ is 90°. The relations valid for Fig. 7.6a are valid for Fig. 7.6c, also.

Completely Graphical Solution

Example. The mechanism chosen for illustration of the completely graphical method is the Atkinson engine, shown in Fig. 7.7a. The crank, link 2, is rotating at an angular speed of 67 rad/sec clockwise. The angular acceleration of the crank is 1200 rad/sec² counterclockwise.

The space scale is 3 in. = 12 in. Thus, 1 in. = 4 in., or 1 in. = $\frac{1}{3}$ ft, or $k_s = \frac{1}{3}$.

The velocity or acceleration scale may be chosen arbitrarily. The acceleration scale is selected, arbitrarily, as 1 in. = 600 ft/sec², or $k_a = 600$. The velocity scale is found from

$$k_v^2 = k_s k_a$$
$$= (\tfrac{1}{3})(600)$$
$$k_v = 14.1, \text{ or } 1 \text{ inch} = 14.1 \text{ ft/sec}.$$

The first step involves, of necessity, a slide rule calculation. Either the normal component of acceleration of point A or the velocity of point A has to be determined, so that the other quantity may be determined graphically. Here, $A_A{}^n$ is found. Also, it is necessary to calculate $A_A{}^t$.

Therefore, for the case where there is an angular acceleration of the first link analyzed, two calculations are necessary in the completely graphical method. If the first link analyzed does not have an angular acceleration, only one calculation is necessary:

$$A_A{}^n = O_2A\omega_2{}^2$$
$$= \left(\frac{4.8}{12}\right)(67)^2$$
$$= 1800 \text{ ft/sec}^2$$
$$A_A{}^t = O_2A\alpha_2$$
$$= \left(\frac{4.8}{12}\right)(1200)$$
$$= 480 \text{ ft/sec}^2$$

The acceleration diagram is started at O_a in Fig. 7.7c. A_A is determined. To determine the remainder of the acceleration diagram the normal components of relative acceleration must be determined, and, to determine these, it is necessary to obtain the velocity diagram. The proper scale for the velocity diagram has been determined already, 1 in. = 14.1 ft/sec. Thus it is possible to calculate V_A, and draw the

88 Application of the Relative Acceleration Equation

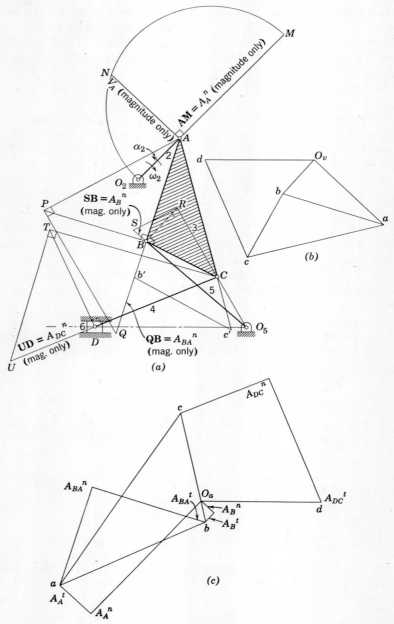

Fig. 7.7. Acceleration analysis for the Atkinson engine. Normal components of acceleration determined graphically.

velocity polygon to the velocity scale. Instead of doing that, however, it is easier to determine the correct vector length of V_A graphically by the method already discussed: locate $A_A{}^n$ along link 2, as shown in Fig. 7.7a, draw the semicircle with O_2M as diameter, erect the perpendicular to O_2A at A to obtain point N. **NA** is then V_A to the scale of $1'' = 14.1$ ft/sec. The complete velocity diagram may be drawn, as shown in Fig. 7.7b, when V_A has been determined.

$A_{BA}{}^n$ may be determined by transferring V_{BA} to Fig. 7.7a, drawing AP, and then making angle $APQ = 90°$. **QB** is then equal to $A_{BA}{}^n$, to scale. **QB** is then transferred to the acceleration diagram. Note that the direction of $A_{BA}{}^n$ is from B to A.

Next, BR is set off equal to V_B, and **SB**, which is $A_B{}^n$, is found. A_B may be found, as shown in Fig. 7.7c, by the simultaneous solution of two vector equations.

Point c in the acceleration diagram is found by similar triangles (triangles A–B–C and A–b'–c' in Fig. 7.7a are similar, with Ab' equal to A_{BA} in magnitude).

$A_{DC}{}^n$ is found by obtaining **UD**. A_D may be found.

Figure 7.7c shows the complete acceleration polygon.

PROBLEMS

7.1. The data given in problem 4.1 are to be used for this problem. Determine the acceleration of the piston and of point C, for the position shown. Draw the acceleration polygon.

What is the angular acceleration of the connecting rod, in magnitude and direction?

Does the direction of rotation of the crank affect the acceleration polygon in any way?

7.2. Refer to the figure of problem 4.3. If the velocity of point A is constant, determine the acceleration of points B, C, and D and the angular acceleration of links 3, 4, and 5.

7.3. Refer to the figure of problem 4.3. If point A is increasing in speed at the rate of 75 ft/sec^2, find the acceleration of points B, C, and D and the angular acceleration of links 3 and 5.

7.4. Refer to the figure of problem 4.3. What acceleration must point A have so that the acceleration of point B is zero?

7.5. Refer to the figure of problem 4.4 for the data. Determine the acceleration of point B for the position shown. What is the angular acceleration of links 3 and 4?

7.6. Refer to the figure of problem 4.5. For the position shown and for the given angular velocity of link 2, the angular velocity being constant, determine the acceleration of links 4 and 6, and the angular acceleration of links 3 and 5.

7.7. Refer to the figure of problem 4.5. If, at the instant shown, link 2 is slowing down at the rate of 1000 rad/sec^2, what is the acceleration of links 4 and 6, and what is the angular acceleration of links 3 and 5?

7.8. Refer to the figure for problem 4.7. Determine the acceleration of points

B and C for the position shown, and determine the angular acceleration of links 3, 4, and 5. Assume link 2 is rotating at a constant angular velocity.

7.9. Refer to the figure for problem 4.7. If, at the instant shown, link 2 is increasing in speed at the rate of 4800 rad/sec^2, determine the acceleration of points B and C, and the angular acceleration of links 4, 5, and 6.

7.10. Refer to the figure for problem 4.8. If link 2 is rotating at a constant angular speed, determine the acceleration of points B, C, and D, and the angular acceleration of links 3, 4, and 5.

7.11. Refer to the figure for problem 4.9. If the relative velocity of A with respect to B is 10 ft/sec, downwards, and if the relative velocity is constant, determine the acceleration of points A, B, and C, and determine the angular acceleration of links 4 and 5.

7.12. Refer to the figure for problem 4.9. What must be the relative acceleration of point A with respect to point B if link 6 is to move with zero acceleration for the position shown? What is the acceleration of points A and B, and the angular acceleration of links 4 and 5.

7.13 (Fig. 7.13). A method available to determine the radius of curvature of a path traced by a moving point on a link uses acceleration analysis. This problem

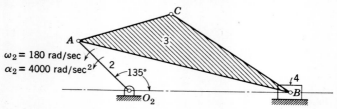

Fig. P-7.13. $O_2A = 3''$; $AB = 9''$; $AC = 4''$; $BC = 6''$.

will deal with the determination of the radius of curvature of the path traced by point C in the figure. Determine:

(1) The velocity and acceleration of point C.

(2) Calculate the radius of curvature, R_c, of the path traced by point C for the instant shown, making use of the relation

$$A_c{}^n = V_c{}^2/R_c$$

where $A_c{}^n$ is the absolute component of acceleration of point C *perpendicular to the velocity* of point C.

(3) Locate the center of curvature, O_c.

(4) To check the location of the center of curvature, plot a portion of the path of C each way from the position shown. Draw an arc, with O_c as center, with a radius equal to the calculated value, R_c. (Note that the radius of curvature for point C changes as point C changes position.)

7.14. Using the method of problem 7.13, find the radius of curvature of the path of the midpoint of link 3 of the figure of problem 4.7, using the dimensions and angular velocity of link 2 given.

7.15 (Fig. 7.15). A cylinder rolls on a flat surface. Show that the x- and y-coordinates of point P, as point P moves from P to P', may be given by

$$x = R\theta - R\sin\theta$$
$$y = R - R\cos\theta$$

Problems

Determine the general equations for the components of velocity in the x- and y-directions as a function of θ. Show that the velocity of point P in its initial position where it is in contact with the horizontal surface is zero, ($\theta = 0$).

Determine the general equations for the components of acceleration of point P in the x- and y-directions and show that the acceleration of point P when $\theta = 0$ is $R\omega^2$ directed from point P to the center of the cylinder, irrespective of the angular acceleration of the cylinder.

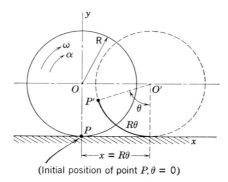

(Initial position of point P, $\theta = 0$)

Fig. P–7.15.

Show that the velocity of the center of the wheel is $R\omega$, where ω is the angular velocity of the cylinder. Show, also, that the acceleration of the center of the wheel is $R\alpha$, where α is the angular acceleration of the cylinder.

Using the relative acceleration equation, show that the acceleration of point P is $R\omega^2$, directed from point P to the center of the cylinder.

7.16. An automobile traveling at 44 ft/sec is increasing in speed at the rate of 15 ft/sec². What are the angular speed and the angular acceleration of the wheels if the tires are 28 in. in diameter? What are the velocity and the acceleration of the point of the tire in contact with the ground, and what are the velocity and the acceleration of the top of the tire?

7.17 (Fig. 7.17). Link 4 is a gear which rolls on a fixed rack. For the given angular velocity of link 2, determine the acceleration of point C, the center of the gear, and the angular acceleration of the gear.

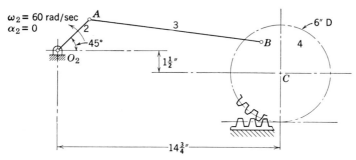

Fig. P–7.17. $O_2A = 3''$; $AB = 11\frac{1}{2}''$; $BC = 2\frac{1}{2}''$.

92 Application of the Relative Acceleration Equation

7.18 (Fig. 7.18). A gear, link 3, rolls around a fixed gear, link 2. With no slip, the arcs of contact must be equal. Or $R\theta = r\phi$. With the axes taken as shown, show that the x- and y-coordinates of point P on link 3 for a general position may be given by the following:

$$x = (r + R) \sin \theta - r \sin (\theta + \phi)$$
$$y = -(r + R) + (r + R) \cos \theta - r \cos (\theta + \phi)$$

Determine the general expressions for the x- and y-components of acceleration of point P for a general position, expressed as a function of r, R, θ, and $\dfrac{d\theta}{dt}$.

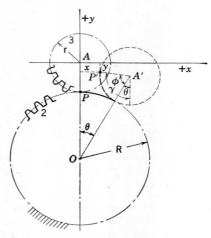

Fig. P-7.18. Note change of angle of line A–P is $(\theta + \phi) = \gamma$, where $R\theta = r\phi$.

Determine the total acceleration of point P for the particular position: $\theta = \phi = 0$, and show that it is equal to

$$A_P = -(r + R)\omega^2 + r\omega^2 \left(1 + \frac{R}{r}\right)^2 = A_A + r\omega_3{}^2 = A_A + A_{PA}$$

where the total acceleration of point P is directed from point P to point A, the center of the gear. (Note that the acceleration of point P is the same irrespective of the angular acceleration of link 3, $\omega = \dfrac{d\theta}{dt}$, the rate of change of angle of the line joining the centers of the two gears, and $\omega_3 =$ the angular speed of link 3.)

The *change* of angle, γ, of a line on link 3 is given by

$$\gamma = \theta + \phi = \theta + \frac{R}{r}\theta$$

Consequently, the angular speed of link 3 is given by $\dfrac{d\gamma}{dt} = \omega + \dfrac{R}{r}\omega$. Also, the angular acceleration of link 3 is given by

$$\frac{d^2\gamma}{dt^2} = \alpha + \frac{R}{r}\alpha$$

where $\alpha = \dfrac{d\omega}{dt}$.

Using the above information, apply the relative acceleration equation to points A and P, and show that the equation given above results.

7.19 (Fig. 7.19). A gear, link 4, rolls around a gear, link 3, while gear 3 is rotating about its own center. In the figure, the motion is broken up into two

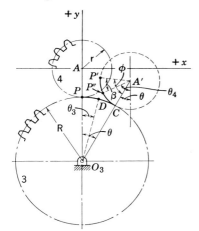

Fig. P-7.19. (a) $\theta c = \theta + \phi - \beta$
(b) arc PC = arc $P'C$
(c) arc PD = arc $P'P''$
(d) $R\theta = r\phi$
(e) $R\theta_3 = r\beta$

parts: point P on link 4 moves to the position shown by P' as the result of gear 4 rolling about gear 3, while gear 3 is held stationary. As a result of the motion of gear 3 rotating clockwise about its own center, gear 4 rotates so that P' moves to P'', with the center of each gear considered stationary. If the angle θ defines the change of angle of the line joining the centers of the gears, $R\theta = r\phi$, with pure rolling and no slip. If gear 3 rotates through an angle of θ_3 radians, gear 4 rotates β radians, with the arcs being equal: $R\theta_3 = r\beta$. The total change of angle of the radial line **AP** is $\theta_4 = \theta + \phi - \beta$.

Show that the x- and y-coordinates for point P in the position P'' may be given by

$$x = (R+r)\sin\theta - r\sin(\theta + \phi - \beta)$$

$$y = -(R+r) + (R+r)\cos\theta - r\cos(\theta + \phi - \beta)$$

Show that the acceleration of point P in the x-direction may be given, for $\theta = \phi = \beta = 0$, by

$$A_P{}^x = (R+r)\alpha - r\alpha_4 = R\alpha_3$$

where α is the angular acceleration of the line joining the centers of the gears, α_4 is the angular acceleration of gear 4, and α_3 is the angular acceleration of gear 3.

Show that the acceleration of point P in the y-direction may be given, for $\theta = \phi = \beta = 0$, by

$$A_P{}^y = -(R+r)\omega^2 + r\omega_4{}^2$$

94 Application of the Relative Acceleration Equation

where ω is the angular speed of the line joining the centers of the gears and ω_4 is the angular speed of the gear 4.

Show that the angular velocity of link 4, expressed as a function of R, r, ω, and ω_3, may be given by

$$\omega_4 = \omega\left(1 + \frac{R}{r}\right) - \frac{R}{r}\omega_3$$

7.20. (It is suggested that problem 7.18 be worked before this problem is attempted.) Refer to the figure of problem 4.21. If the arm A is rotating counterclockwise at 60 rad/sec, and the speed is constant, draw the acceleration polygon for the system. (*Hint.* Proceed in the same order as in the velocity analysis.) What is the acceleration of the point s on gear D, the acceleration of point r on gear D, of point r on gear C, of point t on gear C, and of point t on gear B? Gear B is fixed.

7.21. (It is suggested that problem 7.19 be worked before this problem is attempted.) Refer to the figure of problem 4.21. If the arm A is rotating at 60 rad/sec counter-clockwise, and the speed is increasing at the rate of 1000 rad/sec^2, draw the acceleration polygon. What is the angular acceleration of links 3, 4, and 5? Gear E is fixed.

7.22. Refer to the figure of problem 4.7. Using a completely graphical solution, determine the acceleration of points B and C for the position shown, and determine the angular acceleration of links 3, 4, and 5. Assume link 2 is rotating at a constant angular speed. If the space scale and the acceleration scale are selected, determine the velocity scale.

7.23. Refer to the figure for problem 4.7. Using a completely graphical solution, draw the acceleration polygon if link 2 is increasing in speed at the rate of 4800 rad/sec^2 for the position shown. Specify all scales used

7.24. Refer to the figure for problem 4.8. Draw the acceleration polygon for the position shown if link 2 is rotating at a constant speed. Specify all scales used. Use a completely graphical solution.

7.25 (Fig. 7.25). What must be the angular speed and the angular acceleration of link 2 to have link 6 move with a velocity of 20 ft/sec, as shown, with zero acceleration for the position shown. Use a completely graphical solution. Specify all scales used.

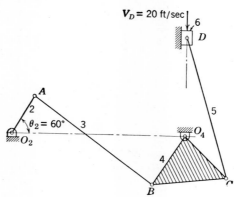

Fig. P-7.25. $O_2A = 3''$; $AB = 10''$; $O_4B = 4''$; $O_4C = 4''$; $BC = 5''$; $CD = 10''$

Problems

7.26. Refer to the figure for problem 4.18. If link 2 is rotating at 30 rad/sec clockwise, determine the accelerations of points A, B, C, and D, and the angular accelerations of links 3, 4, and 5.

Use the semi-analytical method for the determination of the acceleration polygon, and check the solution by using the completely graphical method.

7.27. Refer to the figure for problem 4.18. For the position shown, link 2 is just starting to rotate (that is, $\omega_2 = 0$) with an angular acceleration of 80 rad/sec^2 clockwise. Draw the acceleration polygon. What is the acceleration of point D? The angular acceleration of links 3, 4, and 5?

7.28. Refer to the figure for problem 4.20. Link 2 is rotating at 45 rad/sec clockwise at the instant shown. Determine the acceleration of the midpoint of link 4. What is the angular acceleration of links 3 and 4?

7.29. Refer to the figure for problem 4.20. Link 2 is rotating at 45 rad/sec clockwise at the instant shown, and is increasing in speed at the rate of 900 rad/sec^2. What is the acceleration of points A and B, and the angular acceleration of links 3 and 4?

CHAPTER 8

Acceleration Equation for Two Coincident Points

The relative acceleration equation as discussed in the two previous chapters is limited in application to two points on a rigid link. If a point is moving with respect to a moving body, the analyses of the preceding two chapters do not apply. Coriolis has shown that there is an additional component which must be taken into account.

Two analyses are offered so that the student may appreciate the full significance of the relations which exist.

8.1 Coriolis' component of acceleration. Analytical method

To determine the magnitude of the total acceleration of a point moving with respect to a moving body, consider a point B which is moving with respect to body M at the same time that body M is

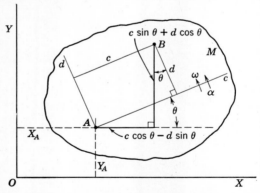

Fig. 8.1. The coordinate system for a point B moving with respect to body M.

moving in a plane, as shown in Fig. 8.1. A coordinate system of axes, X and Y, are fixed and will be used to determine the absolute position of any point in the plane of X and Y. A second system of axes, c and d, are *fixed on body M* and move in the same way that body M moves. The angle θ gives the angular position of the c-axis with the X-axis. Inspection of Fig. 8.1 shows that the X and Y displacements

Coriolis' Component of Acceleration, Analytically

of point B may be expressed by the following, where point A is any arbitrary fixed point on M:

$$X_B = X_A + c \cos \theta - d \sin \theta$$
$$Y_B = Y_A + c \sin \theta + d \cos \theta$$

Differentiate the equations above, remembering that no quantities are constant and that $\dfrac{d\theta}{dt}$ is defined as ω, the *angular speed of body* M:

$$\frac{dX_B}{dt} = V_B{}^x = \frac{dX_A}{dt} - c\omega \sin \theta + \frac{dc}{dt} \cos \theta - d\omega \cos \theta - \frac{dd}{dt} \sin \theta$$

and

$$\frac{dY_B}{dt} = V_B{}^y = \frac{dY_A}{dt} + c\omega \cos \theta + \frac{dc}{dt} \sin \theta - d\omega \sin \theta + \frac{dd}{dt} \cos \theta$$

Set $\dfrac{dc}{dt} = u_c$, the time rate of change of displacement along the c-axis, as if body M were stationary, which is defined as a relative velocity.

$\dfrac{dd}{dt} = u_d$, the time rate of change of displacement along the d-axis, as if body M were stationary, which is defined as a relative velocity.

$\dfrac{dX_A}{dt} = V_A{}^x$, the velocity of point A in the X-direction.

$\dfrac{dY_A}{dt} = V_A{}^y$, the velocity of point A in the Y-direction.

Therefore, the components of velocity of B may be expressed by the following equations, with the above substitutions, and with terms rearranged:

$$V_B{}^x = V_A{}^x - \omega(c \sin \theta + d \cos \theta) + u_c \cos \theta - u_d \sin \theta$$
$$V_B{}^y = V_A{}^y + \omega(c \cos \theta - d \sin \theta) + u_c \sin \theta + u_d \cos \theta$$

Differentiate the expressions above with respect to time:

$$\frac{dV_B{}^x}{dt} = A_B{}^x = \frac{dV_A{}^x}{dt} - \omega(c\omega \cos \theta + u_c \sin \theta - d\omega \sin \theta + u_d \cos \theta)$$
$$+ (c \sin \theta + d \cos \theta) \frac{d\omega}{dt} - u_c \omega \sin \theta + \frac{du_c}{dt} \cos \theta - u_d \cos \theta$$
$$- \frac{du_d}{dt} \sin \theta$$

Acceleration Equation for Two Coincident Points

and

$$\frac{dV_B{}^y}{dt} = A_B{}^y = \frac{dV_A{}^y}{dt} + \omega(-c\omega \sin\theta + u_c \cos\theta - d\omega \cos\theta - u_d \sin\theta)$$

$$+ (c \sin\theta - d \sin\theta)\frac{d\omega}{dt} + u_c\omega \cos\theta + \frac{du_c}{dt}\sin\theta - u_d\omega \sin\theta$$

$$+ \frac{du_d}{dt}\cos\theta$$

Set $\dfrac{du_c}{dt} = a_c$, the time rate of change of relative velocity along the c-axis, as if body M were stationary, which is defined as a relative acceleration.

$\dfrac{du_d}{dt} = a_d$, the time rate of change of relative velocity along the d-axis, as if body M were stationary, which is defined as a relative acceleration.

$\dfrac{dV_A{}^x}{dt} = A_A{}^x$, the acceleration of point A in the X-direction.

$\dfrac{dV_A{}^y}{dt} = A_A{}^y$, the acceleration of point A in the Y-direction.

$\dfrac{d\omega}{dt} = \alpha$, the angular acceleration *of body M*.

The components of acceleration of B may be expressed by the following, with the above substitutions, and with terms rearranged:

$$A_B{}^x = A_A{}^x - \omega^2(c \cos\theta - d \sin\theta) + \alpha(c \sin\theta + d \cos\theta)$$
$$+ (a_c \cos\theta - a_d \sin\theta) - 2u_c\omega \sin\theta - 2u_d\omega \cos\theta$$

$$A_B{}^y = A_A{}^y - \omega^2(c \sin\theta + d \cos\theta) + \alpha(c \cos\theta - d \sin\theta)$$
$$+ (a_c \sin\theta + a_d \cos\theta) + 2u_c\omega \cos\theta - 2u_d\omega \sin\theta$$

The order of vector addition is immaterial. The vectors will be added in a particular way to give a simplified final result:

(a) $A_A{}^x \nrightarrow A_A{}^y = A_A$, the absolute acceleration of point A.

(b) $\omega^2(c \cos\theta - d \sin\theta) \nrightarrow \omega^2(c \sin\theta + d \cos\theta) = \omega^2(c^2 + d^2)^{1/2}$ inasmuch as the components are perpendicular to each other. Since the instantaneous distance from point B to point A is given by $(c^2 + d^2)^{1/2}$, which may be called R, the vector sum is $R\omega^2$. It can be easily shown that $R\omega^2$ is directed from B to A and is along the line from B to A, since $\phi = (\theta + \gamma)$ in Fig. 8.2:

$$\tan(\theta + \gamma) = \frac{c \sin \theta + d \cos \theta}{c \cos \theta - d \sin \theta}$$

$$\tan \phi = \frac{\omega^2(c \sin \theta + d \cos \theta)}{\omega^2(c \cos \theta - d \cos \theta)} = \tan(\theta + \gamma)$$

(c) $\alpha(c \sin \theta + d \cos \theta) \nrightarrow \alpha(c \cos \theta - d \sin \theta) = \alpha(c^2 + d^2)^{1/2} = R\alpha$, inasmuch as the components are perpendicular to each other. The

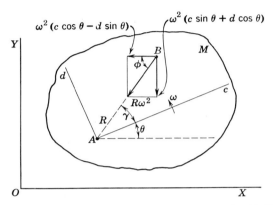

Fig. 8.2. One component of acceleration of point B, $R\omega^2$.

vector, $R\alpha$, is perpendicular to the line from B to A as shown by an investigation of Fig. 8.3:

$$\tan \delta = \frac{\alpha(c \sin \theta + d \cos \theta)}{\alpha(c \cos \theta - d \sin \theta)}$$

$$\tan(\theta + \gamma) = \frac{c \sin \theta + d \sin \theta}{c \cos \theta - d \cos \theta}$$

or $\qquad (\theta + \gamma) = \delta$

Consequently, $R\alpha$ makes an angle of $90° - (\theta + \gamma)$ with the X-axis. Or the angle between $R\alpha$ and the line from B to A is $[90° - (\theta + \gamma) + (\theta + \gamma)]$ or $90°$. It is to be noted that $R\alpha$ is directed in a sense to correspond with the direction of the angular acceleration of body M.

(d) $(a_c \cos \theta - a_d \sin \theta) \nrightarrow (a_c \sin \theta + a_d \cos \theta) = (a_c^2 + a_d^2)^{1/2}$, inasmuch as the components are perpendicular to each other. We will recognize the result as an alternate form of the total relative acceleration point B would have as if body M were stationary. The total relative acceleration is designated by a_r in Fig. 8.4.

(e) $(-2u_c\omega \sin \theta - 2u_d\omega \cos \theta) \nrightarrow (2u_c\omega \cos \theta - 2u_d\omega \sin \theta) = 2\omega(u_c^2$

100 Acceleration Equation for Two Coincident Points

$+ u_d{}^2)^{1/2}$, inasmuch as the components are perpendicular to each other. However, $(u_c{}^2 + u_d{}^2)^{1/2}$ is nothing more than an alternate expression for the total relative velocity of point B as if body M were stationary. Designating the relative velocity of point B on body M by u, we obtain

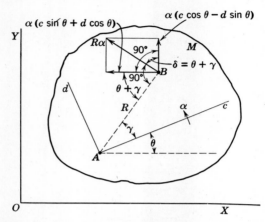

Fig. 8.3. A second component of acceleration of point B, $R\alpha$.

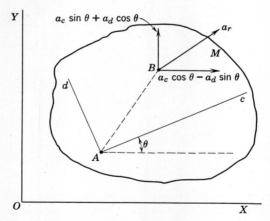

Fig. 8.4. A third component of acceleration of point B, a_r.

for the result $2u\omega$. The direction of $2u\omega$ is perpendicular to the relative velocity u as may be seen from Fig. 8.5: $2u_c\omega \sin\theta \leftrightarrow 2u_c\omega \cos\theta$ is equal to $2u_c\omega$ and is directed at an angle of θ from the vertical axis, as shown; and $2u_d\omega \sin\theta \leftrightarrow 2u_d\omega \cos\theta$ is equal to $2u_d\omega$ and is directed at an angle of θ from the horizontal axis, as shown. Further inspection of Fig. 8.5 reveals that the angle between $2u_c\omega$ and $2u_d\omega$

Coriolis' Component of Acceleration, Analytically

is 90°. The supplementary component of acceleration, $2u\omega$, is perpendicular to u, the relative velocity, since the angle which $2u\omega$ makes with $2u_c\omega$ is the same as the angle which u makes with u_c, ($\rho = \sigma$), and $2u_c\omega$ is perpendicular to u_c.

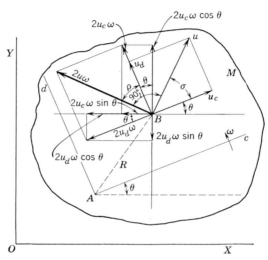

Fig. 8.5. A fourth component of acceleration of point B, $2u\omega$, the Coriolis component which is perpendicular to the relative velocity u. ($\rho = \sigma$.)

An analysis of $2u\omega$ for various directions of angular speeds of the body on which the point is moving, that is, body M in this case, will reveal the following method for determining the sense of direction of acceleration of the supplementary component of acceleration which

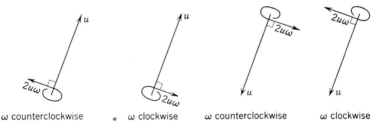

Fig. 8.6. A method to determine the direction of the Coriolis component, knowing the direction of the relative velocity and the direction of the angular velocity of the link on which the point is considered moving.

is always perpendicular to the relative velocity. Set off the proper direction of the relative velocity of the moving point and draw at the tail of the relative velocity a clockwise or counterclockwise curve to represent the angular speed ω of the moving body so that it intersects the relative velocity vector in one place. The arrow for the tip will indicate the direction of $2u\omega$, as shown in Fig. 8.6.

Summary. The acceleration of a point B, moving with respect to moving body M, which has an angular speed of ω and an angular acceleration of α, is given by

$$A_B = A_A \leftrightarrow R\omega^2 \leftrightarrow R\alpha \leftrightarrow a_r \leftrightarrow 2u\omega \qquad \text{(see note*)} \qquad (1)$$

But since $A_A \leftrightarrow R\omega^2 \leftrightarrow R\alpha$ is the acceleration of a point P on body M coincident with point B, which may be called A_P, and since a_r is the relative acceleration of point B with respect to the body M, which is arbitrarily called A_{BP}, the above equation may be written:

$$A_B = A_P \leftrightarrow A_{BP} \leftrightarrow 2u\omega \qquad (2)$$

Attention is called to the similarity between the equation above and that for the acceleration of two points on a rigid link, alike except for the additional Coriolis component, $2u\omega$.

8.2 Coriolis' component of acceleration. Graphical method

A second method of derivation for the acceleration of a point moving on a moving body is presented for further understanding of the concepts involved.

Consider body M, Fig. 8.7a, which is moving in a plane with an angular velocity ω and an angular acceleration α, as shown. At the same time, a point B is moving with respect to M. The velocity of B relative to its path on body M is given as u for the instant shown. u is the velocity determined as if body M were stationary.

In a small time element, Δt, the body M moves to a new position, M', shown by dotted lines in Fig. 8.7a, while B has moved to B' on body M. When B has reached B', it has achieved a new relative velocity represented by u'.

For the initial point of time, the velocity of B may be expressed in relation to that of a coincident point P by the relative velocity equation, as shown in Chapter 4:

$$V_B = V_P \leftrightarrow u \qquad (1)$$

After a time Δt, the velocity of point B is given by $V_{B'}$. Its relation to the velocity of a coincident point, C', on body M is expressed by the relative velocity equation:

$$V_{B'} = V_{C'} \leftrightarrow u' \qquad (2)$$

* *Note.* In determining a_r, the relative angular velocity and acceleration of the line joining the point and the center of curvature must be used.

Coriolis' Component of Acceleration, Graphically 103

Subtract Eq. 1 from Eq. 2:

$$(V_{B'} \to V_B) = (V_{C'} \to V_P) \leftrightarrow (u' \to u) \qquad (3)$$

Figure 8.7b shows the relative path of B on body M, where the relative path is PC. (M is shown as if it were stationary for the picturing

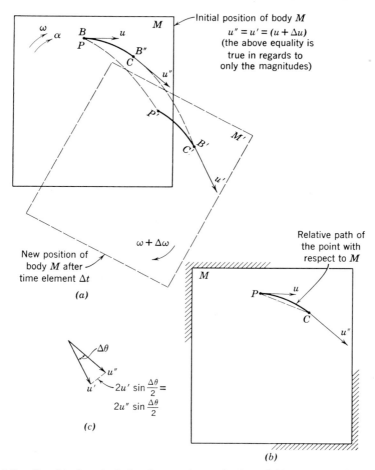

Fig. 8.7. Graphical analysis for change of magnitude and direction of velocity of a point moving on a moving body.

of the relative path.) The relative velocity of B after Δt time is given as u''. Note that u' and u'' are the same in magnitude but have different directions due to the motion of M. Points P and C, which are points on M, are at P' and C' after Δt time, as shown in Fig. 8.7a.

104 Acceleration Equation for Two Coincident Points

A simplification of Eq. 3 may be achieved by adding and subtracting the same vector quantities without affecting the equality, $V_{P'}$ and u''. Also the terms are collected in a particular fashion:

$$(V_{B'} \to V_B) = (V_{C'} \to V_{P'}) \nrightarrow (V_{P'} \to V_P) \nrightarrow (u' \to u'') \nrightarrow (u'' \to u) \tag{4}$$

The rest of the problem is one of proper interpretation of each of the above quantities. Taking each in order, we have:

(a) $(V_{B'} \to V_B)$ is the change of velocity of point $B : \Delta V_B$.

(b) $(V_{C'} \to V_{P'})$ is the relative velocity of C' with respect to $P' : V_{C'P'}$. (This is evident if the relative velocity equation is written for the two points, C' and P', on the rigid body $M : V_{C'} = V_{P'} \nrightarrow V_{C'P'}$). As the time element approaches zero, points P' and C' approach coincidence. If the points are very close to each other, the arc $P'C'$ is practically equal to the chord $P'C'$. The distance along the arc is determined by the distance point B has moved along the relative path. The average velocity is $\dfrac{u + (u + \Delta u)}{2}$. The distance is therefore $\left[\dfrac{u + (u + \Delta u)}{2}\right] \Delta t$. If the angular speed of the body M is $(\omega + \Delta\omega)$, the relative velocity may be expressed by "$R\omega$," or by $\left[\dfrac{u + (u + \Delta u)}{2}\right](\Delta t)(\omega + \Delta\omega)$.

(c) $(V_{P'} \to V_P)$ is the change of velocity of point $P : \Delta V_P$.

(d) $(u' \to u'')$ is the change of direction of relative velocity and is due only to the motion of body M. u' and u'' are the *same in magnitude*. Figure 8.7c shows u' and u'' in position, the angle between the vectors being determined by the angle through which M has turned: $\Delta\theta$. Inspection of Fig. 8.7c shows that $(u' \to u'') = 2u' \sin \dfrac{\Delta\theta}{2} = 2u'' \sin \dfrac{\Delta\theta}{2}$

$= 2(u + \Delta u) \sin \dfrac{\Delta\theta}{2}.$

(e) $(u'' \to u)$ is the change of relative velocity of point B, and is Δu. Consequently, Eq. 4 may be written:

$$\Delta V_B = \left[\dfrac{u + (u + \Delta u)}{2}\right](\Delta t)(\omega + \Delta\omega) \nrightarrow \Delta V_P \nrightarrow 2(u + \Delta u)\sin \dfrac{\Delta\theta}{2} \nrightarrow \Delta u$$

(f) Divide through by Δt, and take the limit of each quantity as Δt approaches zero, disregarding differentials of a higher order:

… Coriolis' Component of Acceleration, Graphically

$$A_B = u\omega \nrightarrow A_P \nrightarrow u\omega \nrightarrow \frac{du}{dt} \qquad (5)$$

Note that in the limit each $u\omega$ component is in the same direction, that is, perpendicular to the relative velocity, and has the same sense. Consequently, the $u\omega$ terms may be added algebraically to give $2u\omega$. Also, $\frac{du}{dt}$ is the relative acceleration of point B with respect to body M, as if body M were stationary. Express $\frac{du}{dt}$ by A_{BP}, the relative acceleration of point B with respect to body M. Rearranging the terms of Eq. 5, we may obtain

$$A_B = A_P \nrightarrow A_{BP} \nrightarrow 2u\omega$$

Note the similarity of the equation above to the equation for the relation of accelerations of two points on a rigid body, except for the additional Coriolis component. Attention is called to the fact that part of the Coriolis component comes from the change of relative velocity between points C' and P', and the other part of it comes from a change of direction of the relative velocity due to the motion of the body on which the point is moving.

An analysis of $2u\omega$ for various directions of angular speeds of the body with respect to which the point is moving, that is, body M in this case, will reveal the following method for determining the sense of direction of acceleration of the Coriolis component of acceleration which is always perpendicular to the relative velocity. Set off the proper direction of the relative velocity of the moving point and draw at the tail of the relative velocity a clockwise or counterclockwise curve to represent the angular speed of the body M so that it intersects the relative velocity vector in one place. The arrow for the tip will indicate the direction of $2u\omega$, as shown in Fig. 8.6.

Example. Certain types of problems can be solved for accelerations in two ways: (1) by the link-to-link method, that is, by the relative acceleration equation for two points on a rigid link; and (2) by use of the Coriolis component. The following problem illustrates the procedure, as well as clarifies the concepts of the terms.

Consider links 2 and 3 shown in Fig. 8.8a. Link 2 is rotating with an absolute angular speed of 12 rad/sec counterclockwise and an absolute angular acceleration of 240 rad/sec^2 counterclockwise. Link 3 is rotating with an absolute angular speed of 18 rad/sec clockwise and an absolute angular acceleration of 360 rad/sec^2 counterclockwise. The acceleration of point B is required, if **O**$_2$**A** is 4 in. and **AB** is 3 in.

106   Acceleration Equation for Two Coincident Points

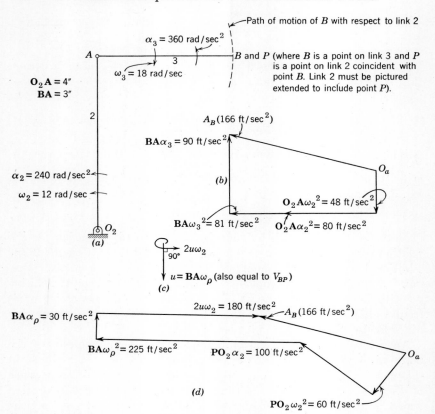

Fig. 8.8. Acceleration of point B found in (b) by the link-to-link method, and in (d) by the relation of accelerations of two coincident points.

The determination of the acceleration of point B by method 1 above is given by

$$A_A = \mathbf{O_2A}\omega_2^2 \nrightarrow \mathbf{O_2A}\alpha_2$$
$$= (\tfrac{4}{12})(12^2) \nrightarrow (\tfrac{4}{12})(240)$$
$$= 48 \text{ ft/sec}^2 \nrightarrow 80 \text{ ft/sec}^2$$
$$A_B = A_A \nrightarrow \mathbf{BA}\omega_3^2 \nrightarrow \mathbf{BA}\alpha_3$$
$$= A_A \nrightarrow (\tfrac{3}{12})(18^2) \nrightarrow (\tfrac{3}{12})(360)$$
$$= A_A \nrightarrow 81 \text{ ft/sec}^2 \nrightarrow 90 \text{ ft/sec}^2$$

Figure 8.8b shows the vector diagram. A_B is 166 ft/sec^2, in the direction shown.

Illustrative Example 107

The determination of the acceleration of point B by method 2 above may be obtained by

$$A_B = A_P \nrightarrow A_{BP} \nrightarrow 2u\omega_2$$

In the equation above, A_P is the acceleration of a point on link 2 which is coincident with the point B on link 3:

$$A_P = \mathbf{PO}_2\omega_2{}^2 \nrightarrow \mathbf{PO}_2\alpha_2$$
$$= \tfrac{5}{12}(12)^2 \nrightarrow \tfrac{5}{12}(240)$$
$$= 60 \text{ ft/sec}^2 \nrightarrow 100 \text{ ft/sec}^2$$

The problem now is to determine the relative acceleration of point B as it moves with respect to link 2. If link 2 is considered stationary, point B rotates about point A, point A being the center of curvature of the path traced by point B on link 2. Consider the line A–B, then, as the rotating radius prescribing the path traced by point B on link 2. The angular velocity of the radius A–B relative to the link 2 is 30 rad/sec clockwise, and the angular acceleration of the radius A–B relative to link 2 is 120 rad/sec^2 counterclockwise, given by ω_ρ and α_ρ, respectively, where the subscript ρ is used to represent the radius of curvature.

Therefore, A_{BP} may be found from

$$A_{BP} = \mathbf{BA}\omega_\rho{}^2 \nrightarrow \mathbf{BA}\alpha_\rho$$
$$= \tfrac{3}{12}(30)^2 \nrightarrow \tfrac{3}{12}(120)$$
$$= 225 \text{ ft/sec}^2 \nrightarrow 30 \text{ ft/sec}^2$$

The Coriolis component is found from

$$2u\omega_2 = 2(\mathbf{BA}\omega_\rho)(\omega_2)$$
$$= 2[(\tfrac{3}{12})(30)]12$$
$$= 180 \text{ ft/sec}^2$$

Figure 8.8c shows the direction of the Coriolis component. Figure 8.8d shows all the vectors in the proper directions. The acceleration of point B is 166 ft/sec^2, the same as is found by a link-to-link analysis.

An exercise left for the student is to draw the velocity polygon, to determine the velocity of point B with respect to point P, and to show that $V_{BP} = \mathbf{BA}\omega_\rho$, in magnitude and direction.

Example of Coriolis' component with relative path a straight line. A numerical example will serve to show the principles involved

108   Acceleration Equation for Two Coincident Points

in determining the acceleration of a point which is moving with respect to a moving body where the relative path is a straight line.

Figure 8.9a shows a point A which is moving on link 2 at the same time that link 2 is moving. The problem is to find the acceleration of a point A_3 on link 3 which is coincident with a point A_2 on link 2. Figure 8.9b shows links 2 and 3 separated to illustrate further the points being discussed. Let us assume that link 2 is rotating clockwise with

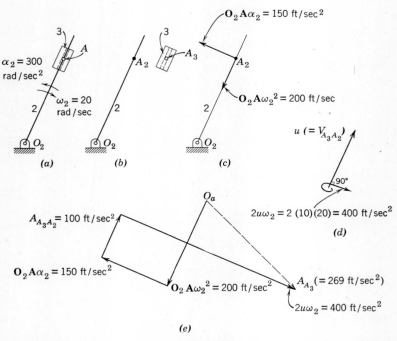

Fig. 8.9. Acceleration of a point moving in a relative straight path.

an angular velocity of 20 rad/sec and is accelerating at the rate of 300 rad/sec² counterclockwise. Let us assume also that the relative velocity of A_3 with respect to the link 2 is 10 ft/sec outward on link 2 away from point O_2; or, $V_{A_3A_2} = u = 10$ ft/sec. Finally, let us assume that the relative velocity is increasing at the rate of 100 ft/sec². The point A_3 is one-half foot from O_2.

The equation which is applicable to the above situation, using the symbols corresponding to the ones given, is

$$A_{A_3} = A_{A_2} \mathbin{+\!\!\!+} A_{A_3A_2} \mathbin{+\!\!\!+} 2u\omega_2$$

Shaper Mechanism Acceleration Analysis   109

The interpretation of each quantity is

(a) A_{A_2} is the acceleration of a point A_2 which is rotating about a fixed center. Its value is determined by $O_2A\omega_2^2 \nrightarrow O_2A\alpha_2 = (\frac{1}{2})(20^2) \nrightarrow (\frac{1}{2})(300) = 200$ ft/sec² $\nrightarrow 150$ ft/sec². These vectors are shown in Fig. 8.9c.

(b) $A_{A_3A_2}$ is the relative acceleration of A_3 with respect to link 2, which is given as 100 ft/sec², and its direction is parallel to the line A–O_2.

(c) $2u\omega_2$ is the Coriolis component. Its magnitude is $(2)(10)(20) = 400$ ft/sec². It is directed perpendicular to the relative velocity in the sense shown in Fig. 8.9d.

The vectors are shown added in Fig. 8.9e. The acceleration of A_3 scales off to 269 ft/sec² in the direction shown.

If the acceleration of any other point on link 3 is desired, we may obtain it readily by noting that the relative acceleration equation for two points on a rigid link applies; and by noting that the angular velocity of link 3 is the same as that of link 2, and the angular acceleration of link 3 is the same as that for link 2.

8.3 Shaper mechanism acceleration analysis

The shaper mechanism discussed in Chapter 4 will now be analyzed for accelerations. The linkage, with its velocity diagram, is shown in Figs. 8.10a and 8.10b. Link 2 is assumed to be rotating at a constant angular velocity (or $\alpha_2 = 0$). The velocity of point A_2 is taken as 40 ft/sec.

The equations for the determination of the acceleration diagram are:

$$A_{A_2} = \frac{V_{A_2}^2}{O_2A} \tag{1}$$

a vector directed from A to O_2. Note that A_2 is a point on link 2, rotating at a fixed radius about O_2. (See Fig. 4.8a.) $A_{A_2} = A_{A_3}$.

Since the slider, link 3, is moving with respect to a moving member, it is necessary to apply Coriolis' equation in order to continue, noting that A_3 is the point which is moving on the moving body, link 4:

$$A_{A_3} = A_{A_4} \nrightarrow A_{A_3A_4} \nrightarrow 2u\omega_4 \tag{2}$$

In Eq. 2, A_{A_3} is known from Eq. 1; the direction of the relative acceleration of A_3 with respect to link 4 is along the line A–O_4; $2u\omega_4$ is, using the terms applying to this case, $2(V_{A_3A_4})(\omega_4)$. The relative velocity of A_3 with respect to A_4 is found from the velocity diagram, as is the angular velocity of link 4. Figure 8.10c shows the direction of Coriolis' component, with ω_4 counterclockwise. As the equation

110 Acceleration Equation for Two Coincident Points

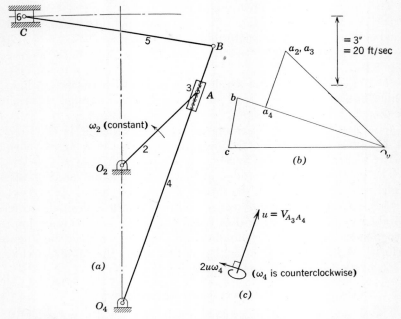

Fig. 8.10a, b, c.   Acceleration analysis of the shaper mechanism.

Given: $V_{A_2} = V_{A_3} = 40$ ft/sec
$O_2A = 4.5''$
$O_4A = 9.8''$
$BC = 8''$

Results from velocity polygon:
$V_{A_4} = 36$ ft/sec
$V_{CB} = 15$ ft/sec
$V_{A_3A_4} = 17.5$ ft/sec

$$\omega_4 = \frac{V_{A_4}}{O_4A} = \frac{36}{\left(\frac{9.8}{12}\right)} = 44.1 \text{ rad/sec ccw}$$

Acceleration components:

$$O_2A\omega_2^2 = \frac{V_{A_2}^2}{O_2A} = \frac{(40)^2}{\left(\frac{4.5}{12}\right)} = 4180 \text{ ft/sec}^2 \qquad \frac{V_{CB}^2}{CB} = \frac{(15)^2}{\left(\frac{8}{12}\right)} = 338 \text{ ft/sec}^2$$

$$A_4O_4\omega_4^2 = \frac{V_{A_4}^2}{O_4A_4} = \frac{(36)^2}{\left(\frac{9.8}{12}\right)} = 1590 \text{ ft/sec}^2 \qquad 2u\omega_4 = 2(V_{A_3A_4})(\omega_4) = 2(17.5)(44.1)$$

$$= 1540 \text{ ft/sec}^2$$

Shaper Mechanism Acceleration Analysis

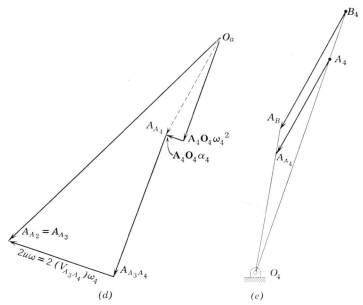

Fig. 8.10d, e. Solution of $A_{A_3} = A_{A_4} \mathrel{+\!\!\!+} A_{A_3 A_4} \mathrel{+\!\!\!+} 2u\omega_4$, where $A_{A_4} = \mathbf{A_4 O_4}\omega_4^2$ $\mathrel{+\!\!\!+} \mathbf{A_4 O_4}\alpha_4$ $\left(\text{or } A'_{A_4} = \dfrac{V_{A_4}^2}{\mathbf{A_4 O_4}} \mathrel{+\!\!\!+} \mathbf{A_4 O_4}\alpha_4 \right.$

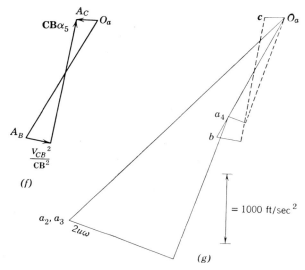

Fig. 8.10f, g. Final acceleration polygon.

112   Acceleration Equation for Two Coincident Points

stands, there are three unknowns, direction and magnitude of A_{A_4}, and magnitude of $A_{A_3A_4}$. One of the unknowns may be eliminated by expressing A_{A_4} as $\mathbf{A_4O_4}\omega_4^2 \nrightarrow \mathbf{A_4O_4}\alpha_4$, since A_4 is a point on link 4.

$$A_{A_3} = \mathbf{A_4O_4}\omega_4^2 \nrightarrow \mathbf{A_4O_4}\alpha_4 \nrightarrow A_{A_3A_4} \nrightarrow 2u\omega_4$$

In the equation above there are two unknowns, magnitude of the tangential acceleration of A_4 and the magnitude of the relative acceleration of A_3 with respect to link 4. Figure 8.10d shows the solution of the equation. The angular acceleration of link 4 is counterclockwise; the slider, link 3, is slowing up, as evidenced by the direction of the relative acceleration of point A_3 with respect to point A_4.

The acceleration of point B is proportional to the acceleration of point A_4, since B and A_4 are rotating about a fixed center O_4. Figure 8.10e shows the graphical solution.

The acceleration of C may be determined by

$$A_C = A_B \nrightarrow \frac{V_{CB}^2}{CB} \nrightarrow \mathbf{CB}\alpha_5 \tag{3}$$

Figure 8.10f shows the solution of Eq. 3. Figure 8.10g shows the final acceleration diagram.

8.4 Oscillating roller follower acceleration analysis

Figure 8.11a shows a cam and an oscillating roller follower. The cam is rotating at a constant speed of 120 rad/sec, counterclockwise. The angular acceleration of link 4 is desired. Consider the two coincident points, B_2 and B_4, which are selected because the center of curvature of the relative path traced by B_4 on link 2 is known: point A, the center of curvature of the cam at the point of contact of the cam and roller. Figure 8.11b shows link 2, where point B_4 is pictured as moving with respect to link 2 in an arc with center at A. Figure 8.11c shows the velocity polygon, with the velocity of point B_4 relative to link 2 ($V_{B_4B_2}$) perpendicular to the line B–A.

The acceleration equation that was used for the acceleration polygon of Fig. 8.11d is

$$A_{B_4} = A_{B_2} \nrightarrow A_{B_4B_2} \nrightarrow 2u\omega$$

or

$$A_{B_4} = A_{B_2} \nrightarrow \rho\omega_p^2 \nrightarrow \rho\alpha_p \nrightarrow 2(V_{B_4B_2})\omega_2$$

where $\rho\omega_p^2 = \dfrac{V_{B_4B_2}^2}{BA}$.

The two unknowns, magnitude of $A_{B_4}{}^t$ and magnitude of $\rho\alpha_p$, are found in the vector solution.

Oscillating Roller Follower Acceleration Analysis 113

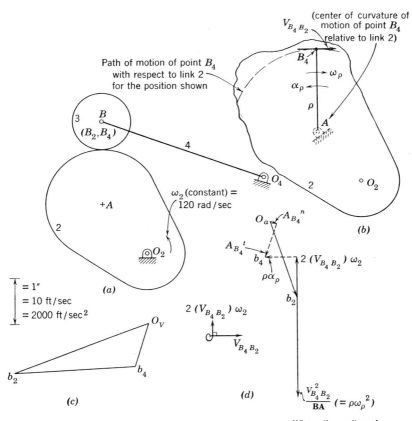

Fig. 8.11. Acceleration analysis of oscillating roller follower cam.

$$A_{B_4} = A_{B_2} \nrightarrow A_{B_4B_2} \nrightarrow 2u\omega$$
$$\text{where } A_{B_4} = A_{B_4}{}^n \nrightarrow A_{B_4}{}^t$$
$$A_{B_4B_2} = \rho\omega_p{}^2 \nrightarrow \rho\alpha_p = \frac{V_{B_4B_2}^2}{BA} \nrightarrow \rho\alpha_p$$

Acceleration components used in the analysis:

$$A_{B_2} = O_2B\omega_2{}^2 = \frac{3.03}{12}(120) = 3640 \text{ ft/sec}^2 \qquad \frac{V_{B_4B_2}^2}{BA} = \frac{(25.5)^2}{\left(\frac{1.8}{12}\right)} = 4330 \text{ ft/sec}^2$$

$$A_{B_4}{}^n = \frac{V_B{}^2}{O_4B} = \frac{(10)^2}{\left(\frac{3.6}{12}\right)} = 333 \text{ ft/sec}^2 \qquad 2u\omega = 2(V_{B_4B_2})\omega_2 = 2(25.5)(120)$$

$$= 6120 \text{ ft/sec}^2$$

114 Acceleration Equation for Two Coincident Points

PROBLEMS

8.1 (Fig. 8.1). A body, link 2, is moving on link 3 at the same time that link 3 is rotating, which indicates the use of Coriolis' component of acceleration. This problem is concerned with the determination of the total acceleration of a point

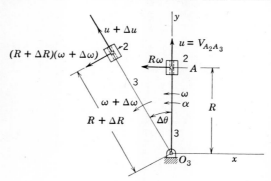

Fig. P–8.1.

A_2 on link 2. Link 3 is taken for convenience as vertical, initially. After a time of Δt link 2 has moved to a new position, whereas link 3 has turned through $\Delta\theta$ radians. The velocity of point A_2 initially is given by

$$V_{A_2} = R\omega \rightarrowtail u$$

where u is defined as $V_{A_2A_3}$. The velocity of point A_2 after link 3 has turned through $\Delta\theta$ radians is given by

$$V_{A_2'} = (R + \Delta R)(\omega + \Delta\omega) \rightarrowtail (u + \Delta u)$$

Show that the change of velocity in the x-direction may be given by

$$\Delta V_x = -(R + \Delta R)(\omega + \Delta\omega)\cos\Delta\theta - (u + \Delta u)\sin\Delta\theta + R\omega$$

and the change of velocity in the y-direction may be given by

$$\Delta V_y = -(R + \Delta R)(\omega + \Delta\omega)\sin\Delta\theta + (u + \Delta u)\cos\Delta\theta - u$$

Expand each expression, divide through by Δt, take the limit as Δt approaches zero and show that the acceleration of point A_2 may be expressed by

$$A_{A_2} = A_{A_3} \rightarrowtail \frac{du}{dt} \rightarrowtail 2u\omega_3$$

Compare the expression with that derived in this chapter, and show that the expressions are the same. Note that $\dfrac{du}{dt}$ may be expressed as $A_{A_2A_3}$, the relative acceleration of point A_2 with respect to link 3.

8.2 (Fig. 8.2). A link, member 2, moves on a rotating link, link 3. Set up expressions for the x- and y-coordinates of a point A_2 as a function of R and θ, differentiate twice with respect to time, and determine the total acceleration of point A_2. Note that R is not a constant quantity. Show that the total acceleration may be given by

$$A_{A_2} = A_{A_3} + \frac{du}{dt} + 2u\omega_3$$

where $u = \frac{dR}{dt}$ (the relative velocity of point A_2 with respect to link 3) and where $\frac{du}{dt}$ is the relative acceleration of point A_2 with respect to link 3.

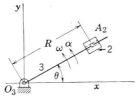

Fig. P-8.2.

8.3 (Fig. 8.3). For the information given on the figure, determine the acceleration of point A_2 on link 2, for the position shown.

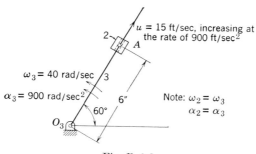

Fig. P-8.3.

8.4. Refer to the figure for problem 4.10. If link 2 is rotating at a constant angular velocity, as shown, determine the angular acceleration of link 4 and the acceleration of point B. Compare the angular acceleration of links 3 and 4.

8.5. Refer to the figure for problem 4.11. If the frame is moving at the given constant speed, determine the acceleration of the piston B.

8.6. Refer to the figure for problem 4.13. If point A is moving at 40 ft/sec with constant angular speed of link 2 in a counterclockwise direction, determine the acceleration of point C on link 5 (or link 6) and determine the angular acceleration of links 3, 4, 5, and 6.

8.7. Refer to the figure for problem 4.15. Link 2 is rotating at a constant angular speed of 20 rad/sec counterclockwise. Determine the acceleration of point C on link 6, and the angular acceleration of links 3, 4, 5, and 6.

8.8. Refer to the figure for problem 4.16. Link 4 is rotating at a constant angular speed of 40 rad/sec counterclockwise for the position shown. Determine

116 Acceleration Equation for Two Coincident Points

the angular acceleration of links 2, 3, and 5. What is the relative acceleration of link 3 on link 4?

8.9. Refer to the figure for problem 4.17. Link 2 is rotating at a constant angular speed of 30 rad/sec. Determine the angular acceleration of each link.

8.10. Refer to the figure for problem 4.17. Link 5 has an angular speed of 20 rad/sec clockwise, and is increasing in speed at the rate of 400 rad/sec^2. What must be the angular speed and the angular acceleration of link 2 to give the prescribed motion?

8.11. Refer to the figure for problem 4.11. If the frame is increasing in speed at the rate of 10 ft/sec^2, determine the acceleration of the piston B and the relative acceleration of the piston B with respect to the frame.

8.12. Refer to the figure for problem 4.23. Determine, for the two cases shown, the acceleration of point B if the crank, link 2, is rotating at a constant speed of 40 rad/sec counterclockwise.

8.13. Refer to the figure for problem 4.23. Determine, for the two cases shown, the acceleration of point B if the crank is rotating counterclockwise at 40 rad/sec and decreasing in speed at the rate of 300 rad/sec^2.

8.14 (Fig. 8.14). Link 2 is rotating at a constant angular velocity of 25 rad/sec, counterclockwise, as shown. Determine the angular acceleration of link 3 in two ways:

(1) By considering the two coincident points, B_2 and B_3. Note that the center of curvature of the path point B_2 traces on link 3 is at A.

(2) By considering the two coincident points A_2 and A_3. Note that the center of curvature of the path point A_3 traces on link 2 has a center of curvature at B.

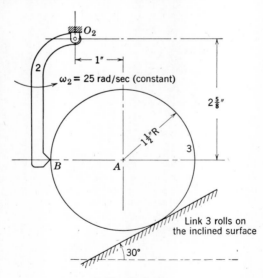

Fig. P–8.14.

8.15 (Fig. 8.15). Link 2 rolls on the flat surface, with the velocity of the center, point A, being 10 ft/sec. The angular velocity of link 2 is constant. Find the angular acceleration of link 3 for the position shown.

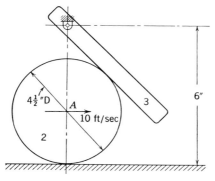

Fig. P–8.15.

8.16 (Fig. 8.16). The pin, the center of which is P, can move relative to the slot of link 2 and relative to the slot of link 4. The angular velocity of link 2 is 1 rad/sec clockwise, whereas the angular acceleration of link 3 is zero. Determine the following:
(a) Velocity of point P.
(b) The angular acceleration of link 2.
(c) The angular acceleration of link 4.
(d) The acceleration of point P.

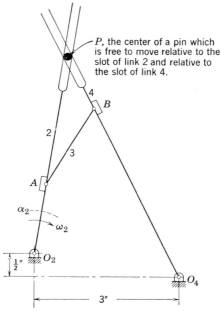

Fig. P–8.16. $O_2A = 1\frac{1}{2}''$; $O_4B = 4''$; $O_2P = 4\frac{1}{4}''$; $O_4P = 5\frac{1}{4}''$.

CHAPTER 9

Special Methods of Acceleration Solution

It was pointed out in Chapter 5 that special methods had to be used with certain types of mechanisms to obtain velocity solutions since the ordinary procedures were not sufficient for complete analyses. So it is with acceleration analysis for certain types of linkages. However, if the student has mastered the concept of the "auxiliary point," he should have no difficulty in applying the same reasoning to acceleration analyses. Some of the mechanisms treated in Chapter 5 will be discussed, and the same auxiliary points will be used. Only problems involving the use of one auxiliary point will be taken up inasmuch as those mechanisms which require the use of two auxiliary points may be analyzed in a similar fashion.

9.1 Watt "walking beam" mechanism

The Watt "walking beam" mechanism is shown in Fig. 9.1a, and the velocity diagram in Fig. 9.1b. Link 2 is assumed to be rotating at constant angular velocity, ω_2 rad/sec counterclockwise. The equations for solution of the acceleration diagram are

$$A_A = \frac{V_A{}^2}{\mathbf{O_2A}} \tag{1}$$

The next point which may be considered is B. However, there are three unknowns in the equation

$$A_B = A_A \mathrel{+\mkern-10mu+} \frac{V_{BA}{}^2}{\mathbf{BA}} \mathrel{+\mkern-10mu+} \mathbf{BA}\alpha_3 \tag{2}$$

magnitude and direction of A_B and magnitude of $\mathbf{BA}\alpha_3$. Since no further information is available, we use an auxiliary point, x, for velocity analysis. Note that points B, C, D, and x are all points on a rigid link, 4. (See Chapter 5, page 45.)

Express the relation between the acceleration of point x and point B, two points on a rigid link:

Watt "Walking Beam" Mechanism

$$A_x = A_B \mapsto \frac{V_{xB}^2}{xB} \mapsto xB\alpha_4 \tag{3}$$

Substitute the value of A_B from Eq. 2 into Eq. 3:

$$A_x = \left(A_A \mapsto \frac{V_{BA}^2}{BA} \mapsto BA\alpha_3\right) \mapsto \frac{V_{xB}^2}{xB} \mapsto xB\alpha_4 \tag{4}$$

There are four unknowns in the equation above: magnitude and direction of A_x, magnitude of $BA\alpha_3$, and magnitude of $xB\alpha_4$.

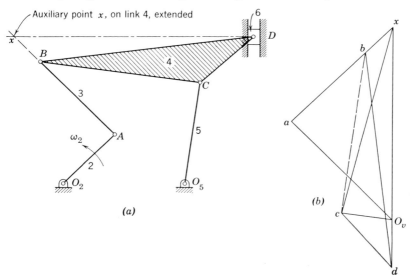

Fig. 9.1a, b. Watt "walking beam" mechanism, with the velocity polygon. An auxiliary point x on link 4 is used.

However, an equation may be written for the relation of accelerations of point x and point D, two points on a rigid link:

$$A_x = A_D \mapsto \frac{V_{xD}^2}{xD} \mapsto xD\alpha_4 \tag{5}$$

wherein two additional unknowns are introduced, magnitude of A_D and magnitude of $xD\alpha_4$.

Or Eqs. 4 and 5 may be set equal to each other:

$$\left(A_A \mapsto \frac{V_{BA}^2}{BA} \mapsto BA\alpha_3\right) \mapsto \frac{V_{xB}^2}{xB} \mapsto xB\alpha_4 = A_D \mapsto \frac{V_{xD}^2}{xD} \mapsto xD\alpha_4 \tag{6}$$

Note that $BA\alpha_3$ and $xB\alpha_4$ are in the same direction, perpendicular to the line B–A (or x–B), and consequently may be replaced by a

Special Methods of Acceleration Solution

single vector which is perpendicular to line B–A. We are not interested in the magnitude of each vector at this point, but are interested in simplifying Eq. 6. Also, A_D and $\mathbf{xD}\alpha_4$ are in the same direction, and may be replaced by a single vector, in the direction of acceleration of point D (or perpendicular to the line x–D, which is the same as A_D). Rewriting Eq. 6, we obtain

$$A_A \twoheadrightarrow \frac{V_{BA}{}^2}{\mathbf{BA}} \twoheadrightarrow \frac{V_{xB}{}^2}{\mathbf{xB}} \twoheadrightarrow \text{(vector perpendicular to line } B\text{–}A\text{)}$$

$$= \frac{V_{xD}{}^2}{\mathbf{xD}} \twoheadrightarrow \text{(vector perpendicular to line } x\text{–}D\text{)} \quad (7)$$

The equation above may be solved since there are only two unknown quantities. Figure 9.1c shows the graphical solution. The acceleration of point x is thus determined.

The remainder of the analysis follows the procedure used in velocity analysis, that is, the acceleration of point C may be found next, then the acceleration of point B, and finally that of point D:

$$A_C = A_x \twoheadrightarrow \frac{V_{Cx}{}^2}{\mathbf{Cx}} \twoheadrightarrow \mathbf{Cx}\alpha_4$$

$$A_C = \frac{V_C{}^2}{\mathbf{CO}_5} \twoheadrightarrow \mathbf{CO}_5\alpha_5 \quad (8)$$

The two equations above solved simultaneously will give A_C, $\mathbf{Cx}\alpha_4$, and $\mathbf{CO}_5\alpha_5$.

$$A_B = A_C \twoheadrightarrow \frac{V_{BC}{}^2}{\mathbf{BC}} \twoheadrightarrow \mathbf{BC}\alpha_4$$

$$A_B = A_A \twoheadrightarrow \frac{V_{BA}{}^2}{\mathbf{BA}} \twoheadrightarrow \mathbf{BA}\alpha_4 \quad (9)$$

Note that the two equations of (9) are only one method for obtaining A_B. Another method for determining A_B is to use only one of the equations, solving for α_4 from Eq. 8, and calculating, say $\mathbf{BA}\alpha_4$.

$$A_D = A_C \twoheadrightarrow \frac{V_{DC}{}^2}{\mathbf{DC}} \twoheadrightarrow \mathbf{DC}\alpha_4 \quad (10)$$

or the following may be used:

$$A_D = A_B \twoheadrightarrow \frac{V_{DB}{}^2}{\mathbf{DB}} \twoheadrightarrow \mathbf{DB}\alpha_4 \quad (11)$$

Watt "Walking Beam" Mechanism

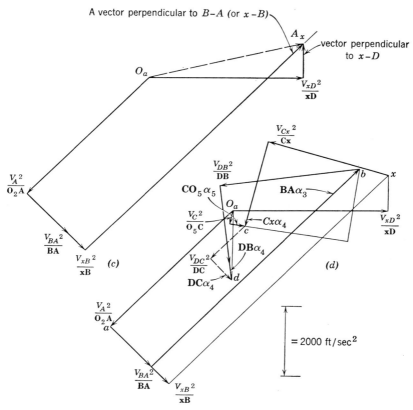

Fig. 9.1c, d. Acceleration polygon. Auxiliary point x is used. Fig. 9.1c shows acceleration solution for point x.

Acceleration components:

$$\frac{V_A{}^2}{O_2A} = \frac{(40)^2}{\left(\frac{4}{12}\right)} = 4800 \text{ ft/sec}^2$$

$$\frac{V_{BA}{}^2}{BA} = \frac{(28.6)^2}{\left(\frac{6}{12}\right)} = 1340 \text{ ft/sec}^2$$

$$\frac{V_{xB}{}^2}{xB} = \frac{(10.9)^2}{\left(\frac{2.1}{12}\right)} = 680 \text{ ft/sec}^2$$

$$\frac{V_{xD}{}^2}{xD} = \frac{(69.9)^2}{\left(\frac{13.35}{12}\right)} = 4390 \text{ ft/sec}^2$$

$$\frac{V_{Cx}{}^2}{Cx} = \frac{(56)^2}{\left(\frac{10.75}{12}\right)} = 3500 \text{ ft/sec}^2$$

$$\frac{V_C{}^2}{O_5C} = \frac{(14.2)^2}{\left(\frac{6}{12}\right)} = 403 \text{ ft/sec}^2$$

$$\frac{V_{BC}{}^2}{BC} = \frac{(47.0)^2}{\left(\frac{9}{12}\right)} = 2945 \text{ ft/sec}^2$$

$$\frac{V_{DC}{}^2}{DC} = \frac{(21.0)^2}{\left(\frac{4}{12}\right)} = 1325 \text{ ft/sec}^2$$

$$\frac{V_{DB}{}^2}{DB} = \frac{(62.6)^2}{\left(\frac{6}{12}\right)} = 3920 \text{ ft/sec}^2$$

122   Special Methods of Acceleration Solution

Either Eq. 10 or Eq. 11 may be used to find A_D. The student is afforded an opportunity to check his work by using both equations. A final check would be to determine the similarity of figures B–x–D–C and b–x–d–c.

The complete acceleration diagram is shown in Fig. 9.1d.

9.2 Modified shaper mechanism

The modified shaper mechanism, already analyzed for velocities in Chapter 5, will now be analyzed for accelerations. The mechanism is shown in Fig. 9.2a and the velocity diagram is shown in Fig.

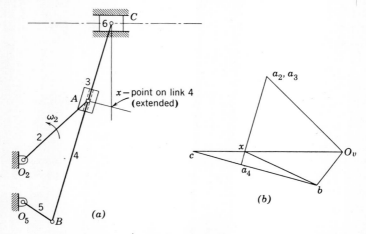

Fig. 9.2a, b. Velocity polygon for modified shaper mechanism. Auxiliary point x on link 4 used.

9.2b. The link 2 is assumed to be rotating at a constant angular velocity of ω_2 rad/sec counterclockwise. The complete acceleration diagram is desired. The solution follows:

$$A_{A_2} = A_{A_3} = \frac{V_A{}^2}{O_2 A} \tag{1}$$

where A_2 and A_3 are the common points on links 2 and 3. Since the slider is moving with respect to a moving body, the Coriolis component is indicated. The relation of the acceleration of two coincident points, A_3 and A_4, is given by

$$A_{A_3} = A_{A_4} \nrightarrow A_{A_3 A_4} \nrightarrow 2u\omega_4 \tag{2}$$

where A_{A_3} is known from Eq. 1; A_{A_4} cannot be determined now since A_4 is not rotating about a fixed point; $A_{A_3 A_4}$ is known in direction,

Modified Shaper Mechanism

since the relative acceleration is along link 4, but is unknown in magnitude; $2u\omega_4 = 2(V_{A_3A_4})\omega_4$, which can be calculated and whose direction can be found. Consequently, there are three unknowns in Eq. 2. We use an auxiliary point, x, in order that the analysis may be continued. Note that x is a point on link 4 extended. (See Chapter 5, page 50.)

$$A_x = A_{A_4} \nrightarrow \frac{V_{xA_4}^2}{xA_4} \nrightarrow \mathbf{xA}_4\alpha_4 \tag{3}$$

Since the substitution of A_{A_4} from Eq. 2 into Eq. 3 will involve the subtraction of vectors, let us rewrite Eq. 2 by considering point A_4 as the point which is moving with respect to the moving body, link 3, noting that $\omega_3 = \omega_4$:

$$A_{A_4} = A_{A_3} \nrightarrow A_{A_4A_3} \nrightarrow 2(V_{A_4A_3})\omega_3 \tag{4}$$

Substitution of Eq. 4 into Eq. 3 gives

$$A_x = [A_{A_3} \nrightarrow A_{A_4A_3} \nrightarrow 2(V_{A_4A_3})\omega_3] \nrightarrow \frac{V_{xA_4}^2}{xA_4} \nrightarrow \mathbf{xA}_4\alpha_4 \tag{5}$$

Rearranging the terms, and recognizing that $A_{A_4A_3}$ and $\mathbf{xA}_4\alpha_4$ are both unknown in magnitude but have the same direction (parallel to link 4), we may simplify Eq. 5 by

$$A_x = A_{A_3} \nrightarrow \frac{V_{xA_4}^2}{xA_4} \nrightarrow 2(V_{A_4A_3})\omega_3 \nrightarrow \text{(vector parallel to link 4)} \tag{6}$$

An additional equation to allow for solution of A_x is

$$A_x = A_C \nrightarrow \frac{V_{xC}^2}{xC} \nrightarrow \mathbf{xC}\alpha_4 \tag{7}$$

Since A_C and $\mathbf{xC}\alpha_4$ are in the same direction, Eq. 7 may be expressed by

$$A_x = \frac{V_{xC}^2}{xC}$$

$\nrightarrow$ (vector perpendicular to the line x–C, or in the direction of A_C) (8)

Equations 6 and 8 may be solved simultaneously for A_x. Figure 9.2c shows the vector solution for A_x, and Fig. 9.2d shows the direction of the Coriolis component.

The equations which may be used for one of the procedures for the remainder of the solution to give the complete acceleration diagram

Special Methods of Acceleration Solution

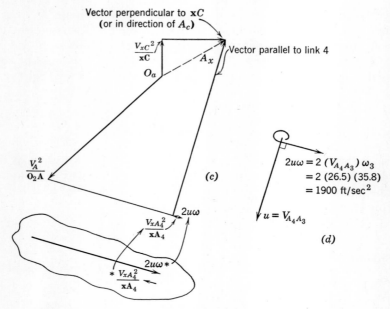

Fig. 9.2c, d. Acceleration of auxiliary point x determined.

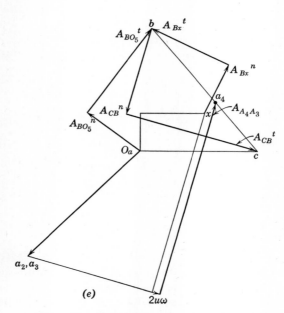

Fig. 9.2e. Complete acceleration polygon.

shown in Fig. 9.2e are

$$A_B = A_x \mapsto \frac{V_{Bx}^2}{Bx} \mapsto Bx\alpha_4$$

$$A_B = \frac{V_B^2}{BO_5} \mapsto BO_5\alpha_5 \tag{9}$$

The acceleration of points A_4 and C may be found readily since the accelerations of two points, B and x, on link 4 are known.
The construction actually used in Fig. 9.2e was to find A_C by

$$A_C = A_B \mapsto \frac{V_{CB}^2}{CB} \mapsto CB\alpha_4$$

Point a_4 in the acceleration diagram was found by using

$$A_{A_4} = A_{A_3} \mapsto 2(V_{A_4A_3})\omega_3 \mapsto A_{A_4A_3}$$

and noting that point a_4 should fall on b–c of the acceleration diagram. A check of the accuracy of the work is that

$$\frac{a_4 b}{cb} = \frac{A_4 B}{CB}$$

PROBLEMS

9.1. Refer to the figure for problem 5.1. If point D is moving at a constant velocity, determine the acceleration of points A, B, and C and the angular acceleration of links 2, 3, 4, and 5.

9.2. Refer to the figure for problem 5.1. If point D is increasing in speed at the rate of 20 ft/sec, determine the acceleration of points A, B, and C and the angular acceleration of links 2, 3, 4, and 5.

9.3. Refer to the figure for problem 5.2. If link 6 is moving at a constant speed, determine the angular acceleration of each link.

9.4. Refer to the figure for problem 5.3. If link 2 is rotating at a constant speed of 1200 rpm clockwise, determine the angular acceleration of each link.

9.5. Refer to the figure for problem 5.4. If the velocity of the piston D is 30 ft/sec as shown, and if the acceleration of point D is to be zero for the position shown, what is the angular acceleration of links 2, 3, 4, and the acceleration of point A?

9.6. Refer to the figure of problem 5.5. Determine the angular acceleration of each link if link 5 is rotating at a constant speed.

9.7. Refer to the figure of problem 5.5. Determine the angular acceleration of each link if the point A is moving with a velocity of 8 ft/sec and if the angular acceleration of link 5 is 120 rad/sec^2 in a counterclockwise direction.

9.8. Refer to the figure for problem 5.6. If link 2 is rotating at 10 rad/sec counterclockwise, determine the acceleration of point C and the angular acceleration of link 3.

9.9. Refer to the figure for problem 4.3. Link 2 is moving with a velocity of 10 ft/sec at the instant shown. What must be the acceleration of link 2 to have the acceleration of link 6 zero?

9.10 (Fig. 9.10). Link 2 is rotating at 120 rad/sec clockwise at a constant rate. Determine the angular acceleration of link 5.

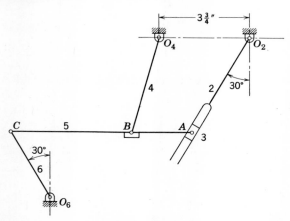

Fig. P–9.10. $O_2A = 4\frac{3}{4}''$; $AB = 2\frac{1}{2}''$; $O_4B = 4\frac{1}{4}''$; $CA = 7\frac{1}{2}''$; $O_6C = 3\frac{1}{4}''$.

CHAPTER 10

Equivalent Mechanisms

As has been pointed out in the acceleration analysis of a point which is moving with respect to a body, it is necessary to know the path which is traced on the body to determine the relative acceleration. Mechanisms which have rolling or sliding surfaces, as in cams, for instance, may become rather involved in the solution if a point-to-point analysis is attempted, because of the difficulty of determining the analytical expressions for the path of relative motion. In such cases, the use of an equivalent mechanism may permit the by-passing of the

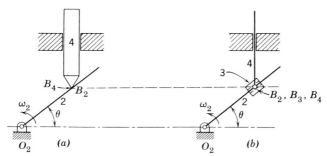

Fig. 10.1. A kinematically equivalent mechanism for (a) is (b).

difficulty and afford the necessary solution. Such equivalent mechanisms are defined as those which give identical motion to the part being analyzed, as found in the original mechanisms. The equivalent mechanism may change in configuration and proportions during a cycle, and it is necessary to restrict our thinking to an equivalent mechanism at an instant of time.

The illustrative examples to be given may not, in some cases, be necessary for the analysis, but will serve to better our picture of the system for a quicker appreciation of the motions.

Case I. Figure 10.1a shows link 2 rotating counterclockwise at a given angular velocity while link 4 is being raised with pure translation. Figure 10.1b shows an equivalent mechanism in that the motion of link 4 is identical to the motion of link 4 in the original mechanism, even though link 3 has been added. The acceleration of link 4 may be found from

$$A_{B_3} = A_{B_2} \nrightarrow A_{B_3 B_2} \nrightarrow 2u\omega_2$$

where B_3 is a point on the slider (and also a point on link 4), and B_2 is a point on link 2 coincident with B_3. The magnitude A_{B_3} is not known, and the magnitude of $A_{B_3 B_2}$ is not known. The other quantities are known. A vector solution can therefore be obtained.

Note that this problem may be solved directly without the use of an equivalent mechanism, which does, however, simplify the concept of motion.

Case II. Figure 10.2a shows a modification of Case I. Figure 10.2b shows an equivalent mechanism. The use of Coriolis' component of acceleration is indicated in the solution of this problem. The procedure of analysis is comparable to Case I.

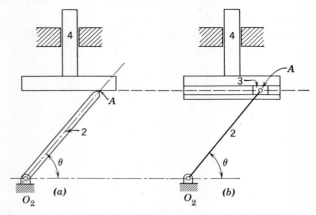

Fig. 10.2. A kinematically equivalent mechanism for (a) is (b).

Case III. Figure 10.3a shows a mechanism where link 4 is raised with pure translation by the action of link 2, at the end of which is a roller. Since the center of the roller, A, is always at a fixed distance from the bottom surface of link 4, the same effect is achieved by the linkage shown in Fig. 10.3b. The analysis for acceleration is comparable to that for Fig. 10.2b.

Case IV. Figure 10.4a shows link 4 being raised with pure translation by a cam, link 2. The radius of curvature of the cam at the point of contact with link 4 is given by R. The equivalent mechanism, shown in Fig. 10.4b, has the pin connection of the added link 3 at the center of curvature of the cam in contact with the follower, link 4.

Equivalent Mechanisms

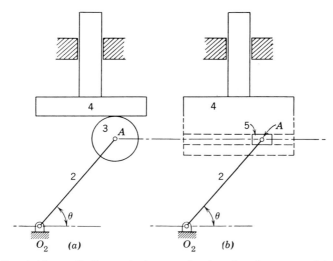

Fig. 10.3. A kinematically equivalent mechanism for the cam and follower in (a) is (b).

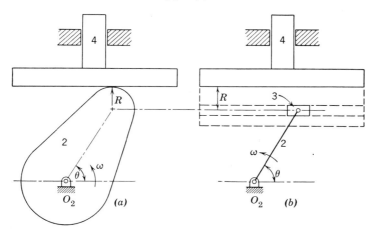

Fig. 10.4. Kinematically equivalent mechanisms.

Note that the equivalent mechanism of this case is the same as for the preceding case.

Case V. Figure 10.5a shows a cam raising a roller follower. The center of curvature of link 2 is at A. Since the distance between A and B is constant as long as the portion of the cam in contact with the roller has a center of curvature at A, a rigid link, 5, may be used to represent that fact in the equivalent linkage in Fig. 10.5b. Point B

130 Equivalent Mechanisms

has pure translation. The equivalent mechanism is the familiar slider-crank mechanism, which may be analyzed for accelerations very easily.

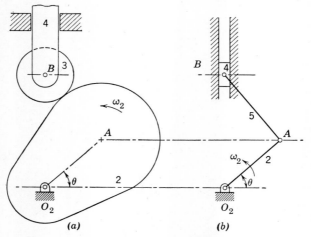

Fig. 10.5. Kinematically equivalent mechanisms

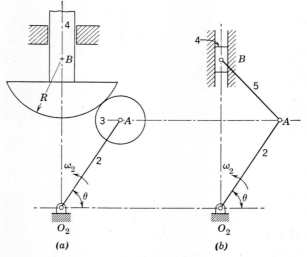

Fig. 10.6. Kinematically equivalent mechanisms.

Case VI. Figure 10.6a is a system comparable to the system of the previous case. Again, since the distance between A and the center of curvature of the button head follower, B, is constant, a link 5 may be used in the equivalent mechanism of Fig. 10.6b to represent the fact that the distance between A and B is constant. The familiar slider-crank mechanism results.

Case VII. A modification of Case VI is shown in Fig. 10.7a, where the center of curvature of the cam is given at A. Figure 10.7b shows

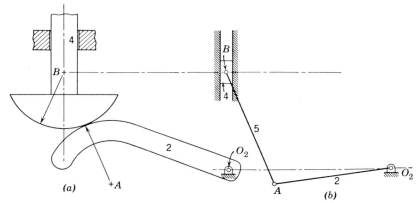

Fig. 10.7. Kinematically equivalent mechanisms.

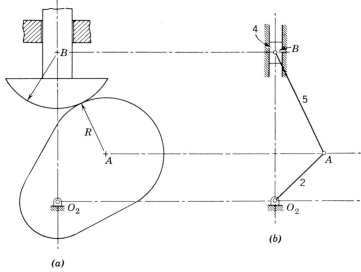

Fig. 10.8. Kinematically equivalent mechanisms.

the equivalent mechanism where link 5 will give the same distance between A and B.

Case VIII. A further modification of the two previous cases is shown in Fig. 10.8a. A represents the center of curvature of the portion of the cam in contact with the follower. The equivalent mechanism is shown in Fig. 10.8b.

Case IX. Figure 10.9a shows a linkage wherein point A is always at a fixed distance from the contact surface of link 4. Figure 10.9b shows the equivalent mechanism.

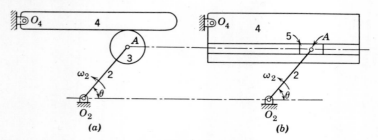

Fig. 10.9. Kinematically equivalent mechanisms.

Case X. Figure 10.10a shows a modification of Case IX. A represents the center of curvature of the portion of cam in contact with the follower, link 4. Since A is at a fixed distance from the under side of link 4, the same motion may be achieved in the equivalent mechanism shown in Fig. 10.10b.

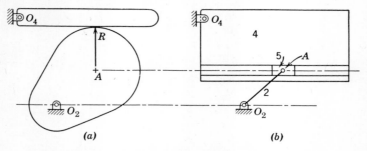

Fig. 10.10. Kinematically equivalent mechanisms.

The few illustrative examples given do not make for a complete story of equivalent mechanisms. However, the principles discussed should give the student sufficient background to enable him to see the procedure of solution. The reader is referred to textbooks in kinematics for further illustrative problems in the expansion of pairs of elements and equivalent mechanisms.

PROBLEMS

10.1 (Fig. 10.1). A cam and follower of the in-line type is made up from circular arcs, with the dimensions shown. For the position shown, determine the acceleration of the follower and the relative acceleration at the point of contact, for a constant angular speed of the cam of 1800 rpm, counterclockwise.

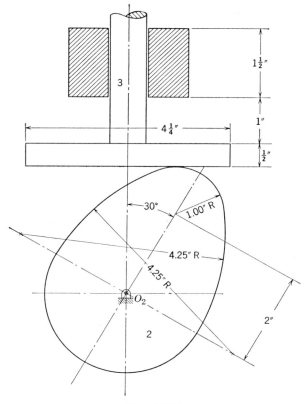

Fig. P-10.1.

10.2 (Fig. 10.2). An application of acceleration, as pointed out in problem 7.13 is the determination of radius of curvature. This problem is concerned with the determination of radius of curvature for a cam, such information being necessary in determining the contact stress (using the Hertz equation for contact stress of curved surfaces under a load). If cams are designed on the basis of use of circular arcs, such information is already available. Cams may be designed on the basis of prescribed motions and lifts of the follower, however, and in such cases the radius of curvature would have to be determined. This problem indicates a procedure of handling such a case, for one position of a cam, such procedure being valid for any cam profile inasmuch as any curve may be considered as being composed of a series of circular arcs.

A follower is shown in the figure in a given position when it is known that the velocity of the cam follower is 2 ft/sec as shown, and the acceleration of the follower is 3000 ft/sec². What is the angular speed of the cam and the radius of curvature of the cam in contact with the follower. The cam is rotating at a constant speed, counterclockwise.

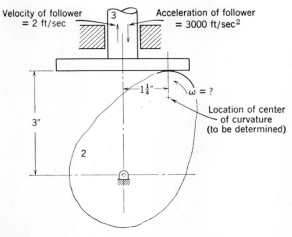

Fig. P–10.2.

(Use a kinematically equivalent system in part and a graphical velocity and acceleration solution, taking into account the fact that the center of curvature of the cam profile is unknown.)*

* A series of various cams are analyzed for radius of cam curvature in "Cam Curvature," by A. R. Holowenko and A. S. Hall, *Machine Design*, August, September, November, 1953.

CHAPTER 11

Review of Static Forces and Graphic Statics

The study of forces in machines involves two types, which may be classified as static forces and dynamic forces. It will be shown in a later chapter that the latter can be handled as a static force system. Consequently, a brief review of the basic principles of static force analysis will be given. Graphical analysis of machine forces will be used in this book because of the simplification it offers to a problem, especially for the more complex machines. It is to be noted that the graphical analysis of forces is a *direct* application of the equations of equilibrium.

11.1 Equations of equilibrium

A machine is a three-dimensional object, with forces acting in three dimensions. In some machines forces can be considered applied in a plane, as a slider-crank mechanism. In some machines forces are applied in parallel planes, as might be illustrated by a four-link mechanism where a link (or links) is offset to avoid interference of parts. In some machines forces are acting in various planes, as might be illustrated by a worm and worm gear speed reducer transmitting power to a bevel gear. In any event, a complete analysis should show that for equilibrium the forces in any plane should be balanced and the moment of the forces about any axis should be balanced. If forces are acting in parallel planes, the force analysis may be made by projecting the forces into a single reference plane without any error; similarly, the moment analysis may be made by projecting the forces into several planes. In general, then, for a three-dimensional system, a complete analysis may be made by projecting the forces into three mutually perpendicular planes.

For each reference plane, it is necessary that the vector sum of the applied forces is zero and that the moment of the forces about any axis perpendicular to the reference plane, or about any point in the plane, is zero for equilibrium. These two conditions are expressed by

$$\Sigma \vec{F} = 0$$
$$\Sigma M = 0 \tag{1}$$

These equations are Newton's equations for the special case of a body at rest, or moving at a constant velocity, which define equilibrium.

An alternate way of expressing the same relations for the forces in a plane is

$$\Sigma F_x = 0$$
$$\Sigma F_y = 0 \tag{2}$$
$$\Sigma M_o = 0$$

where rectangular components of the forces in any x- and y-directions are considered instead of the resultant forces.

There are still other ways of expressing the equations of equilibrium, but the forms given above will be the ones used in this book because they are readily applied to a graphical solution.

11.2   Forces as vectors

A force is a vector quantity and as such can be handled as velocities and accelerations were handled. As a vector quantity, there are three inherent properties to define a force completely: (1) <u>magnitude</u>; (2) a <u>point on the line of action of the force</u>; and (3) <u>direction of the force</u>.

11.3   Couples

A couple is defined as two equal, opposite, and parallel forces, as shown in Fig. 11.1. The resultant force is zero. However, *the*

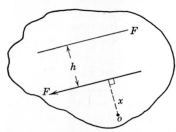

Fig. 11.1.   A couple defined as two equal, opposite, and parallel forces.

moment of the two forces is of constant magnitude, regardless of the point about which the moment of the forces is taken, and is equal to the product of the force F and the distance h between the forces. This may be shown by taking the moments of the forces about any point, o, which is at a distance x from one of the forces, as shown in Fig. 11.1. The

Three Non-parallel Forces in Equilibrium   137

clockwise moment is $F(h + x)$, and the counterclockwise moment is $F(x)$. The sum of the moments is $F(h + x) - F(x)$, clockwise, and is equal to $F(h)$. Or the moment of the couple does not depend on the center of moments, since the distance x does not appear in the final expression for the magnitude of the couple.

11.4 Three non-parallel forces in equilibrium

Consider three forces, F_1, F_2, and F_3, acting on a link, as shown in Fig. 11.2a. *The graphical solution of the relation that the forces must have a zero resultant is obtained when the force polygon*, as shown in Fig. 11.2b, *gives a closed figure.* However, even though the resultant force

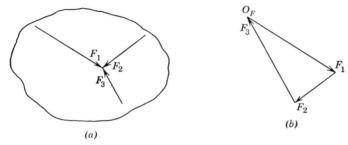

Fig. 11.2. To satisfy equations of equilibrium, three forces must give a closed force polygon and must intersect at a common point.

is zero, it is still possible that the moment equation is not satisfied because the resultant of the forces acting on the body may be a couple, that is, two equal and opposite parallel forces for which the resultant force would still be zero. It is to be noted that the moment of a couple is of constant magnitude, regardless of the point about which moments are taken. *If the moment about any point is taken, and the moment is zero, there cannot be a couple acting on the body.* Consequently, if three forces, whose resultant is zero, intersect at a common point, the moment of each force about a particular point, the point of intersection, is zero, which precludes the possibility of a couple. Therefore, a body on which are acting three non-parallel forces is in equilibrium if the resultant force is zero, as determined by the force polygon closing, and if the forces intersect at a common point.

As an illustration of the type of problem encountered, consider Fig. 11.3a. Force F_1 is known in direction and in magnitude, force F_2 is known in direction only, and a point m on the line of action of a force F_3 is given. It is required to find the magnitude of F_2 and magnitude and direction of F_3. The direction of F_3 is obtained first in that it must pass through the intersection of F_1 and F_2; or a moment equation is applied first. The force polygon is shown in Fig. 11.3b.

It is to be noted that three unknown quantities were determined: magnitude of F_2, and magnitude and direction of F_3. In any statically

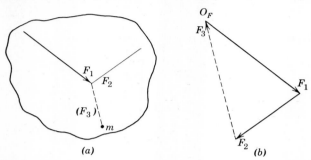

Fig. 11.3. Force F_2 found in magnitude, and force F_3 found in magnitude and direction. Or three unknowns determined.

determinate force problem, three unknown quantities may be determined, and only three. Thus an analytic or graphical solution allows the determination of the same number of unknowns in a given problem.

11.5 Four non-parallel forces in equilibrium

Two cases are discussed here for the situation where four non-parallel forces acting on a body are in equilibrium: (1) three unknown magnitudes; and (2) two unknown magnitudes and one unknown direction.

Case 1. Figure 11.4a shows a link in equilibrium on which are acting four forces, F_1, F_2, F_3, and F_4. F_1 is known completely, whereas F_2, F_3, and F_4 are known in direction only. The graphical solution may be analyzed in either of two ways, both of which give the same result.

Method a. If moments of forces are taken about point m, the intersection of the forces F_3 and F_4, the following equation may be written:

$$(F_1)(a) = (F_2)(b)$$

where a and b represent the distance from point m to the corresponding force. (F_2 must be directed upward to the left as a result of the moment analysis.) This equation may be solved by similar triangles if the equation is rewritten:

$$\frac{F_1}{F_2} = \frac{b}{a}$$

Figure 11.4b shows the similar triangles which permit finding the

Four Non-parallel Forces in Equilibrium 139

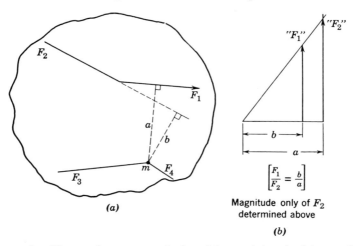

Fig. 11.4a, b. Three unknown magnitudes of forces determined by application initially of a moment equation in a four-force system.

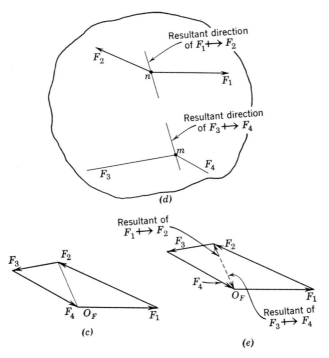

Fig. 11.4c, d, e. Alternate application of a moment equation in a four-force system.

magnitude of F_2. F_3 and F_4 may be determined next from a force polygon as shown in Fig. 11.4c.

Method b. The second method, which requires less work, is to note that if moments are taken about point m, the resultant of F_1 and F_2 must pass through point m so that the moment equation be satisfied. Figure 11.4d shows the direction of the resultant. Figure 11.4e shows the determination of F_2 with the direction of the resultant of F_1 and F_2 known. By similar reasoning, the resultant of F_3 and F_4 must pass through point n, the intersection of F_1 and F_2. The resultant of F_3 and F_4 must be equal and opposite to the resultant of F_1 and F_2 for equilibrium of the body. Figure 11.4e shows the direction and magnitude of F_3 and F_4.

Case 2. Figure 11.5 shows a link in equilibrium under the action of four forces: F_1 and F_2 are known completely; F_3 is known in direction only; a point only on the line of application of F_4 is known.

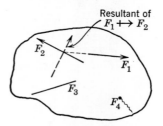

Fig. 11.5. A four-force system which reduces to a three-force system.

This case will reduce to that of a three-force member if F_1 and F_2 are combined to a single resultant force.

11.6 Five or more non-parallel forces in equilibrium

In such problems, where five or more forces are encountered, the problem may be reduced to a three- or a four-force member and may then be handled in the manner described in the previous sections.

11.7 Parallel forces

It is necessary to satisfy the same equations of equilibrium for a system of parallel forces acting on a given body in equilibrium. However, the interpretation of the application of the equations to a graphical solution is slightly different inasmuch as the forces do not intersect at a finite point.

Three parallel forces are shown in Fig. 11.6a. Force P is known completely; forces F_1 and F_2 are unknown in magnitude. A force polygon cannot be used to determine F_1 and F_2 since all the forces

Parallel Forces 141

are parallel. It is necessary to resort to a moment equation for either one or both of the reactions. Take moments about any point on the line of action of F_1, point o, for instance:

$$\Sigma M_o = 0 = +(P)(a) - (F_2)(b)$$

or
$$\frac{P}{F_2} = \frac{b}{a}$$

The foregoing relation may be solved by a similar triangle relationship. The construction to give the above is shown in Fig. 11.6b,

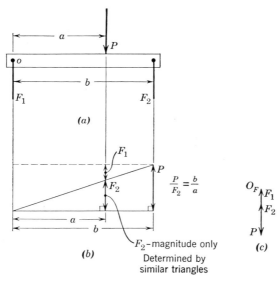

Fig. 11.6. Moment equation applied to a system of three parallel forces results in a solution by similar triangles.

where P is placed to scale at F_2 and the figure drawn as shown. A check of the figure will show that the similar triangle relationship is satisfied. The direction of F_2 is determined by inspection of the moment equation.

F_1 may be determined in one of two ways:

(1) By a force polygon, as shown in Fig. 11.6c.
(2) By applying a moment equation about any point on the line of action of F_2, in a manner comparable to that for finding F_2.

Note that the interpretation of the application of the equations of equilibrium to parallel forces is comparable to that for non-parallel

142 Review of Static Forces and Graphic Statics

forces in that, if the forces have a zero resultant and if the moment about any point is zero, which precludes a couple, the system must be in equilibrium.

11.8 Parallel forces—alternate method

A second method, the method of resolution, for the solution of parallel forces, should be mentioned. From any point on the line of action of the known force, draw lines m and n intersecting F_1 and F_2, as shown in Fig. 11.7a. Draw a force polygon, as shown in Fig. 11.7b,

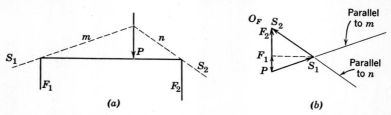

Fig. 11.7. Alternate method of solving a system of three parallel forces by introducing components which cancel each other: the horizontal components of S_1 and S_2.

with the known force, P, and forces parallel to lines m and n. The component of S_1 and S_2 parallel to P will give the forces F_1 and F_2, respectively. Note that essentially two equal, opposite, and collinear forces were added vectorially to the system to obtain a non-parallel system.

11.9 Resultant of two parallel forces

It is sometimes desirable to find the resultant of two parallel forces, P_1 and P_2, as shown in Fig. 11.8a. A resultant force is defined as one which will have the same effect as the components. Consequently, the vector sum, here the algebraic sum since the forces are parallel, is $P_1 + P_2$. Figure 11.8b shows the resultant at an unknown distance x from P_1. Also, since the resultant force will have the same effect as the components, the moment of the resultant force about any point will be the same as the moment of the components. The following equation, with point o in Fig. 11.8b used as the center of moments, results:

$$(P_1 + P_2)(x) = (P_2)(b)$$

or

$$\frac{x}{b} = \frac{P_2}{P_1 + P_2}$$

It is left up to the student to show that the construction in Fig. 11.8c gives x.

Resultant of Two Parallel Forces

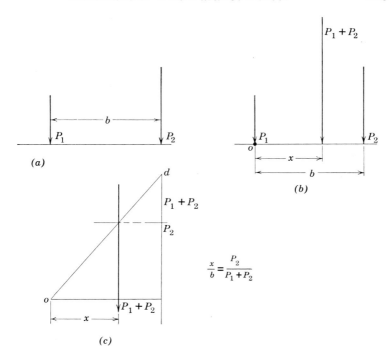

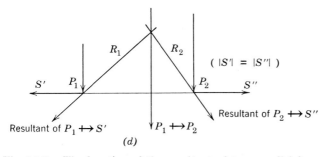

Fig. 11.8. The location of the resultant of two parallel forces found in (c) by a moment equation and found in (d) by adding equal, opposite, and collinear forces which cancel each other.

11.10 Resultant of two parallel forces—alternate method

An alternate method of finding the resultant of two parallel forces is to introduce two *equal, opposite,* and *collinear* forces, S' and S'', which do not affect the relation of forces in any way. The resultant of P_1 and S' is R_1, the resultant of P_2 and S'' is R_2, as shown in Fig. 11.8d. The resultant force, $P_1 + P_2$, must pass through the intersection of R_1 and R_2.

11.11 Two-force members

A type of loading encountered very frequently in force analysis is the so-called two-force member. Figure 11.9a shows a link with two forces, F_1 and F_2, acting at A and B. It is desired to obtain the relation of F_1 and F_2.

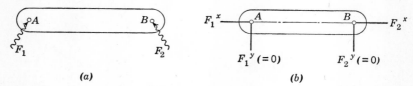

Fig. 11.9. A two-force system requires the forces to be equal, opposite, and collinear to satisfy the equations of equilibrium.

Consider F_1 resolved into two components: $F_1{}^x$ along the line from A to B; $F_1{}^y$ perpendicular to the line from A to B, as shown in Fig. 11.9b. Similarly, $F_2{}^x$ and $F_2{}^y$ are obtained.

The sum of the forces in a given direction must add up to zero. Therefore, $F_1{}^x$ must be equal and opposite to $F_2{}^x$. Also, $F_1{}^y$ must be equal and opposite to $F_2{}^y$. However, if $F_1{}^y$ is equal and opposite to $F_2{}^y$, a couple results for any magnitude of each force. In order to satisfy the moment equation, it is necessary that $F_1{}^y = F_2{}^y = 0$.

Consequently, if only *two* forces act on a body in equilibrium the forces must be equal, opposite, and collinear.

11.12 Special construction

Two special constructions should be mentioned to enable the student to solve more easily certain types of graphical force problems. The constructions are given to provide for greater accuracy and less work.

Case 1. Line of action of a force to pass through a given point x and the intersection of two forces at a point off the paper. Figure

11.10a shows the line of action of two forces, F_1 and F_2. It is desired to have a force, F_3, pass through point x and the intersection of F_1 and F_2. The method of construction and proof are based upon the properties of similar triangles. In Fig. 11.10b, any line, l–l, is drawn through point x; and any line, m–m, is drawn parallel to l-l. Point p

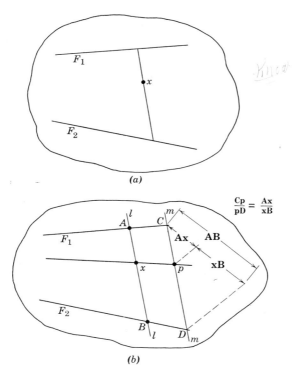

Fig. 11.10. Relation of similar triangles used to determine the direction of a force to pass through a given point and through the intersection of two forces at a point off the paper.

is determined so that $\dfrac{Cp}{pD} = \dfrac{Ax}{xB}$, as shown in the figure. The line through points p and x will pass through the intersection of F_1 and F_2.

Case 2. Resultant of two nearly parallel forces. Figure 11.11a shows two known forces, P_1 and P_2, which are almost parallel. It is desired to obtain their resultant, in magnitude and line of action. If two equal, opposite, and collinear forces S' and S'', are imagined to be acting on the link, no change is made. The resultant of P_1 and S', and P_2 and S'' may be determined. The resultant R will act through point a, with the magnitude and direction determined by a force diagram as shown in Fig. 11.11b.

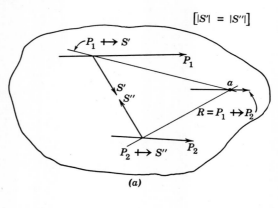

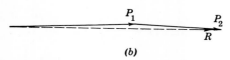

Fig. 11.11. Addition of equal, opposite, and collinear forces to find the location of the resultant of two forces which intersect off the paper. Procedure applicable also to parallel forces.

PROBLEMS

11.1 (Fig. 11.1). (a) The resultant of F_1 and F_2 acts through point P. Determine the magnitude and direction of the resultant force.

(b) Determine the force acting through point P that puts the system shown in equilibrium.

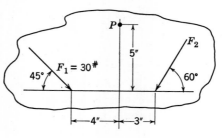

Fig. P–11.1.

Problems

11.2 (Fig. 11.2). Determine the force that puts the system shown in equilibrium.

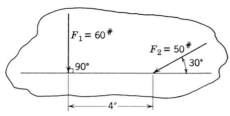

Fig. P-11.2.

11.3 (Fig. 11.3). (a) For the system shown, the resultant of F_1, F_2, F_3 is known to act through point P. Determine the magnitude of the resultant force and the magnitude of F_3.
(b) What is the magnitude of the force acting through point P that puts the system in equilibrium?

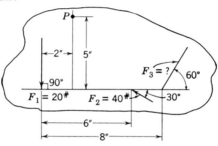

Fig. P-11.3.

11.4 (Fig. 11.4). The system shown is composed of a force and a couple. What is the resultant force, and where is it applied? Show that the system can be put in equilibrium by a single force.

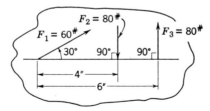

Fig. P-11.4.

11.5. (a) Show that the resultant of any force and any couple is a single force.
(b) Show that a single force may be replaced by a force and couple.

11.6. (Fig. 11.6). Determine the resultant of the two parallel forces shown in a and the resultant of the two parallel forces shown in b.

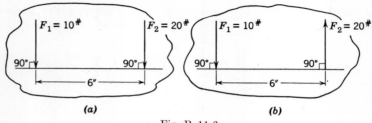

Fig. P–11.6.

11.7 (Fig. 11.7). Determine the resultant of the three parallel forces shown.

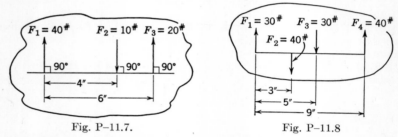

Fig. P–11.7.   Fig. P–11.8.

11.8 (Fig. 11.8). Determine the resultant of the four parallel forces shown. How could the system be put in equilibrium?

11.9 (Fig. 11.9). Forces are applied at the pins A, B, and C of the beam, as shown. Determine the magnitude of F_4 and the magnitude and direction of F_1 for equilibrium.

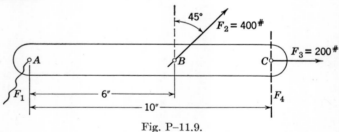

Fig. P–11.9.

11.10 (Fig. 11.10). The forces F_1 and F_2 are known to be perpendicular to the axis of the beam. Find the magnitude of F_2 and the magnitude and direction of F_3 for equilibrium.

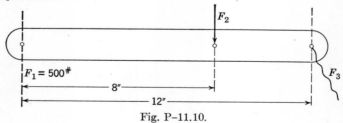

Fig. P–11.10.

Problems

11.11 (Fig. 11.11). The belt forces F_1 and F_2 are applied to a pulley, as shown. Determine the direction and the magnitude of the resultant force.

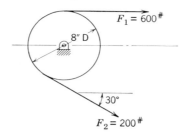

Fig. P–11.11.

11.12 (Fig. 11.12). Determine the magnitude of F_1 and F_3 for equilibrium.

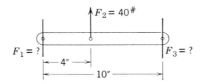

Fig. P–11.12.

11.13 (Fig. 11.13). Determine the magnitude of F_1 and F_2 for equilibrium.

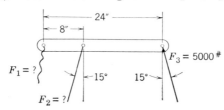

Fig. P–11.13.

11.14 (Fig. 11.14) A force F_1 is applied to a body, as shown. Replace the force F_1 by a force applied at point p and a couple to give the same effect as F_1.

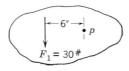

Fig. P–11.14.

CHAPTER 12

Static Forces in Machines

The forces acting upon machine members may arise from many different sources: weights of parts, forces from energy transmitted, forces of assembly, forces from applied loads, friction forces, forces from changes of temperature, impact forces, spring forces, and inertia forces. Each and all of these forces must be considered in the final design of a machine. This book, however, will be restricted to the analysis of forces from weights of parts, forces from energy transmitted, forces from applied loads, inertia forces, friction forces, and spring forces. For consideration of the effect of the other forces, the student is referred to books on machine design.

Static forces only will be considered in this chapter.

12.1 How forces are applied in machines

Forces are applied or transmitted through gears, pins, shafts, sliding members, and the various links which make up a machine.

Case a. Gears. Space does not permit a complete description of all types of gears and the components of forces acting on gears. This book is restricted to the simple straight-tooth spur gears where the force transmitted between two gears is directed along the common normal to the tooth surfaces at the point of contact, if friction is disregarded. The common normal is the so-called pressure line for teeth of involute profile. The commonly used standard pressure angles are $14\frac{1}{2}$ and 20 degrees. Figure 12.1a shows two gears, A and B. Gear A is the driver, and gear B is the driven gear. The pressure angle, ϕ, is shown. Figure 12.1b shows the resultant force, R, acting through the pitch point and the radial and tangential components, F_T and F_R, of the resultant force.

Case b. Pins. If friction and the weight of the pin are disregarded, the forces acting on a pin must pass through the center of the pin. Figure 12.2a shows a pin in a link with each differential force normal to the surface and passing through the center of the pin.

How Forces are Applied in Machines   151

Consequently, the resultant force must pass through the center of the pin, as shown in Fig. 12.2b.

If friction is considered, the resultant force on the pin does not pass through the center of the pin, but is offset from the center an amount to give a torque equal to the friction torque, as shown in

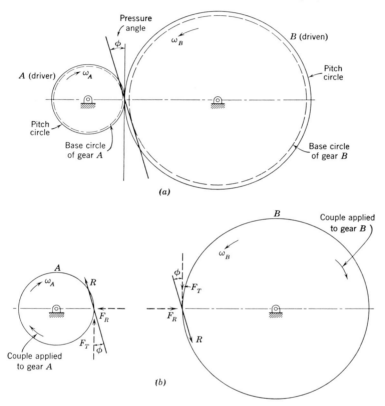

Fig. 12–1. Forces applied through gears.

Fig. 12.2c. The exact location of the resultant force will be discussed later, in the section on friction.

Case c. Sliding members. Another common machine member is the slider, or piston, or cross-head shown schematically in Fig. 12.3a. The reaction forces are perpendicular to the surfaces in contact, if friction is disregarded, as shown in Fig. 12.3a. If friction is considered, the resultant force is not perpendicular to the surface, but is inclined from the vertical by an angle ϕ, as shown in Fig. 12.3b. The angle ϕ is defined by

152   Static Forces in Machines

$$\tan \phi = \frac{\mu N}{N} = \mu$$

where μ is the coefficient of friction. ϕ is called the friction angle.

For simplification, friction will be disregarded for the initial machines analyzed. It will be considered in more detail in later sections.

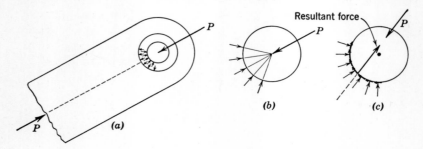

Fig. 12.2.   Forces acting on pins.

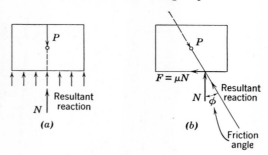

Fig. 12.3.   Forces on sliding members.

Weights of parts will not be discussed as separate forces since they may be considered as external loads and may be treated as such.

12.2 Slider-crank mechanism

Figure 12.4a shows the slider-crank mechanism. A force P, which we may assume to be the resultant of gas pressure, is acting on the piston as shown. The system is kept in equilibrium as the result of a couple applied to link 2 through the shaft at O_2. It is required to find the forces acting on all the links, including the pins, and the couple on link 2. Note that the couple on link 2 may be pictured as applied through the shaft at O_2 or may be pictured as an actual couple applied to link 2 directly.

The procedure of attack for all problems in force analysis is the same: *Isolate each member, making a free body diagram of the forces acting on the member. If the unknowns are not greater than three in number, the problem may be solved by the application of the equations of equilibrium. If there are more than three unknowns for a single body, additional information must be obtained elsewhere by consideration of other members.* The proper isolation of members, or the making of free body diagrams, cannot be over emphasized.

Figure 12.4b shows each member isolated, with the known quantities of the various forces. Link 3 is a two-force member inasmuch

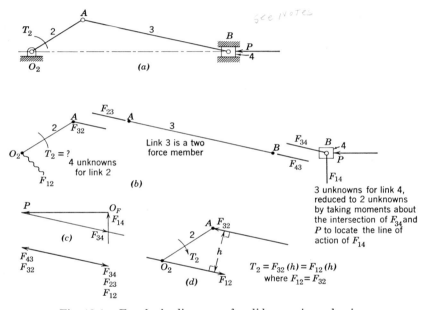

Fig. 12.4. Free body diagrams of a slider-crank mechanism.

as forces are applied at the ends of the bar and no other forces are acting on the link. Whether link 3 is in compression or tension cannot be determined from an analysis of link 3 by itself. Link 4 has three forces acting on it: (1) the known force P; (2) a force F_{34} exerted by link 3 on link 4, which is known in direction because the action of link 4 on link 3 must be along line A–B, since link 3 is a two-force member, and the action and reaction between links 3 and 4 must be equal and opposite; and (3) a force F_{14} perpendicular to the guide surface, which is known in direction, but the magnitude is unknown and a point on the line of action of F_{14} is unknown.

Link 2 has four unknowns: (1) the force F_{32} exerted by link 3 on link 2, known in direction, unknown in magnitude; (2) the force exerted by link 1 on link 2, unknown in direction and magnitude; and (3) the unknown couple applied to link 2, T_2.

Note the nomenclature used for the force representation: F_{14} means the force exerted by link 1 on link 4; F_{41} means the force exerted by link 4 on link 1. This system will be used throughout force analysis.

Link 4, which has only three unknowns, is analyzed first. F_{14} must pass through the intersection of P and F_{34} to satisfy the moment equation. The other two unknowns, magnitude of F_{34} and F_{14}, are found by a force polygon, as shown in Fig. 12.4c. F_{43} is equal and opposite to F_{23}, which puts link 3 in compression for this case.

F_{12} must be equal and opposite to F_{32}, to balance the forces on link 2. However, the two forces, equal, opposite, and parallel give a couple which can be balanced only by another couple. The balancing couple, T_2, is equal to $(F_{32})(h)$, and is clockwise. Figure 12.4d shows the final system for link 2.

12.3 Press mechanism

Figure 12.5a shows a press with a known force P acting on link 7. Links 2 and 3 are gears with 20-degree involute teeth. It is

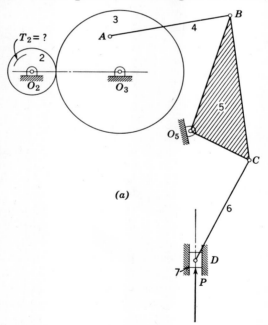

Fig. 12.5a. Given system of forces acting on a press mechanism.

Press Mechanism   155

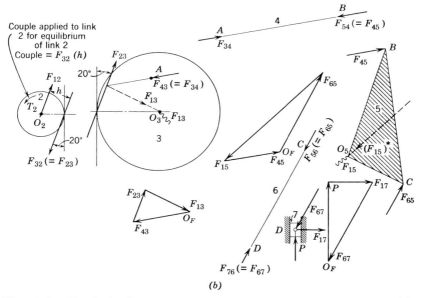

Fig. 12.5b. Free body diagrams for the press mechanism. *Direction found by method of page 145.

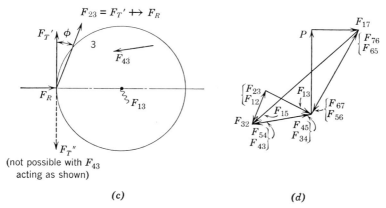

Fig. 12.5c, d. Force analysis for the gear, link 3, and the complete force polygon.

required to find the couple which must be applied to gear 2 to maintain equilibrium.

Figure 12.5b shows free body diagrams of each link. Links 4 and 6 are two force members. The arrows for the forces were put on the free body diagrams *after* the force polygons were drawn. The force analysis is begun with link 7, inasmuch as only three unknowns are present. The force diagram for link 7 is shown. F_{67} is equal and opposite to F_{76}. Link 6 is in compression. F_{15} must pass through the intersection of F_{45} and F_{65}*, since link 5 is a three-force member. The force diagram for link 5 is shown. Link 4 is in compression.

Link 3 has acting on it three forces: F_{43}, F_{13}, F_{23}. F_{23}, the action of gear 2 on gear 3, is inclined at the pressure angle, as shown. It must be acting as shown in order to counteract the moment about O_3 due to F_{43}.† Link 3 is a three-force member, with the three forces intersecting at a point. The direction of F_{13} is thus determined. The force diagram for link 3 is shown in Fig. 12.5b.

An analysis of link 2 shows that F_{12} and F_{32} are the only forces acting on link 2. These two forces must be equal and opposite for equilibrium of forces, but at the same time they will set up a couple which must be balanced by a couple applied externally to link 2. The external couple is equal to $(F_{32})(h)$, where h is the distance between F_{32} and F_{12}, and must be applied in a counterclockwise direction.

The combining of all the individual force polygons gives the complete force polygon shown in Fig. 12.5d.

12.4 Riveter mechanism

Another illustrative example is the riveter mechanism shown in Fig. 12.6a. (A known force, P, is maintained in equilibrium by a force Q.) Links 3, 5, 6, and 7 are two force members. The analysis of the forces acting on link 2, with force P known, is shown in Fig. 12.6b. Link 4, shown in Fig. 12.6c, has four forces acting on it, one of which, F_{34}, is known since link 3 is a two-force member and F_{32} has been found. The resultant R of F_{54} and F_{34}, intersecting at x, must pass through the intersection of F_{64} and F_{74}, that is, point y, to satisfy the moment equation. Note that with three unknowns, it is necessary to apply a moment equation first. The force polygon for link 4 is shown in Fig. 12.6d.

* The construction used for the determination of the point of intersection, when the point of intersection is off the paper, is described on page 145. The construction lines are *not* shown.

† A convenient method for determining the general direction of F_{23} is to consider the resultant gear force broken up into radial and tangential component forces. The radial component for spur gears is *always* directed towards the center of the gear. That leaves only the determination of the direction of the tangential component. Figure 12.5c shows two possibilities: $F_{T'}$ and $F_{T''}$. If moments are taken about the center, it is evident that for the direction of F_{43}, $F_{T'}$ is the only correct direction of tangential force to give a moment about the center of the gear to counteract the moment due to F_{43}. Thus $F_{23} = F_{T'} +\!\!\!\!+ F_R$.

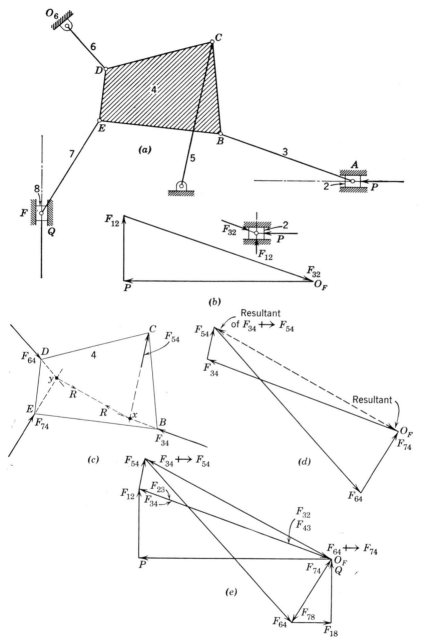

Fig. 12.6. Force analysis of a riveter mechanism.

158 Static Forces in Machines

The complete force polygon for the entire mechanism is shown in Fig. 12.6e.

12.5 Four-link mechanism

The four-link mechanism shown in Fig. 12.7a has acting on it two known forces, P and S, as shown. The system is in equilibrium as the result of a couple T_2 applied to link 2. It is required to find the various pin forces and the couple applied to link 2.

Is it possible to solve for the required quantities by analysis of the entire mechanism as a whole? There are five unknowns if the mechanism is considered as a whole: direction and magnitude of F_{14},

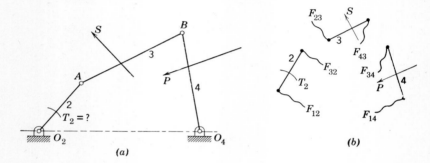

Fig. 12.7a, b. Four-link mechanism, with the forces acting on the isolated links. There are nine unknowns and nine equations of equilibrium.

magnitude and direction of F_{12}, and T_2. With only three equations of equilibrium, we cannot obtain a direct solution.

Consider each link isolated, as shown in Fig. 12.7b. There are nine unknowns, magnitude and direction of each F_{14}, F_{34} (or F_{43}), F_{23} (or F_{32}), F_{12}, and the magnitude of T_2. However, nine equations of equilibrium may be applied (three for each link). The key to the solution, then, is to isolate the links and to consider each separately.

If link 4 is considered, there are four unknowns to be found; if link 3 is considered, there are four unknowns to be found; and if link 2 is considered, there are five unknowns. Therefore, each link cannot be analyzed by itself. Note that if links 3 and 4 are considered, there are six unknowns, with six equations of equilibrium that can be applied, three for each link. Therefore, consider links 3 and 4, as shown in Fig. 12.7c, where F_{34} is broken up into components: F_{34}^{T4}, the component perpendicular to $B-O_4$, and F_{34}^{N4}, the component along $B-O_4$. F_{34}^{T4} may be found by taking moments about point O_4.

By action and reaction, the forces acting at B on link 3 are equal and opposite to the forces acting at B on link 4. The forces acting at B on link 3 are represented by $F_{43}{}^{T4}$ and $F_{43}{}^{N4}$ where $F_{43}{}^{T4}$ has just been found from an analysis of link 4. If the attention is focused on link 3, we will recognize three unknowns: magnitude and direction of F_{23} and magnitude of $F_{43}{}^{N4}$. Link 3 can be reduced to a three-force system by combining S and $F_{43}{}^{T4}$. F_{23} must act through the intersection of the resultant so found and $F_{43}{}^{N4}$ by the application of a moment equation. The force polygon for link 3 is shown in Fig. 12.7d.

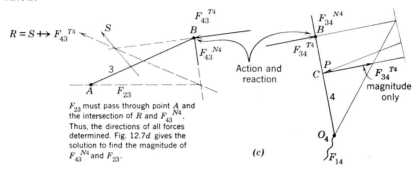

Fig. 12.7c. Analysis of links 3 and 4, with six unknowns and six equations of equilibrium.

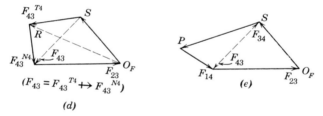

Fig. 12.7d, e. Force polygon for link 3 and the complete force polygon.

The force polygon for the entire system is shown in Fig. 12.7e. The couple applied to link 2 may be readily found since F_{23} has been found.

Note that a moment equation could have been applied to link 3 first, and with the information obtained link 4 could be analyzed. The complete analysis for such a procedure is left to the student.

12.6 Static couples on links

Figure 12.8a shows a mechanism with couples applied to two links: T_4 on link 4 and T_2 on link 2. In addition, a force P is applied to link 3. T_4 and P are known. The value of T_2 is desired.

Static Forces in Machines

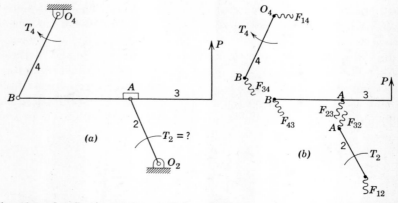

Fig. 12.8a, b. Static couples applied to two links. There are nine unknowns: magnitude and direction of F_{14}, F_{34} (or F_{45}), F_{23} (or F_{32}), F_{12}, and the magnitude of T_2. There are nine equations of equilibrium, so that a solution is possible. If links 3 and 4 are considered, there are six unknowns: magnitude and direction of F_{14}, F_{34}, and F_{23}, with six equations of equilibrium available. Thus, links 3 and 4 are analyzed first.

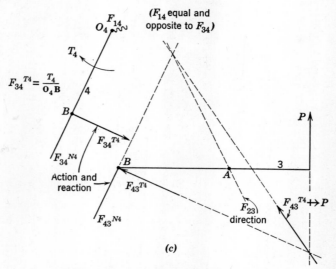

Fig. 12.8c. Analysis of links 3 and 4.

Figure 12.8b shows the links isolated. Note that there are nine unknowns, and nine equations of equilibrium. Link 4 is analyzed first, in part, with a moment equation applied to find $F_{34}{}^{T4}$, as shown in Fig. 12.8c. Link 3 can be handled as a three-force system, as shown in the same figure. Figure 12.8d shows the complete force polygon, whereas link 2 is isolated in Fig. 12.8e to show how T_2 can be found. It is left up to the student to go through each step in the analysis in detail.

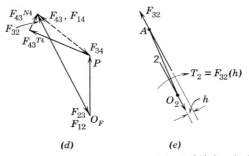

Fig. 12.8d, e. Complete force polygon in (d), and link 2 isolated in (e).

12.7 Zoller double-piston engine

Figure 12.9a shows the Zoller double-piston mechanism which is assumed to be in equilibrium under the action of two external forces, P and S, and a couple applied to link 2. Figure 12.9b shows each link isolated.

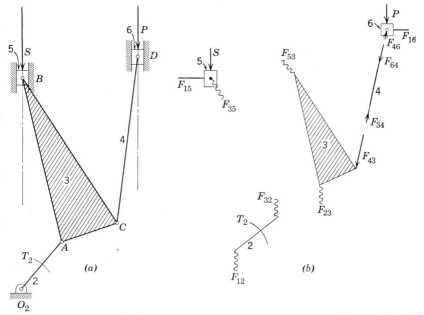

Fig. 12.9a, b. Zoller double-piston engine, with forces shown acting on isolated links.

Link 4 is a two-force member. The forces on link 6 may be determined quickly.

Link 3 cannot be analyzed completely since there are four unknowns, magnitude and direction of F_{53} and F_{23}. Link 5 cannot be analyzed either since there are three unknowns: magnitude of F_{15}, direction and magnitude of F_{35}. (Locating a point on the line of action of F_{15}, the

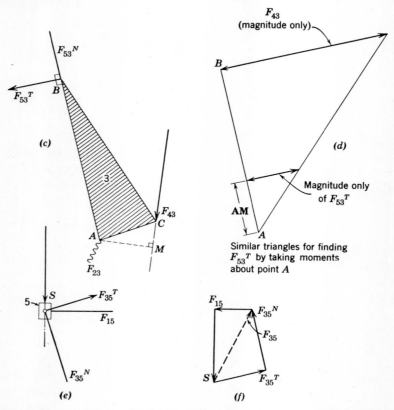

Fig. 12.9c, d, e, f. Determination of $F_{53}{}^T$ by application of a moment equation to link 3 gives the necessary information for analysis of link 5 and then of link 3.

center of the pin of link 5, has used up one equation, the moment equation.) There are five unknowns for links 3 and 5, and five equations available. Consequently, the five unknowns should be determinable. The procedure of solution is comparable to that for the four-link mechanism. Consider F_{53} broken up into components, $F_{53}{}^T$ perpendicular to CA, and $F_{53}{}^N$ along AB, as shown in Fig. 12.9c. By taking moments about A, we may find $F_{53}{}^T$ by the relation: $(F_{43})(\mathbf{AM})$

Friction   163

$= (F_{53}{}^T)(\mathbf{AB})$. Similar triangles as shown in Fig. 12.9d will allow the solution of the equation for the magnitude of $F_{53}{}^T$. Its direction is such as to oppose the moment of F_{43} about point A.

Figure 12.9e shows link 5 redrawn with the force F_{53} replaced with the components, one of which was determined from link 3. Link 5 may be solved now as shown in Fig. 12.9f since there are only two unknowns, $F_{35}{}^N$ and F_{15}.

Figure 12.9g shows link 3 redrawn with the resultant F_{53} shown applied. Two unknowns at this point, magnitude and direction of

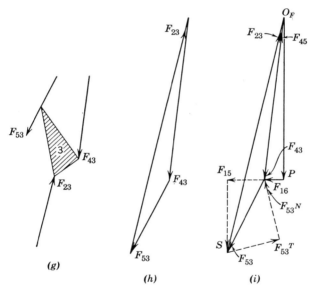

Fig. 12.9g, h, i. Completion of the force polygon.

F_{23}, may be determined by a force polygon, as shown in Fig. 12.9h. A check of the accuracy of the work is made to learn if the direction of F_{23} as found from the force polygon is such that F_{23} will pass through the intersection of F_{53} and F_{43}.

The determination of the couple that has to be applied to link 2 follows the procedure given in the preceding sections. It is found to be counterclockwise.

The complete force polygon is shown in Fig. 12.9i.

12.8 Friction

Friction plays a strange role in that many machines could not operate without it, yet at the same time the same machine must work against it. As an illustration, an automobile could not be started were it not for friction of the tires on the pavement, and could not be stopped, once it had been set in motion, were it not for the brake friction and the friction of the tires on the pavement. On the other hand, there is friction in all the moving parts of the automobile—bearings, tires, gears, pistons, etc.—which requires the use of extra gasoline.

Friction forces are difficult quantities with which to work because of their irregular behavior. The slightest change in condition of the surfaces in contact may introduce considerable change in the friction.

In elementary work, friction of dry surfaces with average pressures is considered independent of speed of sliding and the area of contact of the surfaces and the temperature, but depends only upon the materials. Actually, the friction will in general increase considerably for high pressures as the result of deformation of the material, will decrease for high velocities, and will decrease with an increase in temperature.

If the surfaces are lubricated, the friction depends on many quantities: speed, pressure, viscosity of oil, temperature. The friction forces for "imperfect," or thin film lubrication, are very much different from the friction forces under so-called thick film lubrication, or "perfect" lubrication.

Consequently, once a coefficient of friction has been evaluated, the theory behind the use of the coefficient is definite. However, the evaluation of a coefficient is not as precise. It is beyond the scope of this book to deal with the variations of coefficients of friction. Coefficients of friction will be assumed, and the procedure of solution of problems involving friction will be given.

12.9 Sliding friction

The procedure of handling problems involving sliding friction is best illustrated by several examples.

Slider-crank mechanism. The slider-crank mechanism shown in Fig. 12.10a has three forces, P, Q, and S, acting on it as shown. The problem is to determine the forces on the pins and the couple that must be applied to link 2 to give equilibrium. The coefficient of sliding friction is assumed to be 0.2. Pin friction is neglected in this problem.

The problem as stated so far is not complete in that link 4 may be on the point of moving in either direction along the guide, for which the full friction force would be developed. However, two different answers will be obtained for impending motion in the two directions.

Sliding Friction

This may be seen by visualizing link 4 to be on the point of moving to the right. The friction force on link 4 would then be acting to the left, the same direction as for the load P. The couple applied to link 2 would have to overcome the force P as well as friction. On the other hand, if the impending motion of link 4 is to the left, the friction force on link 4 would be to the right, or, in the opposite direction to P. In which case, P would be partly balanced by the friction force, the rest of the balancing effect coming from the couple on link 2. Consequently, let us assume that link 4 is in impending motion to the right.

Figures 12.10b and 12.10c show link 4 isolated, with F_{34} unknown in magnitude and direction (since link 3 is not a two-force member) but known to pass through the center of the pin. F_{14} must pass through the center of the pin since the other two forces intersect at that point.

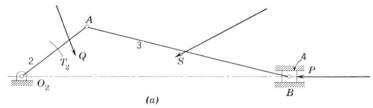

Fig. 12.10a. Slider-crank mechanism, with known forces P, S, and Q.

Whether F_{14} is downward to the left or upward to the left, or, expressing it differently, whether the normal force on link 4 is on the top or bottom, cannot be determined at this point. The procedure of solution is to assume either case, proceed with the solution, and check the final result with the assumed direction. If the directions check, the solution is satisfactory. If the solution does not check, the other possibility of direction must be tried. In some cases, we can visualize the forces and assume the direction properly at first to eliminate the necessity of two solutions.

Before we proceed, we might note that the case of impending motion to the left would result in the direction of forces shown in Figs. 12.10d and 12.10e.

There are two procedures available at this point for solving the problem:

(1) Consider links 3 and 4 isolated, as shown in Fig. 12.10f. The resultant force at B is broken up into components, one perpendicular to A–B and the other parallel to A–B. The component perpendicular to A–B can be found by a moment equation. Link 4 can be analyzed completely when $F_{43}{}^{T3}$ has been found. The solution is then com-

parable to that for the Zoller double-piston engine. The solution by this method is left to the student.

(2) In the second method links 3 and 4 are considered as one body, as shown in Fig. 12.10g. Here we have essentially a three-force system

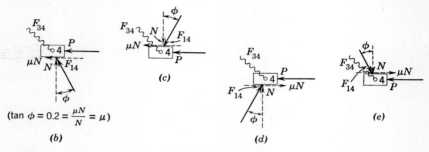

$(\tan \phi = 0.2 = \dfrac{\mu N}{N} = \mu)$

(b)

(c)

(d)

(e)

Fig. 12.10b, c. Possible locations of the normal force on the slider. In (b), the impending motion of link 4 is to the right, with contact assumed on the bottom; in (c) the impending motion is to the right, with contact assumed on the top.

Fig. 12.10d, e. Impending motion to the left, with contact on the bottom gives (d), whereas impending motion to the left with contact on the top gives (e).

with three unknowns: magnitude and direction of F_{23} and magnitude of F_{14}. (F_{14} must pass through the center of the pin as can be seen from application of a moment equation about the center of the pin

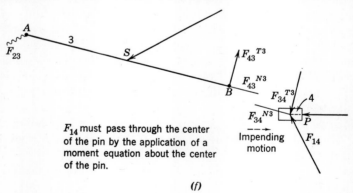

F_{14} must pass through the center of the pin by the application of a moment equation about the center of the pin.

(f)

Fig. 12.10f. Moment equation applied to link 3 to permit analysis of link 4. Link 3 considered isolated from link 4.

for link 4, isolated.) The forces P and S can be combined to give a single resultant, R. Figure 12.10g shows the direction of F_{23}, which must pass through the intersection of R and F_{14}. The forces are

Sliding Friction 167

shown in Fig. 12.10i. Note that the direction of F_{14} is as assumed; contact exists on the bottom surface of link 4.

Q and F_{32} acting on link 2 are combined to give a resultant force acting through point m in Fig. 12.10h. F_{12} must be equal, opposite,

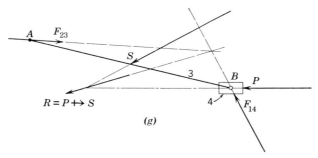

Fig. 12.10g. Links 3 and 4 isolated as one body.

and parallel to the resultant of Q and F_{32}. The resulting couple is opposed by a couple T_2 applied to link 2 in a clockwise direction. $T_2 = (F_{12})(h) = (F_{32} \looparrowright Q)(h)$.

Crank shaper mechanism. Figure 12.11a shows the crank shaper mechanism. A force P from the cutting action on metal acts

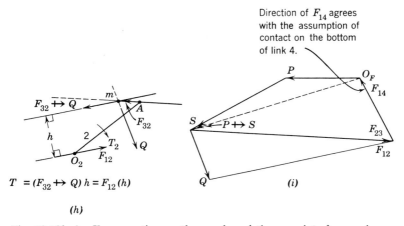

Fig. 12.10h, i. Forces acting on the crank and the complete force polygon.

on the tool attached to link 7 as shown. The system is in equilibrium under the action of a couple applied to link 2, a gear. The pressure angle is 20 degrees. The coefficient of friction for the sliding members 4 and 7 is assumed as 0.1. Link 7 is in impending motion to the left.

It is required to analyze the system completely for all the forces present. Pin friction is to be neglected.

Before proceeding with the force analysis, we must fix upon the direction of impending motion of link 4 in link 5 in order to obtain the direction of the friction force. A velocity analysis, as though link 7 were actually moving to the left, will show that link 4 would be moving

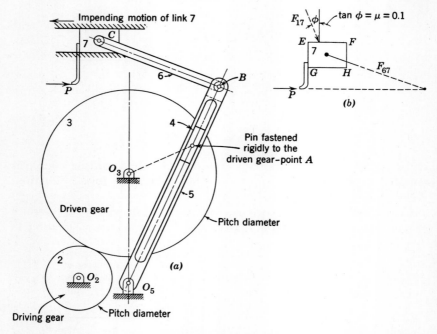

Fig. 12.11a, b. Crank shaper mechanism. F_{17} acting on the top surface of link 7 is impossible, inasmuch as equilibrium cannot be obtained since F_{17} cannot pass through the intersection of F_{67} and P, yet act on the top surface at an angle ϕ from the vertical. (A similar analysis with assumed contact on the bottom gives, also, an impossible system.)

away from O_5, relative to link 5. Also, link 3 is on the point of rotating counterclockwise.

Consider link 7 first, isolated as shown in Fig. 12.11b. Force P is known completely, the line of action of F_{67} is known in direction since link 6 is a two-force member, and the friction force is known to act to the right. However, the location of the normal force is unknown, and the surface on which the normal force acts is unknown. It is not possible at this point to specify whether the normal force is acting on the top or the bottom surface. If a single normal force were acting on link 7, it would be impossible to satisfy the moment equation. That is, if moments were taken about the intersection of P and F_{67}, the assumed reaction F_{17} shown in Fig. 12.11b would have a moment about the intersection. (Note that F_{17} shown is assumed to be acting on the top surface, but the same conclusion would be reached if F_{17} were assumed to be acting on the bottom surface.) The result, therefore, of P and F_{67} is to give a "cocking" action to link 7 which set up forces at points E and H or at points G and F. The same general procedure as used for the slider crank mechanism is necessary in that an assumption must be made as to the location of the forces. It will be assumed that the forces act at F and G, and the forces are shown in

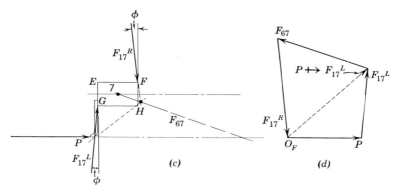

Fig. 12.11c, d. "Cocking" action requires two forces in the analysis. Assumed locations of forces at F and G are verified by the force polygon.

Fig. 12.11c by $F_{17}{}^R$ and $F_{17}{}^L$, with each force inclined from the vertical by the friction angle. The system that results is a four-force member, with the resultant of P and $F_{17}{}^L$, for instance, passing through the intersection of F_{67} and $F_{17}{}^R$. The force polygon for link 7 is shown in Fig. 12.11d. Note that the assumed reactions are in accord with the force polygon. Link 6 is in compression as the result of the analysis of link 7.

Consider next links 4 and 5, shown isolated in Fig. 12.11e. Link 4 has acting on it F_{34}, known to pass through the center of the pin, a friction force whose direction only is known, and a normal reaction. Whether the normal force on link 4 is on the right or the left side has to be determined by an analysis of link 5. Also, whether the reaction is on one side or on the corners can be determined by noting that F_{54} drawn through the center of the pin with the proper inclination will pass through the surface in contact with link 5. Inspection of link 5 shows that the direction of F_{45} must be such as to balance the moment

170　Static Forces in Machines

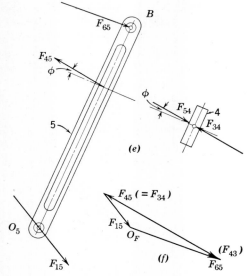

Fig. 12.11e, f.　Link 4 is a two-force member, while link 3 is a three-force member.

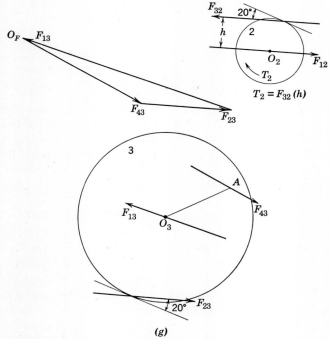

Fig. 12.11g.　Forces acting on the gears, links 2 and 3, with the force polygon for link 3.

Pin Joint Friction

of F_{65} about O_5. The direction of F_{15} may be determined (on the space drawing) by having it pass through the intersection of F_{65} and F_{45}. Figure 12.11f shows the force polygon for links 4 and 5.

Figure 12.11g shows the forces acting on links 2 and 3 and the force polygon for line 3. Link 3 is a three-force member, and link 2 is a member with two equal and opposite couples acting on it.

Figure 12.11h shows the complete force polygon for the entire mechanism.

Note that the couple applied externally to link 2 is clockwise, and is equal to $(F_{32})(h)$.

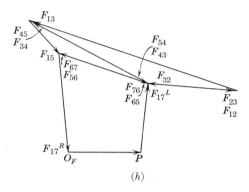

Fig. 12.11h. Complete force polygon.

It is also interesting to note that the couple applied to link 2 could be decreased by lengthening link 7, eliminating the corner reactions, and thus substituting a single force, F_{17}.

12.10 Pin joint friction

Mention was made in the beginning of this chapter that the resultant force on a pin does not pass through the center of the pin if friction is taken into account. The problem is to determine the magnitude and the direction of the forces present on a pin with friction considered.

It is necessary to emphasize the fact that the coefficient of friction will be assumed to be known and to be independent of the load and speed. It was pointed out earlier that the assumptions are not correct, but the procedure of solution with the assumptions give results which are sufficiently accurate for most design work.

Figure 12.12a shows a pin, assumed to be held rigidly, in a clearance hole in link 3. The clearance is exaggerated considerably to allow the showing of the forces acting. (Actually, the clearance is of the order of 0.001 in. per inch of diameter of the pin in the usual case.)

Link 3 is on the point of rotating counterclockwise. The elemental friction forces are as shown because friction *always* opposes the direction of motion. The friction effect is to give a clockwise moment about the center of the hole to counteract the counterclockwise impending motion.

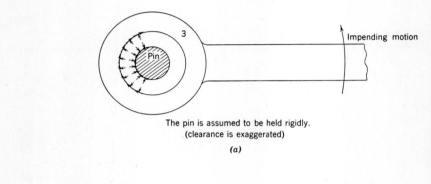

The pin is assumed to be held rigidly.
(clearance is exaggerated)

(a)

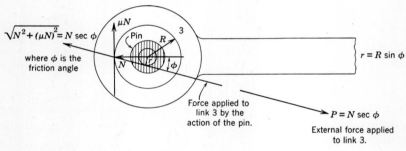

(b)

Fig. 12.12. Forces on a pin with friction considered, giving the "friction circle" radius.

In Fig. 12.12b the resultant of N, the normal force, and μN, the friction force, is

$$(N^2 + \mu N^2)^{1/2} = N(1 + \mu^2)^{1/2} = N(1 + \tan^2 \phi)^{1/2} = N \sec \phi$$

where μ is the coefficient of friction and ϕ is the friction angle.

The moment of the components should equal the moment of the resultant about the center of the pin:

$$(N \sec \phi)(r) = (\mu N)(R)$$

Pin Joint Friction

where r is the distance from the center of the pin to the resultant force and R is the radius of the pin.

The equation above may be reduced to the following equation, noting that $\mu = \tan \phi$:

$$r = R \sin \phi$$

Now, if a single force P were applied to link 3 to give equilibrium, the force P would have to be equal to $N \sec \phi$, and would have to be collinear with $N \sec \phi$, or P would be located at a distance of r from the center of the pin, as shown in Fig. 12.12b.

The distance r is called the radius of the "friction circle," or, expressing it differently, the reaction and applied force are tangent to a friction circle of radius $r = R \sin \phi$. This principle will be used to locate the line of action of forces where pin friction is to be considered.

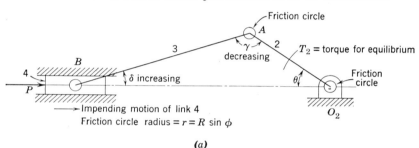

Fig. 12.13a. Slider-crank mechanism, with friction circles at the pins.

If the pin were an integral part of link 3, with the pin in a fixed bearing, the same forces would be present, giving the same results.

Slider-crank mechanism. As an illustration of the application of the friction circle, consider the slider-crank mechanism shown in Fig. 12.13a. A known force P is applied to link 4. For simplification, the friction circles are shown in the schematic line diagram, and are shown in exaggerated size to bring out more clearly the method of solution. It is assumed that the radius of the friction circle is the same at each pin.

Link 4 is assumed to be on the point of moving to the right. A couple is applied to link 2 for equilibrium. Sliding friction is disregarded.

Link 3 is the starting point for this problem. The force acting at B on link 3 is tangent to the friction circle; the force acting at A on link 3 is tangent to the friction circle. Since there are no other forces acting on link 3, the force at B must be equal, opposite, and collinear to the force at A. Link 3 can be seen to be in compression. Four

174   Static Forces in Machines

possible lines of action of the force on link 3 are shown in Fig. 12.13b. Which is the correct one? A procedure that can be used is to determine the angular velocity of each link, as though link 4 were in actual

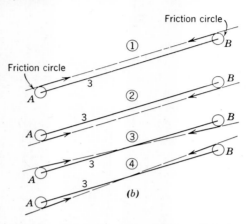

Fig. 12.13b. Possible directions of forces applied to link 3, as a result of pin friction. Link 3 is assumed in compression, which assumption is verified in the force polygon.

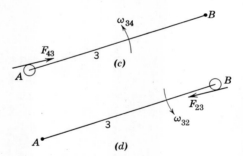

Fig. 12.13c, d. Location of forces at A and B found by consideration of relative angular velocity.

motion to the right with any arbitrary velocity of link 4, and to determine the angular velocity of link 3 with respect to link 4, expressed as ω_{34}, from

$$\omega_3 = \omega_4 + \omega_{34}$$

where the angular velocity of link 3 is the angular velocity of link 4 plus the relative angular velocity of link 3 with respect to link 4. For this case, since $\omega_4 = 0$, $\omega_3 = \omega_{34}$. An arbitrary velocity polygon shows ω_{34} to be counterclockwise. Figure 12.13c shows F_{43} so as to give a clockwise moment to oppose the counterclockwise relative motion.

In a similar manner, from $\omega_3 = \omega_2 + \omega_{32}$, we shall find that ω_{32} is counterclockwise, and that F_{23} must be applied at B as shown in Fig. 12.13d to give a clockwise effect.

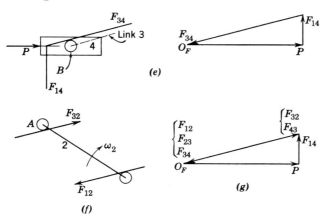

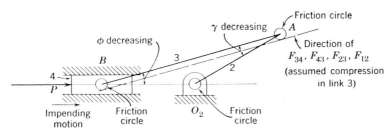

Fig. 12.13e, f, g. Forces acting on the piston and crank as a result of pin friction.

The same results could be obtained by noting that the angle δ is increasing and the angle γ is decreasing, the angles being shown in Fig. 12.13a. The correct force direction, then, is shown in Fig. 12.13b–(3).

Fig. 12.14. Effect of pin friction for the crank in a different position than that of Fig. 12.13a.

Figure 12.13e shows the force analysis for link 4. Figure 12.13f shows the forces applied to link 2, and Fig. 12.13g shows the complete force polygon.

Figure 12.14 shows the mechanism in a different position.

Four-link mechanism. Figure 12.15a shows a four-link mechanism, with a known force P that acts on link 4, and an unknown force

Q that acts on link 2. Link 4 is on the point of rotating clockwise. Link 3 is assumed to be in tension, which assumption is verified by the force polygon. Figure 12.15b shows the forces acting.

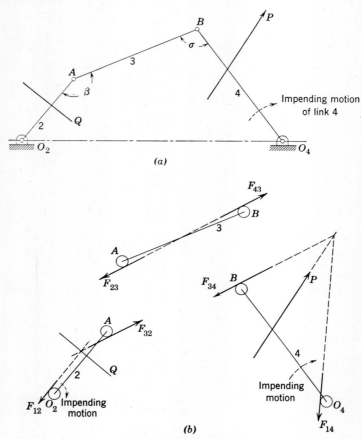

Friction circles are shown at O_4, B, A, and O_2.

Fig. 12.15. Effect of pin friction in a four-link mechanism. Angle σ is decreasing (or ω_{34} is counterclockwise). Angle β is increasing (or ω_{32} is counterclockwise).

Press mechanism. Another illustrative example is the press mechanism shown in Fig. 12.16a. A known force P is applied to link 7, and the impending motion of link 7 is downwards. The mechanism is driven by a couple T_2 applied to link 2. Pin friction and sliding friction are to be taken into account. Friction of the gear teeth is neglected. Figure 12.16b shows the complete force polygon.

Pin Joint Friction

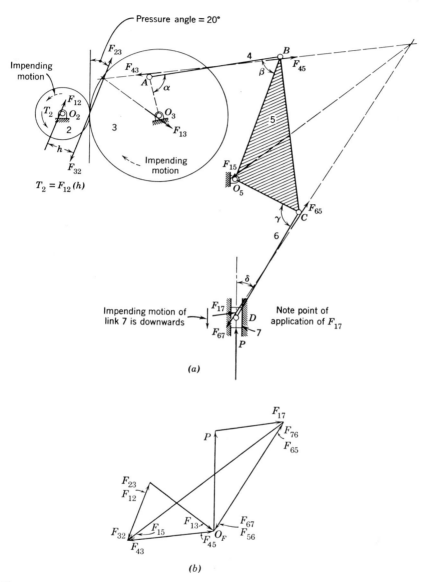

Fig. 12.16. Force analysis of a press mechanism, with sliding friction and pin friction taken into account. As a result of impending motion downwards of link 7, angle α is increasing, or ω_{43} is counterclockwise; angle β is decreasing, or ω_{45} is counterclockwise; angle γ is increasing, or ω_{65} is counterclockwise; angle δ is decreasing, or ω_6 is counterclockwise.

PROBLEMS

Note. Neglect sliding and pin friction in problems 12.1 to 12.24, inclusive.

12.1 (Fig. 12.1). Determine the couple, T_2, applied to link 2, necessary for equilibrium. Determine also the pin forces. The force P is applied to link 4.

Fig. P–12.1. $O_2A = 4''$; $AB = 13''$.

12.2 (Fig. 12.2). A couple is applied to link 2 as shown. Determine the force Q applied to link 4 necessary for equilibrium, and the pin forces. What is the magnitude of F_{14} and F_{12}? If a reaction Q is set up on the cylinder wall, what is the resultant force applied to the structure, that is, what is the resultant of F_{21}, F_{41}, and Q?

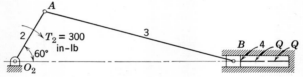

Fig. P–12.2. $O_2A = 4''$; $AB = 13''$.

12.3 (Fig. 12.3). The force P is applied to link 4, and the force S is applied to link 3. Determine the couple that must be applied to link 2 for equilibrium.

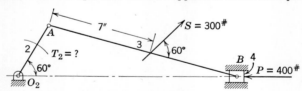

Fig. P–12.3. $O_2A = 4''$; $AB = 13''$.

12.4 (Fig. 12.4). A couple T_2 is applied to link 2, and a force S is applied to link 3. Determine the force Q that must be applied to link 4 for equilibrium. What are the pin forces?

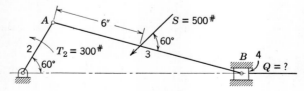

Fig. P–12.4. $O_2A = 4''$; $AB = 13''$.

12.5 (Fig. 12.5). Known forces P and S are applied to links 4 and 3, respectively. What is the magnitude of the force Q applied to link 2 as shown necessary for equilibrium? What are the pin forces?

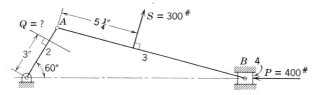

Fig. P–12.5. $O_2A = 4''$; $AB = 13''$.

12.6 (Fig. 12.6). A known force P is applied to link 4 as shown. What is the magnitude of the couple that has to be applied to link 2 for equilibrium? What is the resultant of the forces exerted on the frame, F_{41} and F_{21}?

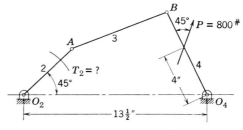

Fig. P–12.6. $O_2A = 5''$; $AB = 7\frac{1}{2}''$; $O_4B = 7''$.

12.7 (Fig. 12.7). Known forces P and S are applied to links 4 and 3, respectively. What is the resultant of the forces exerted on the frame, F_{41} and F_{21}? What is the couple T_2 that must be applied to link 2 for equilibrium?

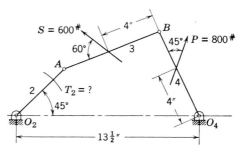

Fig. P–12.7. $O_2A = 5''$; $AB = 7\frac{1}{2}''$; $O_4B = 7''$.

12.8 (Fig. 12.8). A known couple T_4 is applied to link 4, and a known force P is applied to link 3. What couple T_2 must be applied to link 2 for equilibrium?

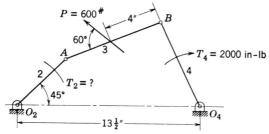

Fig. P–12.8. $O_2A = 5''$; $AB = 7\frac{1}{2}''$; $O_4B = 7''$.

12.9 (Fig. 12.9).   Forces P_1 and P_2 are applied to link 4, and forces S_1 and S_2 are applied to link 3 as shown.   Determine the force Q that must be applied to link 2 for equilibrium.   What are the pin forces?

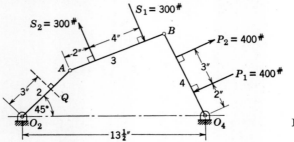

Fig. P–12.9. $O_2A = 5''$; $AB = 7\frac{1}{2}''$; $O_4B = 7''$.

12.10 (Fig. 12.10).   A known couple is applied to link 4, and a known force P is applied to link 3.   What is the couple T_2 applied to link 2 for equilibrium?

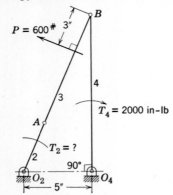

Fig. P–12.10. $O_2A = 4''$; $AB = 9''$; $O_4B = 12''$.

12.11 (Fig. 12.11).   Known forces F_2, F_3, F_4, and F_6 are applied to links 2, 3, 4, and 6, respectively.   What couple T_2 must be applied to link 2 for equilibrium? What are the pin forces?

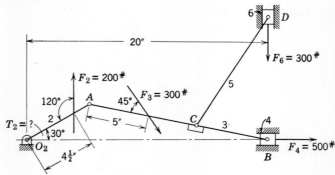

Fig. P–12.11 $O_2A = 6''$; $AB = 15''$; $CD = 11''$.

12.12 (Fig. 12.12). Known forces F_4 and F_6 are applied to links 4 and 6, respectively. What couple must be applied to link 2 for equilibrium?

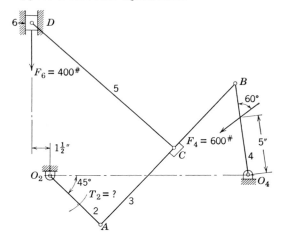

Fig. P–12.12. $O_2A = 6''$; $AB = 16\tfrac{1}{2}''$; $O_4B = 8''$; $AC = 9''$.

12.13 (Fig. 12.13). Known couples T_4 and T_6 are applied to links 4 and 6, respectively. What couple must be applied to link 2 for equilibrium?

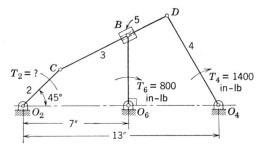

Fig. P–12.13. $O_2C = 3\tfrac{1}{2}''$; $CD = 8''$; $O_4D = 7''$.

12.14 (Fig. 12.14). Known forces P and S are applied to links 4 and 3, respectively. What couple must be applied to link 2 for equilibrium?

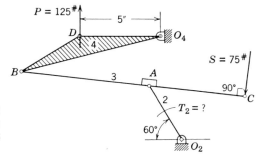

Fig. P–12.14. $O_2A = 4''$; $AB = 8''$; $O_4B = 8\tfrac{3}{4}''$; $BD = 4\tfrac{1}{4}''$; $O_4D = 5''$; $BC = 14''$.

12.15 (Fig. 12.15). A known couple T_2 is applied to the gear, link 2, and a known force P is applied to link 4. What force Q must be applied to link 5 for equilibrium? What are the pin forces?

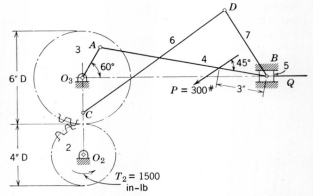

Fig. P–12.15. $O_3A = 2\frac{1}{4}''$; $O_3C = 2\frac{1}{4}''$; $AB = 10\frac{1}{2}''$; $CD = 11''$; $BD = 5''$. Gear pressure angle = 20°.

12.16 (Fig. 12.16). A known couple T_2 is applied to link 2. What is the magnitude of the force Q applied to link 3 for equilibrium?

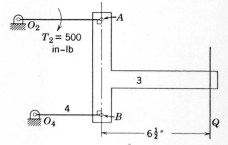

Fig. P–12.16. $O_2A = 5''$; $AB = 6''$; $O_4B = 4''$.

12.17 (Fig. 12.17). Known forces F_3, F_4, F_6 are applied to links 3, 4, and 6 as shown. What couple must be applied to link 2 for equilibrium?

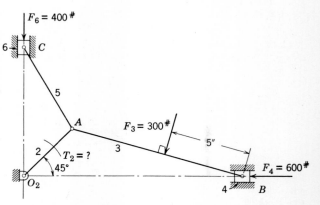

Fig. P–12.17. $O_2A = 4''$; $AB = 11''$; $AC = 6''$.

Problems

12.18 (Fig. 12.18). Known forces F_3, F_4, F_6 are applied to links 3, 4, and 6, respectively. What couple T_2 applied to link 2 is necessary for equilibrium? What are the pin forces?

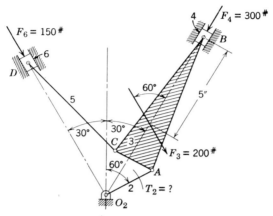

Fig. P–12.18. $O_2A = 2\frac{1}{2}''$, $AB = 7''$; $BC = 7''$; $AC = 2''$; $CD = 6''$.

12.19 (Fig. 12.19). A known force P is applied to the cutting tool attached to link 7. What couple T_2 must be applied to the gear, link 2, for equilibrium? The pin A is fastened rigidly to the gear, link 3. Link 4 is permitted to rotate about the center of pin A, while link 5 is sliding in link 4.

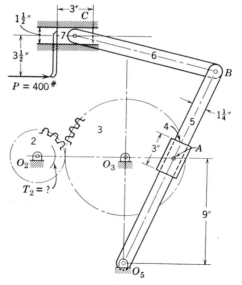

Fig. P–12.19.   Diameter of gear 2 $= 4\frac{1}{2}''$; diameter of gear 3 $= 10''$; $O_3A = 4''$, $O_5B = 18''$; $BC = 12''$. Distance from O_3 to the line of motion of point C is $10\frac{1}{2}''$.

12.20 (Fig. 12.20). A known couple T_2 is applied to link 2. What is the magnitude of the force Q applied to link 6 for equilibrium?

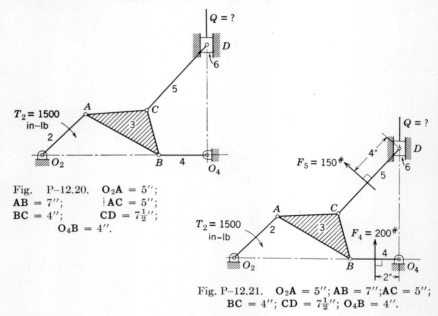

Fig. P–12.20. $O_2A = 5''$; $AB = 7''$; $AC = 5''$; $BC = 4''$; $CD = 7\frac{1}{2}''$; $O_4B = 4''$.

Fig. P–12.21. $O_2A = 5''$; $AB = 7''$; $AC = 5''$; $BC = 4''$; $CD = 7\frac{1}{2}''$; $O_4B = 4''$.

12.21 (Fig. 12.21). A known couple T_2 is applied to link 2, and known forces F_4 and F_5 are applied to links 4 and 5, respectively. What force Q applied to link 6 is necessary for equilibrium? What are the pin forces?

12.22 (Fig. 12.22). A known couple T_2 is applied to link 2. Determine the couple T_6 applied to link 6 for equilibrium, and determine the pin forces.

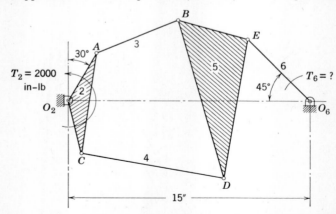

Fig. P–12.22. $O_2A = 3\frac{1}{2}''$; $O_2C = 3\frac{1}{2}''$; $AC = 6\frac{1}{2}''$; $CD = 9''$; $BD = 10\frac{1}{2}''$; $BE = 4\frac{1}{2}''$; $DE = 9''$; $O_6E = 5\frac{1}{2}''$.

Problems 185

12.23 (Fig. 12.23). A known couple T_2 is applied to link 2, and known forces F_3, F_4, and F_5 are applied to links 3, 4, and 5, respectively. Determine the couple T_6 applied to link 6 necessary for equilibrium, and determine the pin forces.

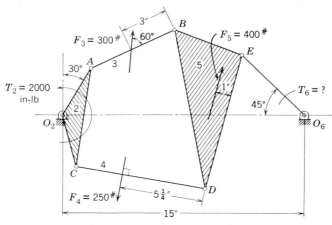

Fig. P–12.23. $O_2A = 3\frac{1}{2}''$; $O_2C = 3\frac{1}{2}''$; $AC = 6\frac{1}{2}''$; $CD = 9''$; $BD = 10\frac{1}{2}''$; $BE = 4\frac{1}{2}''$; $DE = 9''$; $O_6E = 5\frac{1}{2}''$.

12.24 (Fig. 12.24). Figure 12.24 shows a planetary gear system,* where the arm rotates about its axis through the centerline of its bearings. Gears A and B rotate at the same speed. Gear C is integral with the output shaft. Gear D is a stationary gear.

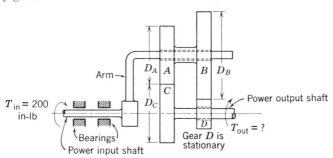

Fig. P–12.24.

The velocity analysis may be made by the conventional kinematic analysis. However the purpose of this problem is to show that force and energy considerations may be used to give the same results. Assume that 200 in.-lb is applied to the input shaft. Assume that the diameter of gear A is 4 in., that the diameter of gear C is 4 in., and that the diameter of gear B is 6 in. Isolate each member

* A portion only of the Model T Ford transmission is shown.

and determine the forces applied to each member. Note that the equations of equilibrium may be applied if each member is rotating at a constant speed. Determine, on the basis of the fact that the energy put into the system is equal to the energy taken out of the system if friction is disregarded, the ratio of angular speeds of the input and output shaft. Determine the couple that must be applied to the gear D if gear D is to be stationary. Show that the couple applied to the input shaft, the couple applied to the output shaft, and the couple applied to the gear D add up to zero.

Note. Neglect pin friction in problems 12.25 to 12.36, inclusive.

12.25. Refer to Fig. 12.1. For a coefficient of sliding friction of 0.15, determine the couple T_2 and the pin forces for
(a) Impending motion of link 4 to the right.
(b) Impending motion of link 4 to the left.

12.26. Refer to Fig. 12.2. For a coefficient of sliding friction of 0.15, determine the force Q and the pin forces for:
(a) Impending motion of link 4 to the right.
(b) Impending motion of link 4 to the left.

12.27. Refer to Fig. 12.3. For a coefficient of sliding friction of 0.15, determine the couple T_2 for impending motion of link 4 to the left.

12.28. Refer to Fig. 12.4. For a coefficient of sliding friction of 0.15, determine the force Q for impending motion of link 2 counterclockwise.

12.29. Refer to Fig. 10.1. A downward force of 60 lb is applied to the follower. If the coefficient of sliding friction is 0.1 (between the follower and its guide, and between the cam and follower) determine the couple that must be applied to the cam for impending motion counterclockwise of the cam.

12.30. Refer to Fig. 12.11. The impending motion of link 4 is to the right. Determine, for a coefficient of sliding friction of 0.15, the couple T_2 that must be applied.

12.31. Refer to Fig. 12.11. The coefficient of sliding friction is 0.15 for links 4 and 6. Determine the magnitude of the couple T_2 applied to link 2 for impending motion of link 2 counterclockwise.

12.32. Refer to Fig. 12.13. Link 4 is in impending motion clockwise. Determine the magnitude of the couple applied to link 2 for the impending motion of link 4. The coefficient of sliding friction is 0.2.

12.33. Refer to Fig. 12.17. The impending motion of link 6 is upwards. Determine the necessary couple T_2 applied to link 2. The coefficient of sliding friction is 0.15.

12.34. Refer to Fig. 12.18. For a sliding coefficient of friction of 0.15, determine the pin forces and the couple T_2 that must be applied to link 2 for clockwise impending motion of link 2.

12.35. Refer to Fig. 12.19. The impending motion of link 7 is to the left. For the known force P, determine the couple T_2 that must be applied to link 2 for a coefficient of sliding friction of 0.1. The pin A is fastened rigidly to the gear, link 3. Link 4 is permitted to rotate about the center of the pin A, while link 5 is sliding in link 4.

12.36 (Fig. 12.36). The schematic sketch of the brake shown, as made by Cutler Hammer, Inc., is released electrically by a magnetic coil (not shown) and actuated by a helical compression spring. The spring force is 200 lb. The coefficient of friction is 0.3 for the brake shoes.

(a) Isolate each member and show on each member the forces applied in magnitude and direction.

(b) What is the horsepower rating at 600 rpm clockwise rotation of the drum?

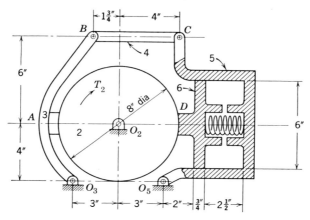

Fig. P-12.36.

12.37. Refer to Fig. 12.1. For a coefficient of sliding friction of 0.15, and for an assumed value of friction in the pins such that the radius of the friction circle is $\frac{1}{8}$ in.* for each pin, determine the couple T_2 and the pin forces for

(a) Impending motion of link 4 to the right.

(b) Impending motion of link 4 to the left.

12.38. Refer to Fig. 12.2. Assuming that the friction in each pin is such that the radius of the friction circle is $\frac{1}{8}$ in.,* and assuming that the coefficient of sliding friction is 0.15, determine the force Q and the pin forces for:

(a) Impending motion of link 4 to the right.

(b) Impending motion of link 4 to the left.

12.39. Refer to Fig. 12.6. Assuming that the radius of the friction circle is $\frac{1}{8}$ in.* for each pin, determine the value of T_2 for:

(a) Impending motion of link 2 clockwise.

(b) Impending motion of link 2 counterclockwise.

12.40. Refer to Fig. 12.16. Assuming that the radius of the friction circle for each pin may be taken as $\frac{1}{8}$ in.,* determine the force Q that must be applied to link 3 for impending motion of link 2 clockwise.

12.41. Refer to Fig. 12.12. Assuming that the radius of the friction circle is known for each pin, and assuming that the coefficient of sliding friction is known, isolate each member. Show on each isolated member the forces acting. Determine the number of unknowns and the number of equation available for the solution. Assume that link 6 is in impending motion upwards. Do not solve for numerical values.

* Radius of the friction circle is given on basis of a full-size drawing.

CHAPTER 13

Inertia Forces

Acceleration analysis has shown that in a mechanism where the links are moving there are definite accelerations which may be determined. Newton's general equations tell us that there must be forces or couples present which cause those accelerations.

This chapter is devoted to the determination of those forces which cause the accelerations, both linear and angular. The general case of plane motion is treated, and links which have only one type of motion, that is, rotation only or translation only, are considered as special cases of the general type of motion. Inertia force concept is introduced after the study is made of the resultant forces causing the motion.

13.1 Forces in plane motion

Consider the body shown in Fig. 13.1a which is moving at the instant with an angular velocity of ω rad/sec counterclockwise and an angular acceleration of α rad/sec^2 counterclockwise. The body is not rotating about any fixed point, but has plane motion. It is desired to determine the resultant force and couple which must be applied to cause the motion at the instant. The x- and y-axes pass through any reference point, A, whose motion is known. The x-axis has been selected, for convenience, as corresponding with the direction of the acceleration of point A. Point G is the location of the center of gravity of the body.

The acceleration of any point, P, by the relative acceleration equation is the vector sum of three quantities:

$$A_P = A_A \mathbin{+\mkern-8mu+} r\omega^2 \mathbin{+\mkern-8mu+} r\alpha$$

The particle at P has a differential mass of dM; consequently, the differential force applied to the particle to give the acceleration is

$$(dM)(A_P) = (dM)(A_A) \mathbin{+\mkern-8mu+} (dM)(r\omega^2) \mathbin{+\mkern-8mu+} (dM)(r\alpha) \qquad (1)$$

Forces in Plane Motion

To simplify the expression, take components of each differential force in the x- and y-directions, and integrate to obtain the total:

$$\int dF_x = \int dM \ A_A - \int dM \ r\omega^2 \cos\theta - \int dM \ r\alpha \sin\alpha \qquad (2)$$

$$\int dF_y = -\int dM \ r\omega^2 \sin\theta + \int dM \ r\alpha \cos\theta \qquad (3)$$

Since A_A is a component which can be applied to every particle, since $r \cos\theta$ is equal to x, since $r \sin\theta$ is equal to y, and since ω and α

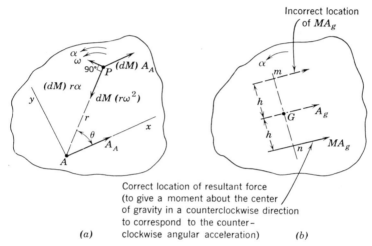

Fig. 13.1. Components of forces applied to a particle, P, to cause a prescribed motion is shown in (a); the resultant force applied to the body is shown in (b).

are fixed quantities for the instant of time, the expressions above may be rewritten as

$$F_x = MA_A - \omega^2 \int dM \ x - \alpha \int dM \ y \qquad (4)$$

$$F_y = -\omega^2 \int dM \ y + \alpha \int dM \ x \qquad (5)$$

Equations 4 and 5 determine the magnitude of the resultant force acting on the body. The moment of the resultant force about point A is determined by taking the moment of each differential component, in Eq. 1 above, about the point A. Note that the moment due to the normal component, $dM \ r\omega^2$, is zero about point A for every point on the body.

$$T_A = -\int (dM \ A_A)y + \int (dM \ r\alpha)r$$

or

$$T_A = -A_A \int dM \ y + \alpha \int dM \ r^2 \qquad (6)$$

Equations 4, 5, and 6 are the general equations of the force components and torque acting on a body in plane motion. Considerable simplification can be made by selecting a special point as the reference point, the center of gravity. *If the center of gravity is used as the reference point, $\int dM\, x = 0$, $\int dM\, y = 0$* from the definition of center of gravity. Equations 4, 5, and 6 may then be rewritten as

$$F_x = MA_g \qquad (7)$$

$$F_y = 0 \qquad (8)$$

$$\mathbf{T} = \mathbf{I}\alpha \qquad (9)$$

where $\int dM\, r^2$ is defined as $\mathbf{I}$, the mass moment of inertia of the body about the center of gravity of the body, and $\mathbf{T}$ is the moment about the center of gravity.

The interpretation of the equations above is that the resultant force is equal to the product of the mass times the acceleration of the center of gravity, and is located in such a position that the moment of the resultant force about the center of gravity is $\mathbf{I}\alpha$. The direction of the resultant force is in the direction of the acceleration of the center of gravity of the body, since Eq. 4 (with the center of gravity as the reference point) shows that $F_x = MA_g$ and Eq. 5 shows that $F_y = 0$. It is to be noted that with the center of gravity taken as the reference point, the x-axis corresponds with the direction of the acceleration of the center of gravity. The position of the resultant force should be such that the direction of the moment of the resultant force about the center of gravity will correspond to the direction of the angular acceleration.

Figure 13.1b shows a body with a known acceleration of the center of gravity. The resultant force is located at a distance h from the center of gravity. At first glance, we might say that there are two possible positions for the resultant force, at m and n in the figure. If the angular acceleration is counterclockwise, the resultant force must pass through n to give a counterclockwise direction of moment of the resultant force about the center of gravity to correspond with the direction of angular acceleration. The distance h is determined from the fact that the moment of the resultant force about the center of gravity is $(MA_g)(h)$, but the moment can be expressed by $\mathbf{I}\alpha$. Therefore, equating the two, we have the following equations:

$$MA_g h = \mathbf{I}\alpha$$

or

$$h = \frac{\mathbf{I}\alpha}{MA_g} = \frac{Mk^2\alpha}{MA_g} = \frac{k^2\alpha}{A_g} \qquad (10)$$

Inertia Force

where k is the radius of gyration of the body about the center of gravity.

Attention is called to the units. If M is measured in lb-sec^2/ft units, α is measured in rad/sec^2 units, I is measured in lb-sec^2-ft units, and A_g is measured in ft/sec^2 units, h will be in foot units.

13.2 Inertia force

The resultant force on a link is shown in Fig. 13.2a. The resultant force is the resultant of the forces applied to the link through the

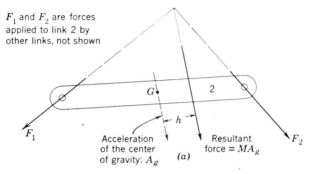

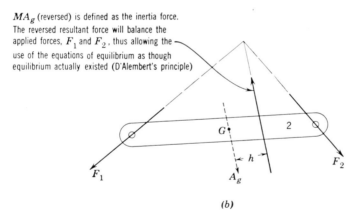

Fig. 13.2. The resultant force, MA_g, is shown in (a); the reversed resultant force, defined as the inertia force, is shown in (b).

pins, F_1 and F_2. Or F_1 and F_2 may be considered to be the components of the resultant force. Or, if the resultant force is reversed in direction, as shown in Fig. 13.2b, the link may be considered to be in equilibrium. The reversed resultant force, called the inertia force, sets up a system to which the equations of equilibrium may be applied

192 Inertia Forces

(known as d'Alembert's principle). Note that the system in not actually in equilibrium since the part is accelerating, but the introduction of the inertia force will allow us to use the method of solution used in static force analysis.

13.3 Link rotating about a fixed point

The same method of attack may be used for a link rotating about a fixed point inasmuch as this case is a special case of the general case. (The resultant force for this case will pass through a fixed point on the link, the so called center of percussion.) Links rotating about a fixed point will be handled by applying the general equations derived.

13.4 Link in translation only

This is another special case of the general case. If the link has pure translation, the angular acceleration is zero. Consequently, the result-

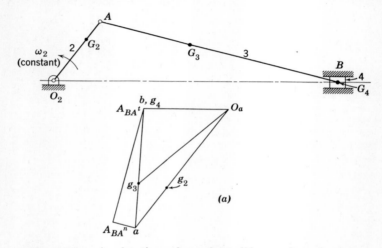

Fig. 13.3a. Acceleration polygon for a slider-crank mechanism.

ant force must pass through the center of gravity inasmuch as $I\alpha$ must be zero, or the torque about the center of gravity must be zero. The direction of the resultant force would be in the direction of the acceleration of the body.

Several examples will be given to illustrate the method of locating inertia forces in mechanisms. In the examples to follow, the acceleration diagrams will be given without discussion.

Slider-crank mechanism. Figure 13.3a shows a slider-crank mechanism and the acceleration diagram for the position shown. The

Slider-Crank Mechanism

location of the center of gravity of each link is given by G_2, G_3, and G_4. Link 2 is assumed to be rotating with a constant angular speed.

The magnitude and direction of the acceleration of the centers of gravity of the links are shown by g_2, g_3, and g_4 in the acceleration diagram.

Consider link 2, isolated in Fig. 13.3b. For this case, since link 2 is rotating at a constant angular speed, the acceleration of G_2 is directed from G_2 to O_2. Since α_2 is zero, the resultant force must pass through G_2, with $I_2\alpha = 0$. The resultant force acting on link 2 is $M_2 A_{g_2}$, and is directed from G_2 to O_2, as may be seen from the acceleration diagram. The inertia force is reversed in direction from the resultant force, and is directed from O_2 to G_2, as shown. The inertia force, represented by f_2, is equal to $M_2 A_{g_2}$.

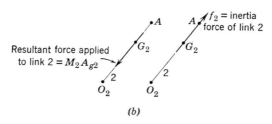

Fig. 13.3b. Resultant and inertia forces of the crank.

Consider link 3 next. The direction of the resultant force corresponds to the direction of the acceleration of G_3. Since the angular acceleration of link 3 is counterclockwise, the moment of the resultant force about G_3 is counterclockwise. To satisfy the necessary relations of direction of acceleration and direction of moment, the resultant force is as shown in Fig. 13.3c. The distance of the resultant force from the center of gravity, h_3, is $I_3\alpha_3/M_3 A_{g_3}$. The inertia force, which is the reversed resultant force, is as shown in Fig. 13.3d.

A second method of interpretation is offered as a means of obtaining a better understanding of locating the inertia forces properly. Consider link 3 again, shown in Fig. 13.3e. Locate the resultant force through the center of gravity and parallel to the acceleration of G_3. This force gives the linear acceleration of G_3. At the same time, a couple may be considered to be acting on the body to give the angular acceleration, the couple being equal to $I_3\alpha_3$, or is equal to $(M_3 A_{g_3})(h)$. The couple may be represented by any two equal and opposite and parallel forces. Consider one of the forces of the couple, $M_3 A_{g_3}$, applied at the center of gravity to balance the force giving the linear acceleration. The other force of the couple must be located to give

194   Inertia Forces

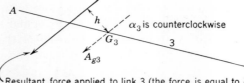

Resultant force applied to link 3 (the force is equal to $M_3 A_{g3}$ $h = I\alpha/(M_3 A_{g3})$). Note that the resultant force acts in the same direction as the direction of the acceleration of G_3 and that the moment about the center of gravity is in the same direction (counterclockwise) as the angular acceleration.

(c)

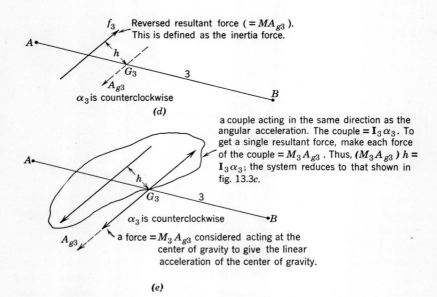

Reversed resultant force ($= MA_{g3}$). This is defined as the inertia force.

(d)

a couple acting in the same direction as the angular acceleration. The couple $= I_3 \alpha_3$. To get a single resultant force, make each force of the couple $= M_3 A_{g3}$. Thus, $(M_3 A_{g3})h = I_3 \alpha_3$; the system reduces to that shown in fig. 13.3c.

a force $= M_3 A_{g3}$ considered acting at the center of gravity to give the linear acceleration of the center of gravity.

(e)

Fig. 13.3c, d, e.   Resultant and inertia forces of the connecting rod.

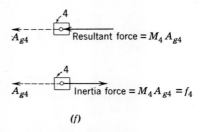

(f)

Fig. 13.3f.   Resultant and inertia forces of the piston.

the counterclockwise angular acceleration of the link, or for this case is as shown. The two forces at the center of gravity cancel each other, giving the same resultant force as found by the first method. The inertia force is again the reversed resultant force.

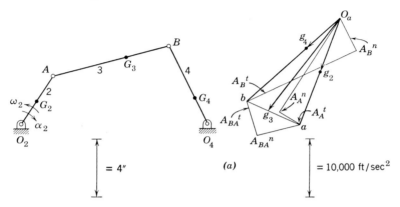

Fig. 13.4a. Acceleration polygon of a four-link mechanism.

$$\alpha_2 = 12{,}000 \text{ rad/sec}^2 \text{ cw}$$
$$\alpha_3 = 9{,}300 \text{ rad/sec}^2 \text{ ccw}$$
$$\alpha_4 = 40{,}000 \text{ rad/sec}^2 \text{ ccw}$$
$$A_{g_2} = 9{,}800 \text{ ft/sec}^2$$
$$A_{g_3} = 20{,}000 \text{ ft/sec}^2$$
$$A_{g_4} = 7{,}100 \text{ ft/sec}^2$$

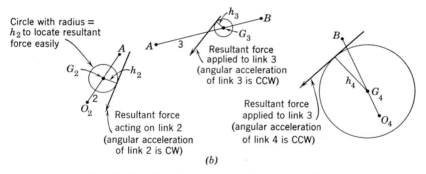

Fig. 13.4b. Resultant force acting on each link.

Consider link 4 next in Fig. 13.3f. Link 4 is moving in translation only. The resultant force must then pass through the center of gravity and must be in the same direction as the acceleration. The resultant force is found from $M_4 A_{g_4}$, or $M_4 A_B$. The inertia force of link 4, f_4, is as shown.

Inertia Forces

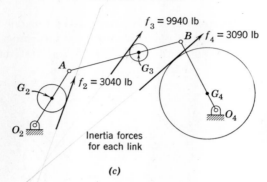

Inertia forces for each link

(c)

Fig. 13.4c.

Weight of each link:
$W_2 = 10$ lb
$W_3 = 16$ lb
$W_4 = 14$ lb

Mass moment of inertia of each link about its center of gravity:
$I_2 = 0.020$ lb-sec^2-ft
$I_3 = 0.050$ lb-sec^2-ft
$I_4 = 0.025$ lb-sec^2-ft

Inertia force of each link:

$$f_2 = M_2 A_{g2} = \frac{10}{32.2}(9800) = 3040 \text{ lb}$$

$$f_3 = M_3 A_{g3} = \frac{16}{32.2}(20{,}000) = 9940 \text{ lb}$$

$$f_4 = M_4 A_{g4} = \frac{14}{32.2}(7100) = 3090 \text{ lb}$$

Distance of inertia force from the center of gravity:

$$h_2 = \frac{I_2 \alpha_2}{M_2 A_{g2}} = \frac{(0.020)(12{,}000)}{\left(\frac{10}{32.2}\right)(9800)} = 0.0788 \text{ ft}$$

$$h_3 = \frac{I_3 \alpha_3}{M_3 A_{g3}} = \frac{(0.050)(9300)}{\left(\frac{16}{32.2}\right)(20{,}000)} = 0.0493 \text{ ft}$$

$$h_4 = \frac{I_4 \alpha_4}{M_4 A_{g4}} = \frac{(0.025)(40{,}000)}{\left(\frac{14}{32.2}\right)(7100)} = 0.324 \text{ ft}$$

Four-link mechanism. A four-link mechanism is shown in Fig. 13.4a, together with the acceleration diagram. The inertia forces will be located directly on the space diagram. The angular acceleration of link 2 is clockwise, that for link 3 is counterclockwise, and that for link 4 is also counterclockwise.

Figure 13.4b shows the resultant force acting on each link to satisfy the direction of the acceleration of the center of gravity of each link and the direction of the angular acceleration. The inertia forces are shown in Fig. 13.4c. The necessary calculations are given with the figures.

Shaper quick-return mechanism. Figure 13.5a shows the shaper quick-return mechanism, and the acceleration diagram for the

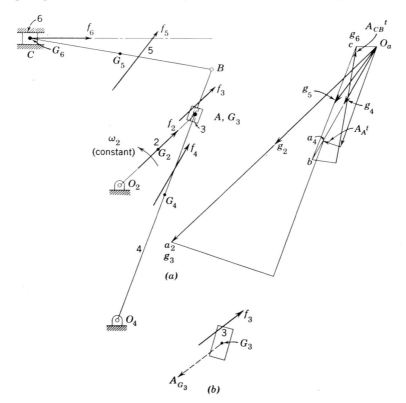

Fig. 13.5. Inertia forces of the shaper quick-return mechanism. (a) $\alpha_2 = 0$; $\alpha_3 = \alpha_4$ (ccw); α_5 ccw; $\alpha_6 = 0$. (b) $\alpha_3 = \alpha_4$ (ccw); $f_3 = M_3 A_{G3}$;

$$h_3 = \frac{I_3 \alpha_3}{M_3 A_{G3}} = \frac{I_3 \alpha_4}{M_3 A_{G3}}.$$

position shown. The inertia forces for links 2, 4, 5, and 6 need no explanation. Link 3, however, is a bit different from the usual link, and a brief discussion of the action is appropriate. The acceleration, both magnitude and direction, of G_3 is found from the acceleration diagram. The direction of the resultant force acting on link 3 is thus known. Link 3 has an angular acceleration, which is the same as that of link 4. α_4 is counterclockwise. Therefore, the resultant force acting on link 3 must be located to give a counterclockwise moment about the center of gravity of link 3: $h_3 = \dfrac{\mathbf{I}_3\alpha_3}{M_3 A_{G_3}} = \dfrac{\mathbf{I}_3\alpha_4}{M_3 A_{G_3}}$.

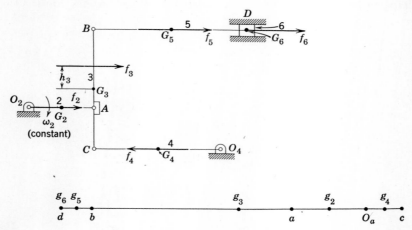

Fig. 13.6. Inertia forces for a special position.

The location and direction of the inertia force of link 3 is shown in Fig. 13.5b. All the inertia forces are shown in the space diagram.

Special position. Figure 13.6 shows a mechanism and the acceleration diagram for a constant angular velocity of link 2. It is left

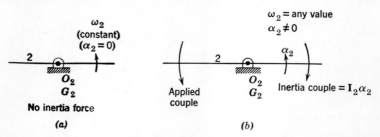

Fig. 13.7. Inertia effects for a counterbalanced link rotating at a constant angular velocity (a), or with an angular acceleration (b).

up to the student to verify the location and the direction of each inertia force.

Counterbalanced link. To modify or eliminate "shaking forces," which will be discussed in detail in later chapters, counterbalancing weights are sometimes added. Figure 13.7a shows a link rotating

Determination of the Mass Moment of Inertia

about O_2, and the center of gravity is at the center of rotation. If the link is rotating at a constant angular velocity, the inertia force is zero, since the acceleration of the center of gravity is zero.

If the link has any angular velocity and an acceleration of α_2, as shown in Fig. 13.7b, the inertia force is still zero, since the acceleration of the center of gravity is zero. However, the moment about the center of gravity is not zero, but is $I_2\alpha_2$. This means that a couple has to be applied to the link to obtain the angular acceleration. For equilibrium, we may picture an inertia couple applied to the link in a direction opposite to the angular acceleration.

13.5 Determination of the mass moment of inertia

The determination of the mass moment of inertia of a body about the center of gravity may be obtained in either of two ways: (1)

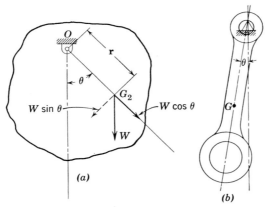

Fig. 13.8. Mass moment of inertia determined experimentally by oscillation.

mathematically or (2) experimentally, applying mathematical principles, if the body is available either as it will be used or to scale. In many cases, the part would not be available until the machine was designed, in which case, mathematical analyses using the basic relation $\mathbf{I} = \int dM\ r^2$* would have to be used. If the part to be used were available, or if the machine warranted making models, or if the mathematical calculations were to be checked by experimental work, the moment of inertia could be found by the equations derived as follows.

Consider the body shown in Fig. 13.8a, which is suspended from

---

* Mass moment of inertia is a purely mathematical quantity and has no physical interpretation.

any point, O, other than the center of gravity. The body is displaced a small angle θ from the vertical, and then released. The problem is to determine the frequency of oscillation of the body, and ultimately the moment of inertia about the center of gravity.

The relation of the moment about point O and the angular acceleration is
$$T_o = I_o \alpha$$
or, for this case:
$$-(W \sin \theta)\mathbf{r} = I_o \frac{d^2\theta}{dt^2}$$

where $W \sin \theta$ is the force component of the weight which causes the torque, $\mathbf{r}$ is the distance from the center of gravity to the point of support, and I_o is the moment of inertia about the axis through the center of support. The minus sign is used because the angle θ decreases as a function of time, or because the torque is in the opposite direction to the positive angle θ.

If the angle θ is small, the above expression may be rewritten, without appreciable error, by
$$-W\theta\mathbf{r} = I_o \frac{d^2\theta}{dt^2}$$
or
$$\frac{d^2\theta}{dt^2} = -\frac{W\mathbf{r}}{I_o}\theta$$

The solution to the equation above is
$$\theta = A\left(\sin\sqrt{\frac{W\mathbf{r}}{I_o}}\,t\right) + B\left(\cos\sqrt{\frac{W\mathbf{r}}{I_o}}\,t\right)$$

The boundary conditions specified are that at $t = 0$, $\theta = \theta_{max}$, and $\frac{d\theta}{dt} = 0$ (or, the angular speed is zero). Solving for the constants of integration, we find that $A = 0$, and $B = \theta_{max}$. Therefore,
$$\theta = \theta_{max}\left(\cos\sqrt{\frac{W\mathbf{r}}{I_o}}\,t\right)$$

The function is a cosine wave which goes through a complete cycle when $\sqrt{\frac{W\mathbf{r}}{I_o}}\,T = 2\pi$.

Mass Moment of Inertia Found by Torsion

Or the time T for a complete cycle is

$$T = 2\pi \sqrt{\frac{I_o}{W\mathbf{r}}}$$

Solve for I_o:

$$I_o = W\mathbf{r}\left(\frac{T}{2\pi}\right)^2$$

The moment of inertia about the center of gravity may then be found from the transfer theorem:

$$I_o = I + M\mathbf{r}^2$$

or

$$I = W\mathbf{r}\left(\frac{T}{2\pi}\right)^2 - \frac{W\mathbf{r}^2}{g}$$

The weight may be determined by weighing the part; $\mathbf{r}$ may be determined by balancing the part horizontally on a knife edge; T may be measured by suspending the part vertically, displacing it a small amount from the vertical, and timing the oscillations. Figure 13.8b shows a connecting rod suspended on a knife edge, which would permit the determination of the time of oscillation.

Example. If a connecting rod weighs 3 lb, the distance from the center of gravity to the knife edge is 5.2 in., and the time for 24 complete oscillations is 19 sec (making the time for one complete oscillation 19/24 sec), the moment of inertia of the connecting rod about the center of gravity is:

$$I = W\mathbf{r}\left(\frac{T}{2\pi}\right)^2 - \frac{W\mathbf{r}^2}{g}$$

$$= 3\left(\frac{5.2}{12}\right)\left(\frac{19/24}{2\pi}\right)^2 - \left(\frac{3}{32.2}\right)\left(\frac{5.2}{12}\right)^2$$

$$= 0.031 \text{ lb-sec}^2\text{-ft}$$

13.6 Alternate method of determining the mass moment of inertia

An alternate method of determining the mass moment of inertia of a body about the center of gravity is that using a torsionally vibrating system. The system, which allows the determination of moment inertia of parts that cannot be conveniently suspended (the connecting rod of the previous section was easily suspended) and can be used with connecting rods as well, is shown in Fig. 13.9. A frame which holds the part to be analyzed is suspended by a wire or steel rod from a stationary surface. The frame is displaced angularly, and the moment of inertia of the part can be found from the time for one oscillation by the following.

Consider the frame displaced θ radians. The torque applied to the frame by the wire is $k\theta$, where k is the spring constant of the wire

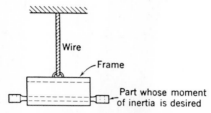

Fig. 13.9   Mass' moment of inertia determined experimentally by torsional vibration.

(inch-pounds of torque per radian). The equation for the motion of the frame is

$$-k\theta = I\frac{d^2\theta}{dt^2}$$

The minus sign is used because θ decreases as a function of time, or because the torque is in the opposite direction to the positive angle θ.

The equation above may be rewritten as

$$-\frac{k}{I}\theta = \frac{d^2\theta}{dt^2}$$

The solution for the equation is

$$\theta = A(\sin\sqrt{k/I}\,t) + B(\cos\sqrt{k/I}\,t)$$

If $t = 0$ when $\theta = \theta_{max}$ and $\dfrac{d\theta}{dt} = 0$, the constants A and B are $A = 0$, and $B = \theta_{max}$.

The final equation for the above boundary conditions is

$$\theta = \theta_{max}(\cos\sqrt{k/I}\,t)$$

The function is a cosine wave which goes through a complete cycle when $\sqrt{k/I}\,T = 2\pi$.

Or the time T for a complete cycle is:

$$T = 2\pi\sqrt{I/k}$$

Or the moment of inertia I of the supporting frame and part being analyzed is

$$I = \frac{kT^2}{4\pi^2}$$

The constant k for the supporting rod may be determined from strength of materials considerations, that is, from $\theta = M_t l/JG$, where k is defined as M_t/θ, or

$$k = \frac{JG}{l} = \frac{\pi d^4 G}{32l}$$

where d is the diameter of the support rod, G is the modulus of elasticity in shear, and l is the length of the support rod. However, it is simpler, perhaps, to make several experimental runs, as indicated by the following analysis, especially for a series of tests.

Let I_b be the moment of inertia of the part being analyzed;
I_f be the moment of inertia of the frame;
T_f be the time for one complete oscillation of the frame itself;
T_{b+f} be the time for one complete oscillation of the frame and part being analyzed.

Then
$$I_b + I_f = \frac{kT_{b+f}^2}{4\pi^2}$$

But
$$I_f = \frac{kT_f^2}{4\pi^2}$$

Thus
$$I_b = \frac{k}{4\pi^2}(T_{b+f}^2 - T_f^2)$$

The constant k may be eliminated by making a run with another piece whose moment of inertia can be found by calculation: a *solid cylindrical piece*, for instance, whose moment of inertia is known mathematically to be $Wr^2/2g$, which can be calculated numerically.

Let I_s be the moment of inertia of the sample part ($I_s = Wr^2/2g$);
T_{s+f} be the time for one complete oscillation of the sample part and the frame.

Then
$$I_s + I_f = \frac{kT_{s+f}^2}{4\pi^2}$$

Or
$$I_s = \frac{k}{4\pi^2}(T_{s+f}^2 - T_f^2) = \frac{Wr^2}{2g}$$

And
$$k = \frac{2\pi^2 Wr^2}{g(T_{s+f}^2 - T_f^2)}$$
(This equation valid only when a solid cylindrical piece is used in the analysis as a sample piece.)

Inertia Forces

Thus the equation for the moment of inertia of the part being analyzed is

$$I_b = \frac{Wr^2}{2g}\left(\frac{T_{b+f}^2 - T_f^2}{T_{s+f}^2 - T_f^2}\right)$$

Example. The following test data are obtained, using a torsional system described. Determine the mass moment of inertia of the part being analyzed.
Weight of solid cylindrical piece: 4.2 lb.
Outside radius of above: 3.5 in.
Time for one oscillation of frame alone: 2.0 sec.
Time for one oscillation of sample and frame: 2.5 sec.
Time for one oscillation of frame and part being analyzed: 3.0 sec.

$$I_b = \frac{(4.2)(3.5)^2}{(2)(386)}\left(\frac{3^2 - 2^2}{2.5^2 - 2^2}\right)$$

$$= 0.148 \text{ lb-sec}^2\text{-in.}$$

(*Note:* For a frame supported by three equally spaced support wires, the reader is referred to an article, "Determining Moment of Inertia," by C. E. Crede in *Machine Design*, August, 1948.)

13.7 Kinetically equivalent systems

It is sometimes desirable to replace a given system by a kinetically equivalent one. Such a replacement may simplify considerably an analysis of a mechanism, especially where the analysis is made for a number of positions or configurations of the mechanism.

Generally, a kinetically equivalent system is defined as a group of bodies, considered rigidly connected together, which will be given the same accelerations as the actual link or body under the action of the same forces. There are three requirements which must be satisfied to have a kinetically equivalent system:

(*a*) The two systems must have the same mass.
(*b*) The two systems must have the center of gravity located in the same position.
(*c*) The two systems must have the same moment of inertia.

These conditions may be seen necessary from inspection of the basic formulas:

(*a*) may be seen from $\Sigma F = MA_g$, where ΣF and A_g are the same for the two systems. Therefore, the two systems must have the same mass.

Kinetically Equivalent Systems

(b) may be seen from the same equation. The acceleration of the center of gravity must be the same for the two systems.

(c) may be seen from $\Sigma T = I\alpha$, where ΣT and α are the same for the two systems. Therefore, the two systems must have the same moment of inertia about the center of gravity, and therefore about any point as seen from the transfer theorem.

An example of the procedure in determining a kinetically equivalent system follows. Two masses will be used in the equivalent system, although, practically, there is no limit to the number of masses which may be used.

Figure 13.10a shows a link with mass M, and with the center of gravity at G. It is desired to obtain the relations of two concentrated masses, m_1 and m_2, which may be substituted for M. The three requirements, expressed mathematically, to give an equivalent system, are

$$m_1 + m_2 = M \tag{1}$$

$$m_1 h_1 = m_2 h_2 \tag{2}$$

where h_1 is the distance from m_1 to G, and h_2 is the distance from m_2 to G.

$$m_1 h_1^2 + m_2 h_2^2 = I = Mk^2 \tag{3}$$

where k is defined as the radius of gyration, about the center of gravity. Note that m_1 and m_2 are considered point masses so that the moment of inertia of each mass, m_1 and m_2, about G may be written as shown.

There are four unknowns, m_1, m_2, h_1, and h_2, but only three equations. This means that any one of the four may be assumed, and the other three determined.

Elimination of m_1, m_2, and M from the above three equations gives:*

$$h_1 h_2 = k^2$$

* Set $m_1 = m_2 \dfrac{h_2}{h_1}$ from Eq. 2 into Eq. 1 to give

$$m_2 \frac{h_2}{h_1} + m_2 = M$$

Or, rearranging terms:

$$m_2 = M \left(\frac{h_1}{h_1 + h_2} \right)$$

Substitution of $m_1 h_1 = m_2 h_2$ from Eq. 2 into Eq. 3 and substitution of $m_2 = M \left(\dfrac{h_1}{h_1 + h_2} \right)$ found above into Eq. 3 gives

$$h_1 h_2 = k^2$$

206   Inertia Forces

The foregoing equation could be used to locate the equivalent masses to suit the reader. For instance, in the connecting rod of the slider-crank mechanism, m_1 might be located at the piston pin, and the

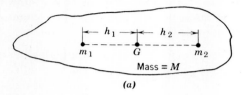

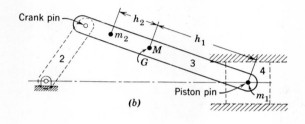

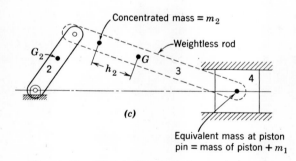

Fig. 13.10. Kinetically equivalent systems.

location of the other mass, m_2, determined by $h_1 h_2 = k^2$. Or

$$h_2 = k^2/h_1$$

h_2 is a fixed quantity, as k is fixed, and h_1 is taken as the distance from the center of gravity to the piston pin, which is a constant. The mass m_2 would always be at a fixed point on the link, regardless of the position of the crank in Fig. 13.10b.

Consequently, by considering the equivalent system, it is possible to obtain for the slider-crank mechanism shown in Fig. 13.10c a simplified system, which would simplify the analysis for a number of positions of the crank. The equivalent system has a mass of piston equal to the original mass of the piston and the mass m_1 at the piston pin. The connecting rod is replaced by a weightless rod, with a concentrated mass m_2 at h_2 from the center of gravity of the original rod. Since m_2 is a concentrated mass, the inertia force of m_2 will always pass through m_2. The direction and magnitude of acceleration of m_2 can be found from the usual acceleration diagram. Consequently, the inertia force of m_2 will pass through a fixed point. Thus the distance from the center of gravity to the inertia force of the connecting rod would not have to be calculated for each position of the connecting rod.

PROBLEMS

13.1 (Fig. 13.1). A link which weights 8.0 lb and has a moment of inertia about the center of gravity of the link equal to 0.020 lb-ft-sec^2 is known at the instant to

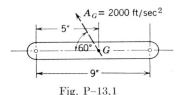

Fig. P-13.1

have an angular speed of 30 rad/sec clockwise and an angular acceleration of 3000 rad/sec^2 counterclockwise. The center of gravity has an acceleration of 2000 ft/sec^2 as shown in the figure.
 (a) What is the radius of gyration, k?
 (b) What is the magnitude of the resultant force applied to the link?
 (c) Where is the resultant force applied to the link?
 (d) What is the magnitude of the inertia force of the link?
 (e) Where is the inertia force located?

13.2 (Fig. 13.2). A link which weighs 6 lb and has a moment of inertia about the center of gravity equal to 0.015 lb-ft-sec^2 is known at the instant to have a resultant

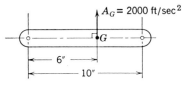

Fig. P-13.2.

acceleration of 2000 ft/sec^2 perpendicular to the link as shown. The angular acceleration of the link is 4000 rad/sec^2 counterclockwise.

(a) What is the magnitude of the resultant force applied to the link, and where is the resultant force applied?

(b) What is the magnitude of the inertia force of the link and where is it applied?

13.3 (Fig. 13.3). The figure shows three different combinations of a link rotating about a fixed center. Determine, for each case:

(a) The magnitude and location of the resultant force applied to each link to give the prescribed motion.

(b) The inertia effect of each link.

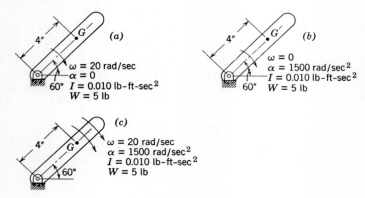

Fig. P-13.3

13.4 (Fig. 13.4). A single-cylinder diesel engine has a 3 in. crank radius and a connecting rod length of $11\frac{1}{4}$ in. If the crank speed is 2500 rpm counterclockwise and is constant, determine the inertia force magnitude and location for each link when the crank makes an angle of 60 degrees with the line through the crank bearing and piston pin. The center of gravity of each link is shown.

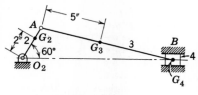

Fig. P-13.4.

$O_2A = 3''$	$W_2 = 5$ lb	$I_2 = 0.004$ lb-ft-sec^2
$AB = 11\frac{1}{4}''$	$W_3 = 8$ lb	$I_3 = 0.030$ lb-ft-sec^2
	$W_4 = 6$ lb	

13.5. Refer to Fig. 13.4. A single-cylinder engine has a stroke of 6 in. and a connecting rod length of $11\frac{1}{4}$ in. Determine the magnitude and location of each inertia force for the position when the crank makes an angle of 60 degrees with the line through the crank bearing and piston pin at which time the crank is rotating at 2500 rpm counterclockwise and decreasing in speed at the rate of 6000 rad/sec^2.

Problems

13.6 (Fig. 13.6). The crank of the single-cylinder engine is rotating at a speed of 2500 rpm counterclockwise and is decreasing in speed at the rate of 6000 rad/sec² for the position shown. The crank is counterbalanced so that the center of gravity of the crank is located at the crank bearing. Determine the inertia effect of each link. Can the inertia effect of link 2 be expressed as a single force? If so, where would the resultant force be considered as applied?

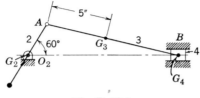

Fig. P-13.6.

$O_2A = 3''$ $W_2 = 10$ lb $I_2 = 0.008$ lb-ft-sec²
$AB = 11\frac{1}{4}''$ $W_3 = 8$ lb $I_3 = 0.030$ lb-ft-sec²
  $W_4 = 6$ lb

13.7 (Fig. 13.7). The four-link mechanism shown has a motion as prescribed by the known acceleration of point A. The center of gravity of each link is at its midpoint. For the information given in the figure, determine the magnitude and location of the inertia force for each link.

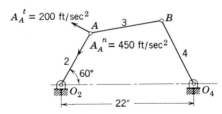

Fig. P-13.7.

$O_2A = 10''$ $W_2 = 4.5$ lb $I_2 = 0.0140$ lb-ft-sec²
$AB = 12''$ $W_3 = 2.0$ lb $I_3 = 0.00820$ lb-ft-sec²
$O_4B = 12''$ $W_4 = 5.3$ lb $I_4 = 0.0205$ lb-ft-sec²

13.8 (Fig. 13.8). The two gears, links 2 and 3, rotate about their own centers, O_2 and O_3, respectively. The angular speed and the angular acceleration of link

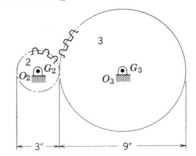

Fig. P-13.8.

$\omega_2 = 20$ rad/sec cw
$\alpha_2 = 3000$ rad/sec² cw
$I_2 = 0.010$ lb-ft-sec²
$I_3 = 0.090$ lb-ft-sec²

210   Inertia Forces

2 are specified: 20 rad/sec clockwise and 3000 rad/sec² clockwise, respectively. Determine the inertia effect of each link, if the center of gravity of each link is at the center of rotation of each link.

13.9 (Fig. 13.9). The angular speed of link 2 is 50 rad/sec clockwise, and the angular acceleration of link 2 is 1500 rad/sec² counterclockwise. Determine the magnitude and the location of the inertia effect of each link.

The center of gravity of links 2, 5, and 6 is as shown. The center of gravity of links 3 and 4 may be considered to be at the center of gravity of each area.

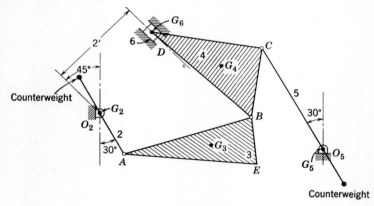

Fig. P–13.9.

$O_2A = 1'$	$W_2 = 8$ lb	$I_2 = 0.050$ lb-ft-sec²
$AB = 2\frac{3}{4}'$	$W_3 = 6$ lb	$I_3 = 0.085$ lb-ft-sec²
$AE = 2\frac{3}{4}'$	$W_4 = 10$ lb	$I_4 = 0.100$ lb-ft-sec²
$BE = 1'$	$W_5 = 10$ lb	$I_5 = 0.060$ lb-ft-sec²
$DB = 2\frac{3}{4}'$	$W_6 = 4$ lb	$I_6 = 0.048$ lb-ft-sec²
$BC = 1\frac{1}{2}'$		
$O_5C = 2\frac{1}{2}'$		

13.10 (Fig. 13.10). Link 2 is rotating at a constant speed of 1200 rpm counterclockwise. Determine the magnitude and location of the inertia effect of each link.

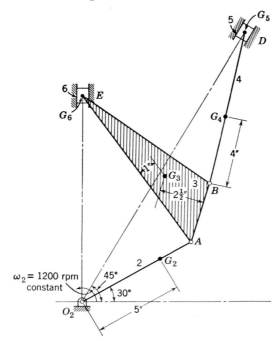

Fig. P–13.10.

$O_2A = 7''$	$W_2 = 18$ lb	$I_2 = 0.05$ lb-ft-sec^2
$AB = 3\frac{1}{2}''$	$W_3 = 15$ lb	$I_3 = 0.04$ lb-ft-sec^2
$AE = 10\frac{1}{2}''$	$W_4 = 10$ lb	$I_4 = 0.15$ lb-ft-sec^2
$BE = 8\frac{3}{4}''$	$W_5 = 12$ lb	$I_5 = 0.20$ lb-ft-sec^2
$BD = 9''$	$W_6 = 12$ lb	$I_6 = 0.20$ lb-ft-sec^2

13.11 (Fig. 13.11). Link 2 is rotating at a constant angular speed of 30 rad/sec counterclockwise. Determine the magnitude and the location of the inertia effect of each link.

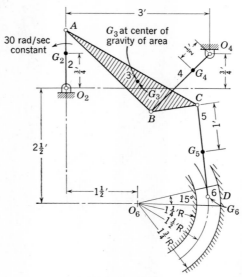

Fig. P-13.11.

$O_2A = 1\frac{1}{4}'$	$W_2 = 5.0$ lb	$I_2 = 0.035$ lb-ft-sec^2
$AB = 2\frac{1}{2}'$	$W_3 = 8.0$ lb	$I_3 = 0.05$ lb-ft-sec^2
$O_4B = 1\frac{3}{4}'$	$W_4 = 6.0$ lb	$I_4 = 0.04$ lb-ft-sec^2
$BC = 1'$	$W_5 = 4.5$ lb	$I_5 = 0.03$ lb-ft-sec^2
$CD = 2'$	$W_6 = 4.0$ lb	$I_6 = 0.3$ lb-ft-sec^2

13.12 (Fig. 13.12). An inclined rod of uniform rectangular cross section is rotating about an axis y–y at a constant angular speed of ω rad/sec.

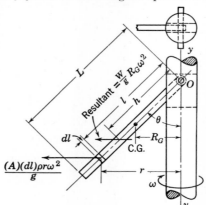

Fig. P-13.12.

(a) Show that the inertia force of an infinitesimal section, dl long, is

$$\frac{(A)(dl)(\rho)(r)(\omega^2)}{g}$$

where A is the area of a cross section, sq ft,
ρ is the weight density, lb/ft³,
ω is the angular speed, rad/sec,
$g = 32.2$ ft/sec².

(b) Show that

$$\text{the resultant inertia force} = \frac{W}{g} R_G \omega^2$$

where W is the total weight of the bar, lb,
R_G is the distance from the axis of rotation to the center of gravity of the bar.

(c) By taking moments of the components of the inertia force about the center of the pin O, and by equating the moment to the moment of the resultant inertia force about the same point:

$$\int_0^L \frac{(A)(dl)(\rho)(\omega^2)l \cos\theta}{g} = \frac{W}{g} R_G \omega^2 h \cos\theta$$

show that the resultant inertia force is located so that $h = \frac{2}{3}L$.

13.13 (Fig. 13.13). (It is suggested that problem 13.12 be worked before this problem is attempted.) An inclined rod of uniform rectangular cross section is rotating about a vertical axis at a constant angular speed of ω rad/sec.

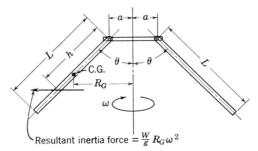

Fig. P–13.13.

(a) Show that

$$\text{the resultant inertia force} = \frac{W}{g} R_G \omega^2$$

where W is the weight of the bar, lb,
R_G is the distance from the axis of rotation to the center of gravity of the bar.

(b) Show that the resultant inertia force is so located that

$$h = \frac{3aL + 2L^2 \sin\theta}{6a + 3L \sin\theta}$$

(c) Sketch the distribution of the inertia force along the link and show that the result in part b may be obtained also by the determination of the center of gravity of a trapezoidal figure.

13.14 (Fig. 13.14). (It is suggested that problems 13.12 and 13.13 be worked before this problem is attempted.) A two-ball weighted governor is shown in Fig. 13.14. Determine, for the position shown, the inertia force of links 3 and 4, the inertia force of the 8-lb weight, and the location of each inertia force. Consider each link to be of uniform cross section.

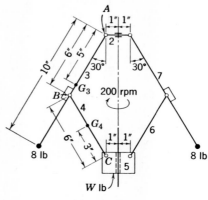

Fig. P-13.14.

$W_3 = W_7 = 3$ lb
$W_4 = W_6 = 2$ lb

13.15 (Fig. 13.15). (It is suggested that problems 13.12 and 13.13 be worked before this problem is attempted.) A two-ball, crossed-arm, weighted governor

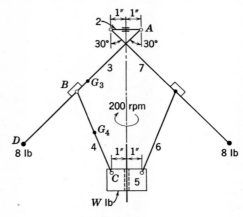

Fig. P-13.15.

$AB = 6''$ $W_3 = W_7 = 3$ lb
$AG_3 = 5''$ $W_4 = W_6 = 2$ lb
$AD = 10''$
$BC = 6''$
$BG_4 = 3''$

is shown in Fig. 13.15. Determine, for the position shown, the inertia force of links 3 and 4, the inertia force of the 8-lb weight, and the location of each inertia force. Consider each link to be of uniform cross section.

13.16 (Fig. 13.16). Link 2 of the Whitworth quick-return mechanism in Fig. 13.16 is rotating at a constant angular speed of 400 rpm clockwise. Determine the inertia effect magnitude and location for each link.

Link 2 is counterbalanced, or the center of gravity of link 2 is at O_2.

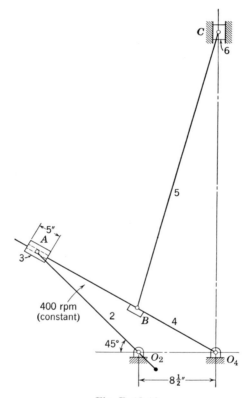

Fig. P–13.16.

$O_2A = 16''$	$W_3 = 3$ lb	$I_2 = 0.03$ lb-ft-sec^2
$O_4B = 10''$	$W_4 = 12$ lb	$I_3 = 0.45$ lb-ft-sec^2
$BC = 33''$	$W_5 = 15$ lb	$I_4 = 0.70$ lb-ft-sec^2
	$W_6 = 10$ lb	$I_5 = 0.80$ lb-ft-sec^2

The center of gravity of link 2 is at the center of pin O_2.
The center of gravity of link 3 is at the center of pin A.
The center of gravity of link 4 is at the center of pin B.
The center of gravity of link 5 is at the midpoint of link 5.
The center of gravity of link 6 is at the center of pin C.

13.17 (Fig. 13.17). Link 2 is horizontal for the position shown. The angular speed of link 2 is zero, but the angular acceleration of link 2 is 1200 rad/sec^2 counterclockwise. The mass of link 2 is negligible.

Determine the magnitude and location of the inertia effect of each link.

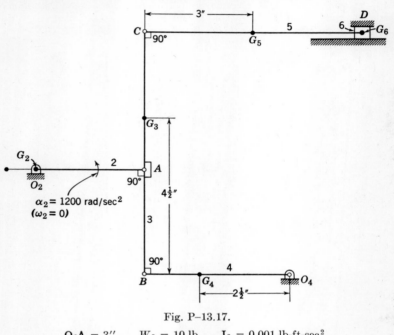

Fig. P-13.17.

$O_2A = 3''$	$W_3 = 10$ lb	$I_3 = 0.001$ lb-ft-sec^2
$AB = 3''$	$W_4 = 8$ lb	$I_4 = 0.005$ lb-ft-sec^2
$O_4B = 4''$	$W_5 = 9$ lb	$I_5 = 0.006$ lb-ft-sec^2
$BC = 7''$	$W_6 = 4$ lb	$I_6 = 0.0015$ lb-ft-sec^2
$CD = 6''$		

13.18 (Fig. 13.18). Link 2 is rotating at a constant speed of 60 rad/sec. For the position shown, link 2 is horizontal.

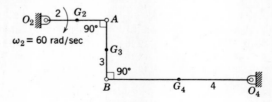

Fig. P-13.18.

$O_2A = 5''$	$W_2 = 4$ lb	$I_2 = 0.004$ lb-ft-sec^2
$AB = 5''$	$W_3 = 3$ lb	$I_3 = 0.001$ lb-ft-sec^2
$O_4B = 12''$	$W_4 = 6$ lb	$I_4 = 0.005$ lb-ft-sec^2

Determine the magnitude and the location of the inertia effect of each link. The center of gravity for each link is located at the center of the link.

13.19 (Fig. 13.19). For the position shown, link 2 has zero angular speed and an angular acceleration of 1800 rad/sec² clockwise.

Determine the magnitude and location of the inertia effect of each link.

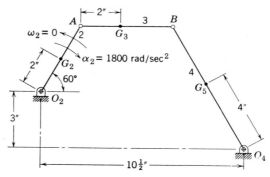

Fig. P–13.19

$O_2A = 4''$	$W_2 = 3$ lb	$I_2 = 0.03$ lb-ft-sec²
$AB = 4\frac{3}{4}''$	$W_3 = 4$ lb	$I_3 = 0.05$ lb-ft-sec²
$O_4B = 7\frac{1}{2}''$	$W_4 = 6$ lb	$I_4 = 0.10$ lb-ft-sec²

13.20. The moment of inertia of a connecting rod is to be found experimentally. The connecting rod is suspended from a knife's edge, displaced a small amount from the vertical, and allowed to swing freely. It is found that the time for 30 complete oscillations is 28.0 sec. The distance from the knife's edge to the center of gravity is found to be 7.25 in. If the connecting rod weighs 6.5 lb., what is the moment of inertia of the connecting rod about:

(a) The point of the connecting rod in contact with the knife's edge?
(b) The center of gravity of the connecting rod?

13.21. The moment of inertia of a body with an irregular shape is to be found experimentally by use of an oscillating torsional system wherein a single wire is used for support for the frame. The test is made with three separate measurements. The frame alone oscillates so that 35 sec are required for 20 oscillations. A solid cylindrical piece, which weighs 5.25 lb and has a diameter of 7.50 in., is placed on the frame, and it is found that the time for 20 complete oscillations is 45 sec. Next, the piece whose moment of inertia is desired is placed on the frame, and it is found that the time for 20 oscillations of the frame and piece is 60 sec. What is the moment of inertia of the irregular shape, about the center of gravity, expressed in pound foot second units?

13.22. Derive an expression for the moment of inertia of a body to be found from experimental tests, using an oscillating torsional system. A single supporting wire is to be used. The sample piece is to be a circular ring, with an outside radius R_o and an inside radius R_i, instead of a solid piece, as used in section 13.6

13.23. (Fig. 13.23) Figure 13.23a shows a body, with dimensions as given. Figure 13.23b shows a system made up from two concentrated weights W_1 and W_2 at points (1) and (2) and connected by a weightless rod.

218   Inertia Forces

Are the two systems kinetically equivalent? Explain carefully, giving reasons for your conclusion.

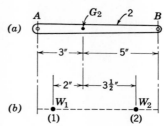

Fig. P–13.23.

$W = 12.0$ lb $W_1 = 4.365$ lb
$I_2 = 0.0259$ lb-ft-sec^2 $W_2 = 7.635$ lb

13.24. A link weighs 12.0 lb and has a moment of inertia about the center of gravity equal to 0.0259 lb-ft-sec^2. It is desired to replace the system by a kinetically equivalent system wherein two concentrated weights of 9 and 3 lb are used. Where should the two weights be located with respect to the center of gravity?

13.25. Refer to Fig. 13.4. (a) Replace the connecting rod by a kinetically equivalent system wherein two concentrated weights are used, one at the crank pin and the other at the proper location.

(b) Replace the connecting rod by a kinetically equivalent system wherein two concentrated weights are used, one at the piston pin and the other at the proper location.

(c) Replace the connecting rod by a kinetically equivalent system wherein two concentrated weights are used, one 2 in. from the crank pin and the other at the proper location.

13.26. Refer to Fig. 13.10. Replace link 3 by a kinetically equivalent system wherein two concentrated weights are used, one at the pin A and the other at the proper location.

13.27. Refer to Fig. 13.19. Replace link 3 by a kinetically equivalent system wherein two concentrated weights are used, one at the pin A and the other at the proper location.

13.28. Refer to Fig. 13.19. Replace link 3 by a kinetically equivalent system wherein *three* concentrated weights are used, one at the pin A, one at pin B, and the third at the proper location. Assume that the concentrated weight at point A is one pound.

CHAPTER 14

Dynamic Analysis

Dynamic analysis is defined as the study of the forces at the pins and guiding surfaces and the forces causing stresses in machine parts, such forces being the result of external forces applied to the machine and inertia forces due to the motion of each part in the machine. This chapter is devoted to the analysis of various machines under the action of external and inertia forces. Analyses using combined inertia and static forces will be discussed, and separate inertia and static force analyses will be described.

Inertia forces in high-speed machines can become very large, and cannot be neglected as might be done in slow-speed engines having light parts. The inertia force of the piston of an automobile traveling at high speed might be a thousand times the weight of the piston. In addition to the necessity of recognizing the magnitude of inertia forces, we must appreciate that inertia forces affect the forces applied to the frame of the machine. With the frame forces varying in magnitude and direction, so-called shaking forces are set up to give unbalance and vibration.

As was pointed out in the previous chapter, inclusion of the inertia force of each link with the forces acting on the link gives a system which can be handled as a static problem. If each link, with its inertia force and forces applied to the link, can be considered to be in equilibrium, the entire machine can be considered to be in equilibrium.

Several examples are given in the following pages.

14.1 Slider-crank mechanism with inertia and applied forces

Figure 14.1a shows a slider crank mechanism with a force P applied to the piston. The crank is assumed to be rotating at a constant angular velocity in a counterclockwise direction. Figure 14.1b shows the acceleration diagram, used in locating the direction of the inertia forces in Fig. 14.1a. The couple T_2 applied to the crank is desired.

Figure 14.1c shows links 3 and 4 isolated together. There are three

220   Dynamic Analysis

unknowns: magnitude and direction of $\mathcal{F}_{23}$,* magnitude of $\mathcal{F}_{14}$. The location of $\mathcal{F}_{14}$ is determined from a consideration of link 4 by itself. The force polygon is shown in Fig. 14.1d. Link 2 is shown isolated in Fig. 14.1e. The couple T_2 applied to link 2 for equilibrium is clockwise, with the couple applied to the driven shaft at O_2 being counterclockwise, the same direction as the rotation of the link. Thus there is a power output from the engine. It is interesting to note that if the applied force P were considerably smaller, it would have been possible to obtain a condition where T_2 applied to link 2 would be

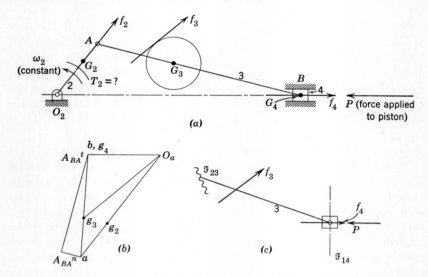

Fig 14.1. Applied and inertia forces in a slider-crank mechanism, with links 3 and 4 isolated.

counterclockwise, or the couple applied to the driven shaft at O_2 would be clockwise, which is in the opposite direction to that of the rotation of the crank. In such a case, there would not be any power output from the engine, but a power input to the engine. Such power could be considered to come from the flywheel attached to the crank.

Another interpretation that can be made for the problem is to think of the original mechanism with the applied force P and the couple T_2 applied to the crank as the only forces acting on the system, as

* The symbols used are: $\mathcal{F}_{23}$ to represent the total force applied by link 2 on link 3; f_{23} to represent the force applied by link 2 on link 3 as a result of the inertia force analysis; F_{23} to represent the force applied by link 2 on link 3 as a result of the static force analysis.

shown in Fig. 14.1f. It would be evident that the system would not be in equilibrium. The force P might be considered to be composed of two parts: one part working against T_2 and the other part going over to accelerate the parts with the prescribed motion. Use of the

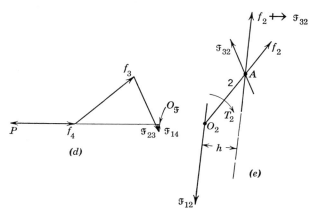

Fig. 14.1d, e. Force polygon and forces acting on the crank.

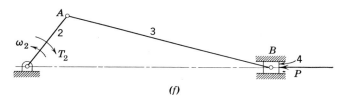

Fig. 14.1f. Mechanism without inertia forces is not in a system of equilibrium

inertia forces in the analysis, however, has simplified the problem to one of analysis of a static system of equilibrium.

14.2 Powell engine

A second illustrative problem, handled exactly as the previous one, is the system shown in Fig. 14.2a, the Powell engine. A known force P is applied to the piston. The velocity of the crank is prescribed, and, from the acceleration polygon shown in Fig. 14.2b, the accelerations are determined for use in calculating the inertia forces. A static force analysis was used to determine the force polygon shown in Fig. 14.2c. The setting up of the free body diagrams and following through the force polygon are left up to the student as an exercise.

222   Dynamic Analysis

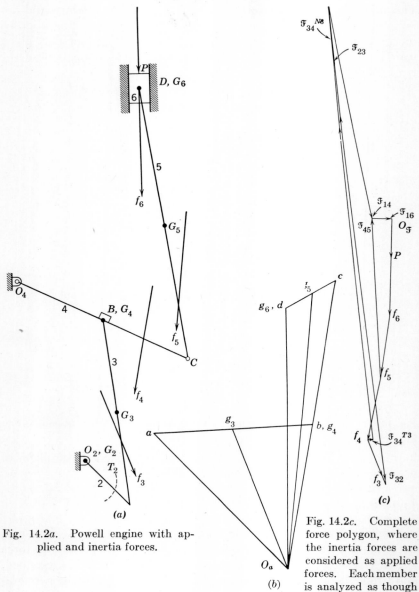

Fig. 14.2a. Powell engine with applied and inertia forces.

Fig. 14.2b. Acceleration polygon.

Fig. 14.2c. Complete force polygon, where the inertia forces are considered as applied forces. Each member is analyzed as though in equilibrium, or the mechanism is treated as a problem in statics.

14.3 Shaper mechanism

Figure 14.3a, the shaper mechanism, has a known couple applied to the crank, link 2, which is rotating at a given constant angular velocity in a counterclockwise direction. The inertia forces are shown to scale in position. The force Q applied to link 6 is desired. The analysis is made as though a static case of the system in equilibrium were being analyzed. Figure 14.3b shows each link isolated. The

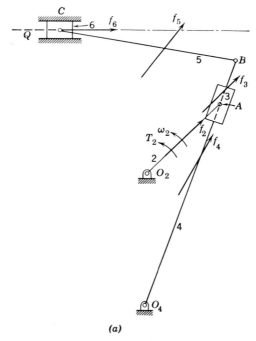

Fig. 14.3a.   Forces in a shaper mechanism.

unknowns for each link are greater than three, so attack cannot be made directly. Links 2 and 3, considered separately, have a total of six unknowns: magnitude and direction of $\mathfrak{F}_{12}$, magnitude and direction of $\mathfrak{F}_{32}$, and magnitude and location of $\mathfrak{F}_{43}$. However, there are six equations that can be applied, three for each link. The force $\mathfrak{F}_{32}$ is broken up into two components: $\mathfrak{F}_{32}{}^{T2}$ and $\mathfrak{F}_{32}{}^{N2}$ components perpendicular and parallel, respectively, to O_2–A, as shown in Fig. 14.3c. $\mathfrak{F}_{32}{}^{T2}$ is found from a moment equation.   Figure 14.3c shows, also, $\mathfrak{F}_{23}{}^{T2}$ and $\mathfrak{F}_{23}{}^{N2}$, by action and reaction, applied to link 3.   Link 3 may be reduced to a three-force system by combining $\mathfrak{F}_{23}{}^{T2}$ and f_3. The intersection of the resultant so found and $\mathfrak{F}_{23}{}^{N2}$ gives a point

224 Dynamic Analysis

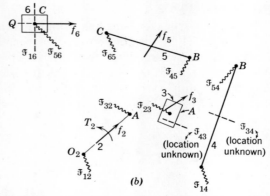

Fig. 14.3b. Free body diagrams for each isolated link.

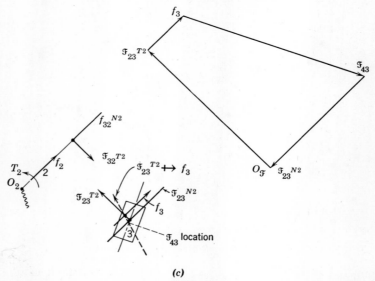

Fig. 14.3c. Force analysis for links 2 and 3, with six unknowns and six equations of equilibrium.

through which $\mathfrak{F}_{43}$ must pass to satisfy a moment equation. The force polygon to find $\mathfrak{F}_{43}$ is shown in Fig. 14.3c.

Links 4 and 5 can be handled next, with six equations and six unknowns. The remainder of the solution is left to the student.

14.4 Analysis of a system for a prescribed motion

This section is devoted to a discussion of the required effort necessary for a prescribed motion of a mechanism. The slider crank, shown

Analysis of a System for a Prescribed Motion

in Fig. 14.4a, will serve the purpose of illustrating the attack for a situation where a force p applied to the piston is to be found for a given motion as defined by the known angular velocity of the crank. The crank is assumed to be rotating at a constant angular velocity.

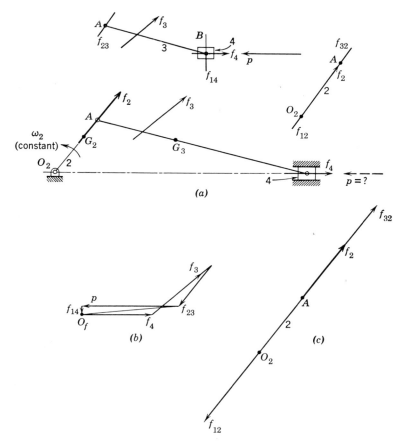

Fig. 14.4. Analysis for the necessary force p to give a prescribed motion, a constant angular velocity of the crank for the position shown. No power taken from the engine. (Note that the crank is a two-force member.)

No power is taken off the crank, the effort p going over to give the acceleration as prescribed by the known angular velocity of the crank. The free body diagram of each link is shown, with the inertia force of each link. The system is handled as a static equilibrium case, with the force polygon for links 3 and 4 shown in Fig. 14.4b. Note that links 3 and 4 considered together give a four-force system, with the resultant of p and f_{14} passing through the intersection of the resultant of f_3 and f_4 with f_{23}. Link 2 is shown separately in Fig. 14.4c.

Figure 14.5a shows the same mechanism as Fig. 14.4a, except that the effort is provided by a couple t_f applied to the crank. No power is taken out of the engine, the couple going over to give the prescribed motion. The free body diagrams are shown, with the force polygon of Fig. 14.5b. The crank is shown in Fig. 14.5c.

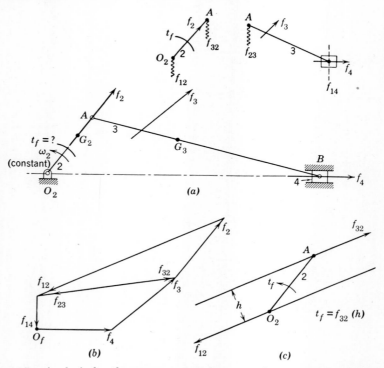

Fig. 14.5. Analysis for the necessary couple, t_f, to give a prescribed motion, a constant angular velocity of the crank.

It is important to note that the same effect, that is, the prescribed motion, is achieved in two different ways, with two different force polygons, and with different pin forces. We cannot analyze such a system for stresses without knowing specifically how the engine is driven.

14.5 Separate static and inertia force analysis

An artifice that is quite frequently used in an attempt to separate the inertia and static effects is the use of separate force polygons for the inertia and static forces. The basic idea underlying the procedure is the principle of superposition of forces, *which is applicable if friction is not considered.* Figure 14.6 illustrates the principle. In (a) is shown a body on which is applied two forces, P and f. (P represents a static force; f represents an inertia force.) The forces putting the system in equilibrium are R_L and R_R. In (b), the force P is considered by itself, and the reactions due to P are determined. In (c), the force f

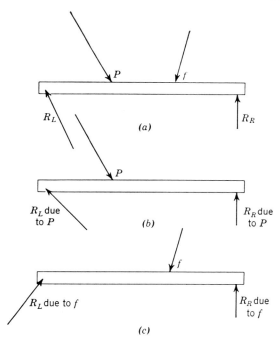

Fig. 14.6. Illustration of the principle of superposition of forces.

is considered by itself, and the reactions due to f are determined. If (b) and (c) are superimposed, with the reactions on the left and right sides of the body found by vector addition, the original case (a) results. Let us examine the significance of such an approach as applied to a slider crank mechanism for illustrative purposes.

Figure 14.7a shows a mechanism with a total force P applied to the piston. The crank is assumed to be rotating at a constant angular velocity. (The mechanism has already been analyzed in Fig. 14.1.) The total couple T_2 applied to the crank is desired.

Various combinations of forces may be selected, in separate force analyses. Figure 14.7b shows a combination where the force P is considered as a static force, which permits the finding of t_s, considered as a couple caused by static effects. Analysis of the system with only the inertia effects permits finding the couple t_f applied to the crank. The sum of t_s and t_f gives T_2, the total couple applied to

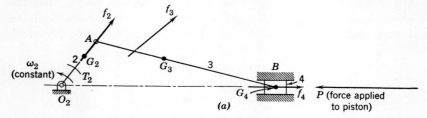

Fig. 14.7a. A system of applied and inertia forces which can be analyzed by different static and inertia force arrangements, as shown in (b) and (c).

the crank. Note that the superposition of the two systems in Fig. 14.7b gives the same total quantities as are given in Fig. 14.7a.

A second possibility in analysis is to consider the system shown in Fig. 14.7c. Here a force p is pictured applied to the piston, with no couple on the crank, to balance the inertia forces. A force $(P-p)$ is

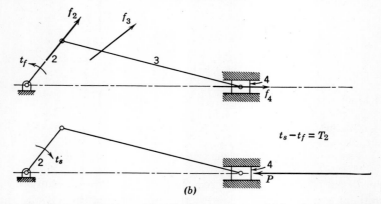

Fig. 14.7b. Separate static and inertia force systems which, when combined, give the original in (a). The total force P applied to the piston is considered a static force.

considered applied to the piston for the determination of T_2. The interpretation here is that, of the total force, a force p is necessary to cause the prescribed motion, only part of the total force going over to cause a couple T_2 on the crank. Again, the two systems in Fig. 14.7c will combine to give the original in Fig. 14.7a.

Now the question arises as to which system is to be used in an analysis. If we are concerned with total effects, the answer is that it makes no difference which system is taken. If we are concerned with effects of inertia forces, for instance, we must be very careful. For a complete cycle in an internal combustion engine, the system changes in that for a portion of the cycle the effort comes from the gas pressure on the piston whereas for the remainder of the cycle the effort comes from the couple applied to the crank through the flywheel. Compare the forces f_{14} and f_{23} in Figs. 14.4b and 14.5b. These forces are

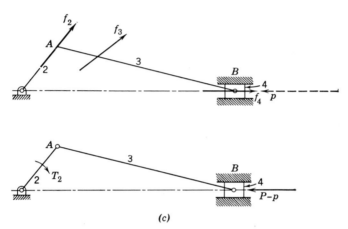

Fig. 14.7c. Separate static and inertia force systems which, when combined, give the original in (a). The total force P applied to the piston minus the force p applied to the piston to cause the prescribed motion is considered a static force. If the force p is directed to the right, the static force would be P plus p.

different as a result of different assumptions as to how the engine is driven. The stresses in the various members of the two systems are different since the pin forces are different.

14.6 Shaking force

A shaking force is defined as the vector sum of the forces exerted on the frame of a machine, with the force varying in direction or magnitude or both. The forces transmitted to the frame of a machine may be considered to be made up of two components: (1) that due to the static forces; and (2) that due to the inertia forces. A combined force analysis will give the various forces exerted on the structure which would have to be considered in the final design of a machine. However, it is desirable to separate the shaking effect of the inertia forces from that of the static forces because in some cases the effect of the inertia forces can be balanced in part, or completely, as will be discussed in later chapters on balancing of engines.

The slider-crank mechanism is chosen again to illustrate the principles. Reference to Figs. 14.4a and 14.4b reveals that, with the effort coming from a gas force p on the piston for the prescribed motion, the force exerted on the stationary frame is the vector sum of a force p applied to the cylinder head (a force to the right, equal and opposite to the force applied to the piston), a force f_{41}, and a force f_{21}, where f_{41} is the force exerted by the piston on the frame and f_{21} is the force exerted by the crank on the frame. If the force polygon, shown in Fig. 14.4b, is analyzed, together with the forces in Fig. 14.4c, it is found that $p \leftrightarrow f_{41} \leftrightarrow f_{21}$ is equal to $f_2 \leftrightarrow f_3 \leftrightarrow f_4$, or the shaking force due to the inertia forces is the vector sum of the inertia forces.

If the engine is driven by a couple applied to the crank, as shown in Fig. 14.5a, the shaking force, $f_{41} \leftrightarrow f_{21}$, is also equal to $f_2 \leftrightarrow f_3 \leftrightarrow f_4$, which can be seen by analysis of Fig. 14.5b. Thus the vector sum of the inertia forces can be determined to give directly the shaking force due to the inertia forces, without a force analysis. It is to be noted, however, that the line of action of the shaking force for Fig. 14.4a is not the same as that for Fig. 14.5a. The line of action of the shaking force has to be determined by a complete force analysis.

Balancing of the shaking force due to the inertia forces does not mean smooth operation of the machine, necessarily. The magnitude of the couple applied to the crank will change even if the total force applied to the piston is constant. If a single-cylinder engine were used, a varying couple applied to the crank would cause varying forces applied to the frame of the machine. In actual practice, a flywheel is used, a couple being set up by the action of the flywheel to counteract the effect of the varying couple. (The effect of a flywheel is discussed in Chapter 16.) Another solution is to use several engines with the cranks displaced so that a more uniform couple could be obtained.

PROBLEMS

14.1 (Fig. 14.1). A block, as shown is accelerating to the right at the rate of 150 ft/sec². The total resistance $Q = 200$ lb is applied as shown. The block

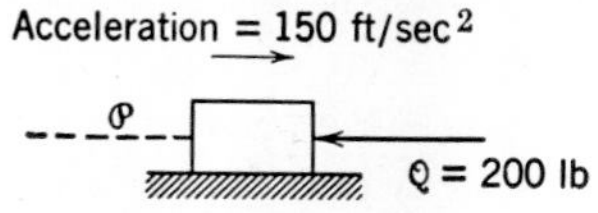

Acceleration = 150 ft/sec²

$Q = 200$ lb

Fig. P-14.1.

weighs 9 lb. Neglecting friction, determine the force $\mathcal{P}$ by each of the following methods:

(a) By a combined static and inertia force analysis.

(b) By a separate static and inertia force analysis, thinking of the total force $\mathcal{P}$ as broken up into components, one component p to accelerate the block and the other component P to balance the force Q considered as a static force:

$$(\mathcal{P} = p \not\mathrel{+\!\!\!+} P)$$

(c) By a separate static and inertia force analysis, thinking of the force $\mathcal{P}$ as a static force, and thinking of Q as broken up into components, one component q to balance the inertia force and the other component Q to balance the static effect:

$$(Q = q \not\mathrel{+\!\!\!+} Q)$$

14.2 (Fig. 14.2). A link, as shown, is counterbalanced so that the center of gravity of the link is at the center of rotation. A couple $T_2 = 40$ ft-lb clockwise

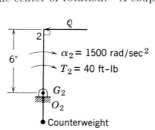

Fig. P–14.2. $I_2 = 0.010$ lb-ft-sec^2.

is applied. A resistance force of Q is applied perpendicular to the link as shown. Determine the force Q by each of the following methods:

(a) By a combined inertia and static force analysis.

(b) By a separate static and inertia force analysis, thinking of the total couple T_2 as broken up into components, one component t_f to balance the inertia effect, and the other component T_s to balance the effect of the resistance force Q.

(c) By a separate static and inertia force analysis, thinking of the total torque T_2 as a static effect, and thinking of the total force Q as broken up into components, one component q to balance the inertia effect and the other component Q to balance the static effect, where $Q = q \not\mathrel{+\!\!\!+} Q$.

14.3 (Fig. 14.13). The purpose of the following problem is to show that the separate analysis of the inertia and static effects cannot be made necessarily correctly if friction is to be taken into account.

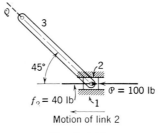

Fig. P–14.3.

Figure 14.3 shows two links, links 2 and 3, with a total force $\mathcal{P} = 100$ lb applied to link 2 and an unknown force Q applied to link 3, along the axis of link 3. The inertia force of link 2 is known to be 40 lb, acting to the right. Link 2 is moving to the left. The inertia force of link 3 is to be neglected.

(a) Using a combined inertia and static force analysis, determine the magnitude of Q if the coefficient of sliding friction is 0.2. Neglect pin friction. Determine the total reaction of link 1 on link 2, $\mathfrak{F}_{12}$.

(b) Using a separate inertia force and static force analysis, determine the reaction of link 1 on link 2 due to the inertia forces only, f_{12}. Determine, also, the reaction of link 1 on link 2 due to the static effects only, F_{12}. Consider that the force $\mathcal{P}$ is a static force, that the accelerating force is applied by q on link 3, and that the force Q on link 3 balances the static forces. The total force $\mathcal{Q} = q \nrightarrow Q$. Assume that the coefficient of sliding friction is 0.2 for the inertia and static force analysis, as used in part a.

Does the vector sum of f_{12} and F_{12} give the same result $\mathfrak{F}_{12}$ as found in part a? Why does the separate analysis with friction taken into account give an erroneous result? Note, also, that $\mathcal{Q}$ as found in part a does not agree with $\mathcal{Q}$, that is, $q \nrightarrow Q$, as found in part b.

(c) If $\mathcal{P}$ and f_2 were in the same direction, would the combined analysis give the same results as the separate analyses? Explain.

14.4 (Fig. 14.4). An air compressor is driven by a couple T_2 applied to the crank, the couple being 500 lb-ft. The suction force Q is 400 lb. The power

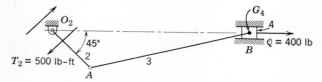

Fig. P-14.4. $O_2A = 6''$; $AB = 18''$.

supplied to the compressor through the crank goes into accelerating the links and drawing in the air. Neglecting the masses of the crank and connecting rod, and neglecting friction, make complete static force and inertia force polygons. Determine the total crank bearing reaction, $\mathfrak{F}_{12}$.

Check the result by making a combined force analysis.

14.5 (Fig. 14.5). In a gear-driven air compressor, shown schematically, the energy required during the working stroke comes from two sources: the prime mover, as represented by the force $\mathcal{P}$ applied to the gear, link 2, and as represented

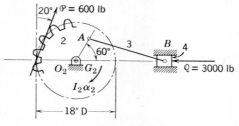

Fig. P-14.5. $O_2A = 6''$; $AB = 18''$.

by the inertia couple of the gear, link 2. $\mathcal{P} = 600$ lb, and $Q = 3000$ lb. Neglect friction and the masses of links 3 and 4.

(a) Using separate inertia and static force analyses, determine the necessary magnitude of the inertia couple of link 2 and the total reaction $\mathfrak{F}_{12}$.

(b) Determine the magnitude of the inertia couple of link 2 and the total reaction $\mathfrak{F}_{12}$ by means of a combined force analysis.

14.6 (Fig. 14.6). A gear, link 2, drives a gear, link 3. The pressure angle is 20 degrees. If a couple, T_2, is applied to link 2, show that the angular acceleration

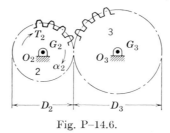

Fig. P–14.6.

of link 2, α_2, may be expressed by

$$\alpha_2 = \frac{T_2}{I_2 + (D_2/D_3)^2 I_3}$$

where I_2 is the moment of inertia of the gear, link 2, about its center of gravity at the center of the pin O_2;

where I_3 is the moment of inertia of the gear, link 3, about its center of gravity at the center of the pin O_3;

where D_2 and D_3 are the diameters of gears 2 and 3, respectively.

Hint. Isolate the two gears and make complete analyses of each.

14.7. Refer to Fig. 13.8. What couple must be applied to the gear, link 2, for the prescribed angular acceleration of link 2?

14.8 (Fig. 14.8). A single-cylinder diesel engine is shown. Determine the couple that must be applied to link 2 to cause the prescribed motion. Consider the inertia force of each link.

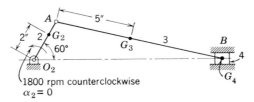

Fig. P–14.8.

$O_2A = 3''$ $W_2 = 5$ lb $I_2 = 0.004$ lb-ft-sec^2
$AB = 11\frac{1}{4}$ $W_3 = 8$ lb $I_3 = 0.030$ lb-ft-sec^2
  $W_4 = 6$ lb

14.9. Refer to Fig. 14.8. Find the force p applied to the piston to cause the prescribed motion.

234 Dynamic Analysis

14.10. A single-cylinder diesel engine is shown in Fig. 14.8. A force $\mathcal{P} = 4000$ lb is applied to the piston. $\mathcal{P}$ is applied to the left and acts through the center of the piston pin.

(a) Using a combined force analysis, find the couple T_2 that must be applied to link 2 for equilibrium, and find $\mathfrak{F}_{12}$ and $\mathfrak{F}_{14}$.

(b) Using separate static and inertia force analyses, considering $\mathcal{P}$ as a static force, determine t_f (the couple applied to the crank to balance the inertia forces) and T_s (the couple applied to the crank to balance the static force $\mathcal{P}$). Determine the total couple applied to the crank. Find, also, $\mathfrak{F}_{12}$ and $\mathfrak{F}_{14}$.

(c) Using separate static and inertia force analyses, find the total couple, T_2 applied to the crank, thinking of T_2 as a static effect, and considering $\mathcal{P}$ as made up of two parts, one part p necessary to balance the inertia effects, and the other part P necessary to balance the static effects. Find, also, $\mathfrak{F}_{12}$ and $\mathfrak{F}_{14}$.

14.11 (Fig. 14.11). The crank of the single-cylinder engine shown is counterbalanced so that the center of gravity of the crank is at the center of the crank

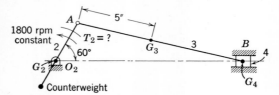

Fig. P–14.11.

$O_2A = 3''$ $W_2 = 5$ lb $I_2 = 0.010$ lb-ft-sec^2
$AB = 11\frac{1}{4}''$ $W_3 = 8$ lb $I_3 = 0.030$ lb-ft-sec^2
$\phantom{AB = 11\frac{1}{4}''}$ $W_4 = 6$ lb

bearing. For the given information, determine the couple applied to the crank to cause the prescribed motion.

14.12. Refer to Fig. 14.11. A force $\mathcal{P} = 4000$ lb is applied to the piston, the force acting to the left and through the center of the piston pin. Using a separate inertia and static force analysis, where $\mathcal{P}$ is considered as a static force, determine t_f (the couple applied to the crank to balance the inertia forces) and T_s (the couple applied to the crank to balance the static force $\mathcal{P}$). Determine the total couple T_2 applied to the crank. Find $\mathfrak{F}_{12}$ and $\mathfrak{F}_{14}$.

14.13. Refer to Fig. 13.7. Determine the couple t_f that must be applied to the crank to give the prescribed motion, as specified by the components of acceleration given for point A. The center of gravity of each link is at its midpoint.

14.14. Refer to Fig. 13.7. A couple $T_4 = 900$ in.-lb is applied to link 4 in a clockwise direction. For the prescribed motion, what couple must be applied to link 2 to accelerate the parts and to work against the couple applied to link 4? The center of gravity of each link is at its midpoint.

(a) Using a combined force analysis.

(b) Using a separate inertia and static force analysis, thinking of T_4 as a static effect, and considering the total couple T_2 applied to link 2 as being made up of two parts, t_f to balance the inertia forces and T_s to balance T_4.

(c) Using a separate inertia and static force analysis, thinking of T_2 as a static effect, and considering T_4 as made up of two parts, one part being t_f, the couple necessary to balance the inertia forces and the other part being T_s, the couple necessary to balance the static effects, with $T_4 = t_f + T_s$.

14.15. A cam follower weighing 0.966 lb is to rise 0.9 in., the first 0.3 in. with a constant acceleration of 1000 ft/sec^2 and the last 0.6 in. with a constant deceleration of 500 ft/sec^2. A spring to back up the follower has been tentatively chosen to have a constant of 30 lb/in. and to be assembled with zero initial force. Determine whether or not the spring is adequate to maintain the follower in contact with the cam.

14.16. Refer to Fig. 13.14. Determine the central weight W pounds which is necessary to have the two-ball governor run at the speed shown in the position shown. Consider the inertia force of each link, and the weight of each link. Consider each link to be of uniform cross section.

14.17. Refer to Fig. 13.15. The two-ball weighted governor is rotating at the speed shown.

(a) Neglecting the weights and inertia forces of links 3, 4, 6, and 7, determine the weight W of the central weight, link 5, to allow the governor to run at the given speed in the position shown.

(b) Considering the effect of the inertia force and weight of each link, determine the weight W of the central weight, link 5, to allow the governor to run at the given speed in the position shown. Consider each link to be of uniform cross section.

14.18. Refer to the data and figure for problem 13.9. A force $\mathcal{P} = 500$ lb is applied to the piston, link 6, and acts through the center of the pin, D, along the line of motion of the piston, in a downwards direction.

(a) Using a combined force analysis, determine the couple T_2 that must be applied to the crank to overcome the inertia effects and the force applied to the piston. Determine the total pin reactions for all the pins. Is power going into or coming from the engine for the prescribed motion for the position shown?

(b) Using separate inertia and static force analyses, determine the couple t_f applied to link 2 to overcome the inertia effects and the couple T_s to balance the static force $\mathcal{P}$.

14.19. Refer to Fig. 13.10. A force $\mathcal{P}_5$ is acting on link 5, directed from D to O_2 and a force $\mathcal{P}_6$ is acting on link 6, directed from E to O_2. $\mathcal{P}_5 = 600$ lb and $\mathcal{P}_6 = 300$ lb. Determine, for the position and motion shown, the turning effort on the crank, link 2. (The turning effort is the force at A perpendicular to the crank.) Use separate inertia and static force analyses, considering $\mathcal{P}_5$ and $\mathcal{P}_6$ as static forces.

14.20. Refer to Fig. 13.11. A clockwise couple $T_4 = 600$ in.-lb is applied to link 4. Using separate inertia and static force analyses, determine the total couple T_2 applied to the crank, link 2, for the prescribed motion. Determine, also, all the pin forces.

14.21. Refer to Fig. 13.16. A downward force $\mathcal{P} = 800$ lb is applied to link 6, the force acting through the center of the pin. What is the required couple T_2 applied to link 2? Use a combined force analysis.

14.22. Refer to Fig. 13.17.

(a) What is the magnitude of the couple T_2 applied to the crank, link 2, for the prescribed motion? What are the pin forces due to the inertia effects?

(b) If a couple is applied to link 4 instead of link 2 for the prescribed motion, determine the necessary magnitude of T_4. How does T_4 compare to T_2 as found in part a? Will the pin forces be the same?

14.23. Refer to Fig. 13.17. A force $\mathcal{P} = 400$ lb is applied to the piston, link 6. The force $\mathcal{P}$ acts to the left and passes through the center of the pin, D. What couple must be applied to link 2 to work against the force and to give the prescribed motion?

Dynamic Analysis

14.24. Refer to Fig. 13.17. A couple of 900 in.-lb is applied to link 4, in a clockwise direction, and a force $\mathcal{P} = 400$ lb is applied to the piston, link 6. The force acts to the left and passes through the center of the pin, D.

(a) Determine the couple applied to link 2.

(b) Determine all the pin forces.

14.25. Refer to Fig. 13.18. What couple applied to link 2 is necessary to obtain the prescribed motion? What are the pin forces due to the prescribed motion? The center of gravity of each link is at the midpoint of the link.

14.26. Refer to Fig. 13.18. A couple $T_4 = 300$ in.-lb is applied to link 4 in a clockwise direction. What couple must be applied to link 2 to balance the couple T_4 and to balance the inertia forces? The center of gravity of each link is at the midpoint of the link.

14.27. Refer to Fig. 13.19. What couple must be applied to link 4 to give the prescribed motion? What are the pin forces?

14.28. Refer to Fig. 13.4. A couple of 800 in.-lb clockwise is applied to the crank. For the position shown, the angular speed of the crank is zero. What is the angular acceleration of the crank and connecting rod, and the linear acceleration of the piston?

14.29. A slider-crank mechanism is in head end dead center position, at which time the angular speed of the crank is zero. A couple T_2 is applied to the crankshaft.

The crank, of length R feet, is counterbalanced and has a moment of inertia about the axis of rotation equal to I_2 lb-ft-sec^2. The connecting rod length is L feet, and the center of gravity of the connecting rod is midway between the crank pin and the piston pin. The weight of the connecting rod is W_3 pounds, and the weight of the piston is W_4 pounds.

Develop an expression for the applied couple, T_2, in terms of the above quantities and the angular acceleration of the crank, α_2. There are no other external forces or couples to be considered.

14.30. Refer to Fig. 13.16.

(a) Determine the force $\mathcal{P}$ which may be applied to the slider, link 6, if 10 horsepower is supplied to the crank, link 2, which is rotating at a constant speed of 400 rpm clockwise. No flywheel is used, and all the energy is supplied to the crank by the motor. Use separate inertia and static force analyses.

(b) What percentage change in the force $\mathcal{P}$ will result if the inertia forces are disregarded, for the position shown.

14.31 (Fig. 14.31). A complete investigation of stresses in a connecting rod involves static and inertia effects. The purpose of this problem is to investigate the distribution of inertia force loading on a connecting rod of a simplified section as the first step in the appreciation of the stress analysis.

Fig. P–14.31. $O_2A = 3\tfrac{3}{4}''$; $AB = 15''$.

Figure 14.31 shows a single-cylinder engine where the crank is assumed to be rotating at a constant angular speed of 3000 rpm counterclockwise. The connecting rod is assumed to be a steel bar of uniform rectangular section, 2 by $1\tfrac{1}{2}$ in., and 15 in. long.

The analysis will be carried out for the one position shown.

Picture breaking up the connecting rod into ten arbitrary sections, each $1\frac{1}{2}$ in. long. Consider, then, ten masses with the weight of each section concentrated at the midpoint of the corresponding section. Determine the inertia force of each of the ten sections and show, to scale, the inertia forces applied to the link, in the proper direction.

Resolve the inertia forces into components, one component perpendicular to the connecting rod and the other component parallel to the connecting rod. These components will show the type of loading encountered in the connecting rod due to the inertia forces.

To compare the accuracy of the approximate procedure outlined above, determine the resultant of the ten components and compare the result with that found from an exact solution, using the methods of Chapter 13.

The density of steel may be taken as 0.283 lb/cu in.

Neglect the effect of the pins and the projection of the connecting rod beyond the pins.

Show that the moment of inertia of the connecting rod, with a uniform cross section, about the center of gravity may be expressed by

$$\mathbf{I}_3 = \frac{1}{12} \frac{W_3}{g} L^2$$

where W_3 is the weight of the connecting rod,
  L is the length of the connecting rod.

14.32 (Fig. 14.32). This problem is concerned with the determination of the inertia force distribution in the floating member of a four-link mechanism, as shown. The analysis is to be carried out for the one position shown, when the crank

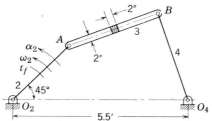

Fig. P–14.32. $O_2A = 2.5'$; $AB = 3'$; $O_4B = 3'$.

link 2 has an angular speed of 600 rpm counterclockwise and an angular acceleration of 800 rad/sec² counterclockwise. The driving couple for the prescribed motion is applied to link 2.

Link 3 is assumed to be of constant cross section, 2 in. by 2 in. The rod is made from steel, with a density of 0.283 lb/cu in.

Picture breaking up link 3 into twelve sections, each 3 in. long. Consider, then, twelve masses, with the weight of each section concentrated at the midpoint of the corresponding section. Determine the inertia force of each of the twelve sections and show, to scale, the inertia forces applied to the link, in the proper direction.

To compare the accuracy of the approximate procedure outlined above, determine the resultant of the twelve components, and compare the result with that found by the methods of Chapter 13.

Neglect the effect of the pins and the projection of the rod beyond the pins.

CHAPTER 15

Analytic Determination of Accelerations in a Slider-Crank Mechanism

Since the slider-crank mechanism is encountered so frequently and since it is desirable to simplify the analysis for a complete cycle to eliminate the tedious graphical solution, an analytic expression will be derived for the acceleration of the piston. Use will be made of the analytic expression in later chapters on balancing of engines.

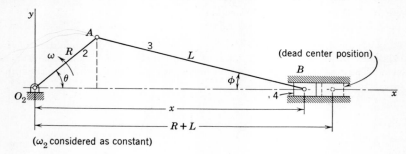

Fig. 15.1.

Consider the in-line slider crank mechanism shown in Fig. 15.1. The position of the point B from O_2, the origin of the coordinate axes x and y, is represented by x. Link 2 is at an angle θ from the x-axis for the position shown. The length of the crank is given by R, and the length of the connecting rod is given by L. The connecting rod makes an angle of ϕ with the x-axis. The distance x may be represented by the sum of the projection of the crank and connecting rod on the x-axis:

$$x = R \cos \theta + L \cos \phi \tag{1}$$

Since it is desirable to represent x as a function of R, L, and θ, ϕ may be eliminated by noting that the vertical projections of R and L are equal, or

$$R \sin \theta = L \sin \phi \tag{2}$$

Accelerations in a Slider-Crank Mechanism

or
$$\sin \phi = \frac{R}{L} \sin \theta \qquad (3)$$

but
$$\cos \phi = \sqrt{1 - (\sin \phi)^2} \qquad (4)$$

Therefore, substitution of Eq. 3 into Eq. 4 gives

$$\cos \phi = \sqrt{1 - \left(\frac{R}{L} \sin \theta\right)^2} \qquad (5)$$

Substitute Eq. 5 into Eq. 1:

$$x = R \cos \theta + L \sqrt{1 - \left(\frac{R}{L} \sin \theta\right)^2} \qquad (6)$$

Equation 6 is exact for the location of the piston from the center of the crank bearing.

The expression for the velocity may be obtained by differentiating x with respect to t, noting that $\frac{d\theta}{dt} = \omega$, the angular velocity of the crank:

$$V = \frac{dx}{dt} = -R\omega \left[\sin \theta + \frac{R}{2L} \frac{\sin 2\theta}{\sqrt{1 - \left(\frac{R}{L} \sin \theta\right)^2}} \right] \qquad (7)$$

The expression for the acceleration of the piston may be obtained by differentiating V with respect to t, considering that ω, the angular speed of the crank, is constant:

$$A = \frac{dV}{dt} = -R\omega^2 \left[\cos \theta + \frac{\frac{R}{L} \cos 2\theta \left[1 - \left(\frac{R}{L}\right)^2\right] + \left(\frac{R}{L}\right)^3 \cos^4 \theta}{\left[1 - \left(\frac{R}{L} \sin \theta\right)^2\right]^{3/2}} \right] \qquad (8)$$

The significance of the negative signs is that if the velocity or acceleration is negative, it is directed from the piston to the crank bearing. The angular velocity of the crank is positive if counterclockwise.

Because Eqs. 6, 7, and 8 are unwieldy in application, the usual practice is to use approximate forms of the equations, which are sufficiently accurate for practical purposes. To obtain the approximate forms, consider expanding the radical in Eq. 6 by the binomial theorem:

$$(a+b)^n = a^n + \frac{na^{n-1}b^1}{1!} + \frac{(n)(n-1)a^{n-2}b^2}{2!}$$
$$+ \frac{(n)(n-1)(n-2)a^{n-3}b^3}{3!} + \cdots$$

Therefore, for $\left[1 - \left(\frac{R}{L}\sin\theta\right)^2\right]^{\frac{1}{2}}$,

$$a = 1$$
$$b = -\left(\frac{R}{L}\sin\theta\right)^2$$
$$n = \frac{1}{2}$$

Substituting the values above into the expression for the binomial theorem, we have

$$\left[1 - \left(\frac{R}{L}\sin\theta\right)^2\right]^{\frac{1}{2}} = 1^{\frac{1}{2}} + \frac{\frac{1}{2}\left(1^{\frac{1}{2}-1}\right)\left[-\left(\frac{R}{L}\sin\theta\right)^2\right]^1}{1}$$
$$+ \frac{\frac{1}{2}\left(\frac{1}{2}-1\right)\left(1^{\frac{1}{2}-2}\right)\left[-\left(\frac{R}{L}\sin\theta\right)^2\right]^2}{(1)(2)}$$
$$+ \frac{\frac{1}{2}\left(\frac{1}{2}-1\right)\left(\frac{1}{2}-2\right)\left(1^{\frac{1}{2}-3}\right)\left[-\left(\frac{R}{L}\sin\theta\right)^2\right]^3}{(1)(2)(3)} + \cdots$$
$$= 1 - \frac{1}{2}\left(\frac{R}{L}\sin\theta\right)^2 - \frac{1}{8}\left(\frac{R}{L}\sin\theta\right)^4 - \frac{1}{16}\left(\frac{R}{L}\sin\theta\right)^6 + \cdots$$

The greatest value of R/L is approximately $1/4$ for practical engines. If the maximum value of $\sin\theta = 1$, and if R/L is taken as $1/4$, the order of magnitude of the terms is

$$1 - \frac{1}{32} - \frac{1}{2,048} - \frac{1}{65,536} + \cdots$$

The series is a rapidly converging one, and little error is introduced if the terms after the second one are neglected. The radical may be written in the approximate form by:

$$\sqrt{1 - \left(\frac{R}{L}\sin\theta\right)^2} \approx 1 - \frac{1}{2}\left(\frac{R}{L}\sin\theta\right)^2$$

The exact equation for displacement, Eq. 6, may be written, in the approximate form, by

$$x = R \cos \theta + L - \frac{1}{2} \frac{R^2}{L} \sin^2 \theta \qquad (9)$$

An expression for the velocity may be obtained by differentiating the displacement with respect to time:

$$V = -R\omega \sin \theta - \frac{R^2}{2L} \omega \sin 2\theta$$

or

$$V = -R\omega \left[\sin \theta + \frac{R}{2L} \sin 2\theta \right] \qquad (10)$$

Differentiate the velocity with respect to time for the acceleration, considering the angular velocity of the crank constant:

$$A = -R\omega^2 \left[\cos \theta + \frac{R}{L} \cos 2\theta \right] \qquad (11)$$

The maximum error in the approximate expressions above is around ½% of the maximum values for $R/L = 1/4$. However, because of the simplification of the equations, the approximate expressions are used in analytical work.

A comparable analysis can be used to determine the acceleration of any point on the connecting rod, if desired.

PROBLEMS

15.1. Refer to Fig. 15.1 of the text. The center of gravity of the connecting rod is at a distance C from the crank pin. Set up expressions for the x- and y-coordinates of the center of gravity, as a function of R, θ, L, and C.

(a) Using exact relations, determine the x- and y-components of the velocity and acceleration of the center of gravity. Consider the angular speed of the crank as constant.

(b) By means of the binomial theorem, simplify the expressions for the x- and y-components of the center of gravity. Using the simplified expressions so obtained, determine simplified expressions for the x- and y-components of the velocity and acceleration of the center of gravity, comparable to Eqs. 10 and 11 above.

15.2. Using an R/L ratio equal to 1/4, calculate and plot the exact velocity and acceleration of the piston of a slider-crank mechanism, in intervals of 15 degrees, for an assumed constant speed of the crank of 1800 rpm. $R = 3$ in. and $L = 12$ in.

Calculate and plot the approximate velocity and acceleration values, as given in Eqs. 10 and 11 above for intervals of 15 degrees, using the values given above.

Determine the percentage error for each 15-degree interval. What is the maximum error, and at what position of the crank does it occur? Determine the ratio of the maximum velocity error to the maximum velocity. Determine, also, the ratio of the maximum acceleration error to the maximum acceleration.

15.3. (Fig. 15.3). The crank of the offset engine shown is rotating at a constant angular speed, ω rad/sec counterclockwise. The crank radius is R, and the connecting rod length is L. The offset is h.

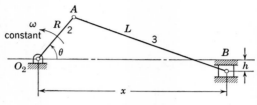

Fig. P–15.3.

(a) The stroke of the piston is not $2R$, as it is for the in-line engine. Show that the stroke S may be expressed by

$$S = [(R + L)^2 - h^2]^{1/2} - [(L - R)^2 - h^2]^{1/2}$$

(b) Show that the approximate displacement of the piston, x, may be expressed by

$$x = R \cos \theta + L \left[1 - \left(\frac{R \sin \theta + h}{L} \right)^2 \right]^{1/2}$$

(c) Show that the approximate velocity of the piston, V, may be expressed by

$$V = -R\omega \left[\sin \theta + \frac{R}{L} \sin 2\theta + \frac{h}{L} \cos \theta \right]$$

(d) Show that the approximate acceleration of the piston, A, may be expressed by

$$A = -R\omega^2 \left[\cos \theta + \frac{R}{L} \cos 2\theta - \frac{h}{L} \sin \theta \right]$$

15.4. An offset engine, as shown in Fig. 15.3, has a crank radius of 3 in. and a connecting rod length of 12 in. The distance $h = 1\frac{1}{2}$ in. The crank rotates at a constant speed of 2400 rpm. Determine the acceleration of the piston for every 15 degrees. Plot the acceleration of the piston *as a function of the stroke of the piston.*

15.5. Refer to Fig. 15.3. The time for the forward and return strokes of an offset engine are not the same. For counterclockwise rotation of the crank, with a constant crank speed, should the offset of the piston be below or above the centerline through the crank bearing to obtain a longer time for the stroke of the piston to the left?

Weight of a Flywheel for a Given Coefficient of Speed 245

where ω_1 is the maximum angular speed of the flywheel,
ω_2 is the minimum angular speed of the flywheel,
ω is the average angular speed of the flywheel,

or by

$$\delta = \frac{V_1 - V_2}{V}$$

where V_1 is the maximum speed of a given point on the flywheel,
V_2 is the minimum speed of the same point on the flywheel,
V is the average speed of the same point on the flywheel.

The maximum permissible coefficients for different applications that have been found to give satisfactory operation vary from around 0.2 for pumps and crushing machinery to around 0.003 for alternating-current generators. Specific values may be obtained in handbooks and in textbooks on machine design.

16.2 Weight of a flywheel for a given coefficient of fluctuation of speed

The kinetic energy of a body rotating about a fixed center is

$$\text{K.E.} = \tfrac{1}{2} I_0 \omega^2$$

If the maximum angular speed of the body is ω_1 and the minimum angular speed is ω_2, the change of kinetic energy, E, is given by

$$E = \tfrac{1}{2} I_0 (\omega_1^2 - \omega_2^2)$$

If the equation above is multiplied and divided by r^2, where r may be taken as the mean radius of the *flywheel rim*, the following is obtained:

$$E = \frac{I_0}{2r^2} [(r\omega_1)^2 - (r\omega_2)^2]$$

Since $r\omega_1 = V_1$, the maximum velocity of a point at the mean radius of the rim, and $r\omega_2 = V_2$, the minimum velocity of the same point, the equation may be further rewritten as

$$E = \frac{I_0}{2r^2} (V_1^2 - V_2^2) = \frac{I_0}{2r^2} (V_1 + V_2)(V_1 - V_2)$$

But $(V_1 + V_2)/2 = V$, the average velocity of the same point at the mean radius of the rim.

And $(V_1 - V_2)/V = \delta$, the defined expression for the coefficient of fluctuation.

Flywheel Analysis

The expression for the change of energy may be written thus:

$$E = \frac{I_0}{2r^2}(2V)(V\delta) = \frac{I_0 V^2 \delta}{r^2}$$

The mass moment of inertia, I_0, may be expressed by $(W/g)k^2$, where k is the radius of gyration. It is usual practice, in flywheel analysis, to consider the weight of the flywheel concentrated at the mean radius of the rim, and to make adjustments later for the fact that the arms and hubs contribute to the flywheel effect. In other words, the radius of gyration is assumed to be equal to the mean radius of the rim, r. With this assumption, the preceding equation may be expressed by:

$$E = \frac{Wr^2 V^2 \delta}{gr^2} = \frac{WV^2 \delta}{g}$$

Or, finally,
$$W = \frac{Eg}{V^2 \delta} *$$

[The expression may be written, in terms of minimum and maximum speed: $W = \dfrac{2Eg}{V_1^2 - V_2^2}.$]

As a result of the above assumption, the actual weight of the rim of the flywheel may be taken as approximately 10% less than that calculated by the above formula to allow for the effect of the arms and hub of the flywheel and other rotating parts, which is sufficient for the usual designs encountered.

For a given engine with a flywheel of a given material, the safe allowable mean rim velocity, V, is determined by the material and the centrifugal stresses set up in the rim. For instance, V for cast iron is usually limited to 4000 to 6000 ft/min maximum, whereas V for steel may be higher, depending upon the construction of the flywheel. The reader is referred to books on machine design and handbooks for actual velocities used for particular types of construction.

Consequently, with a velocity established for a given type of flywheel, the coefficient of fluctuation of speed set by the type of application, the problem now is to determine the maximum excess or deficiency of energy, E, during an energy cycle which causes the speed of the flywheel to change from V_1 to V_2, or vice versa.

Example. The use of a flywheel to decrease the size of motor required for a punching operation is illustrated by the following exam-

---

* If the flywheel is considered as a flat circular plate, the moment of inertia is $\frac{1}{2}(W/g)R^2$. The total weight required is then: $W = 2Eg/V^2\delta$.

ple. A ⅞-in. hole is to be punched in a ¾-in. plate made of annealed SAE 1030 steel. We will assume that 30 holes are to be punched in a minute, or the time between punching operations is 2 sec. We will further assume that the actual punching takes place in one-tenth of the interval between punches, or the actual punching takes place in $(1/10)(2) = \frac{1}{5}$ sec. The driving motor and flywheel runs at 210 rpm, with the necessary velocity reduction through a gear train to give the 30 punching operations per minute. Friction will be neglected completely.

First, let us calculate the approximate size of motor required, without any flywheel, to punch the hole.

Energy for Punching. The determination of the size of the motor depends upon the power required in the punching operation. Although experimental work has been done on the maximum force required, the work done in foot-pounds in the operation is not as specific. The maximum force is a function of several variables: material, treatment of the material, ductility of the material, clearance between the punch and die. The maximum force can be expressed as the product of the area being sheared times a suitable shear stress:

$$P = \pi d t s$$

where d is the diameter of the hole in inches,
t is the thickness of the plate in inches,
s is the resistance to shear in pounds per square inch.*

Thus the maximum force P is approximately

$$P = \pi(7/8)(3/4)(52{,}000)$$
$$= 107{,}000 \text{ lb}$$

The work per stroke is determined by obtaining the area under the force-displacement curve. The force-displacement curve may be illustrated by that shown in Fig. 16.1a, where the maximum force occurs around three-eighths the depth of the plate. It is customary to assume that the area under the force-displacement curve can be approximated by

$$W' = \tfrac{1}{2}Pt$$

where W' is the work done, in foot-pounds, P is the maximum force, and t is the thickness of the plate, in feet.

Thus, $W' = \tfrac{1}{2}(107{,}000)(3/4)/12$
$= 3340$ ft-lb

---

* Marks' *Handbook* (McGraw-Hill Book Co.) gives for s, for SAE 1030 steel, a value of 52,000 psi.

248   Flywheel Analysis

Analysis A—No Flywheel. The average power required, assuming that the force-deflection curve is rectangular in shape, as shown in Fig. 16.1b, is determined by the foot-pounds of work per unit time. In this case, the average power is $3340/(1/5) = 16{,}700$ ft-lb per second. This corresponds to $(16{,}700/550) = 30$ hp. Actually, the instantaneous maximum horsepower would be approximately 60 hp.

Analysis B—A Flywheel Is Used. If a flywheel is used, the motor size could be decreased considerably. The size of motor required can be determined from the condition that the energy taken from the

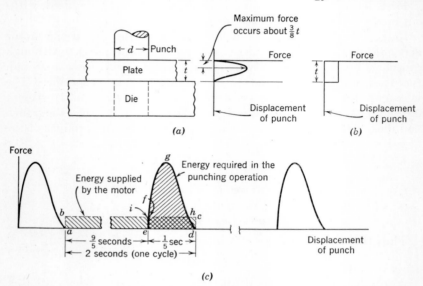

Fig. 16.1. Force variation in a punching operation and energy requirements during a cycle.

flywheel must be returned to it by the motor in a cycle. The energy taken from the flywheel is 3340 ft-lb in $1/5$ second. The 3340 ft-lb must be supplied by a motor, in, however, 2 sec. Or the motor must supply energy at the rate of $3340/2 = 1670$ ft-lb per second. This corresponds to $1670/550 = 3.04$ hp, the size of motor necessary.

The weight of flywheel necessary can be determined by using the curve shown in Fig. 16.1c. The area *abcda* represents the energy supplied by the motor in a cycle. Area *efghde* represents the energy required in the punching operation. Area *eicd* represents the energy supplied by the motor during the actual punching operation. Thus the energy that has to be taken from the flywheel is $3340 - 334 = 3006$ ft-lb. (The energy supplied by the motor during the actual punching

Weight of a Flywheel for a Given Coefficient of Speed   249

operation is $(3.04)(550)(1/5) = 334$ ft-lb, or one-tenth of the energy required in a complete cycle.)

The maximum velocity of the flywheel is, assuming a mean diameter of 2.5 ft:

$$V = \frac{\pi DN}{60}$$

$$= \frac{\pi(2.5)(210)}{60}$$

$$= 27.5 \text{ ft/sec}$$

If a permissible drop in speed of 10% is specified, the minimum speed is 24.8 ft/sec.

The weight of flywheel required is found from

$$W = \frac{2gE}{V_1^2 - V_2^2}$$

$$= \frac{2(32.2)(3006)}{(27.5)^2 - (24.8)^2}$$

$$= 1370 \text{ lb}$$

Actually, only about 90% the weight or about 1230 lb is necessary in the rim of the flywheel, the remainder of the weight effect coming from the spokes and hub. In other words, only about 90% of the flywheel effect comes from the rim. The total weight of the flywheel is about 125% of the rim weight.

The size of rim may be determined from the fact that the weight of the rim is $(b)(h)(\pi D)(12)(0.255)$ lb, where b is the width of the rim in inches, h is the thickness of the rim in inches, D is the mean diameter of the rim in feet, and 0.255 is the density of cast iron in pounds per cubic inch.

Thus, $(b)(h)(\pi D)(12)(0.255) = 1230$

$(b)(h)(\pi 2.5)(12)(0.255) = 1230$

or $(b)(h) = 51.2$

If h is made around $1\frac{1}{4}b$, the dimensions of the rim are approximately $6\frac{1}{2}$ in. by 8 in.

Note that the flywheel rim speed is not at the usual maximum value of around 4000 ft/min. Using a larger diameter flywheel will permit a smaller weight in the rim, with smaller cross-sectional dimensions.

250   Flywheel Analysis

A suggested exercise for the student is the determination of the flywheel weight if the rim is moving with a velocity of 4000 ft/min.

16.3 Procedure in determining flywheel requirements

A single-cylinder, single-acting, four-cycle spark ignition engine is selected as an illustrative engine for the procedure of solution in determining the torque output variation and in determining the flywheel requirements. The procedure to be given is applicable to gas engines, diesel engines, and, with minor variations, to multi-cylinder engines of all types.

Basically, the determination of the torque output of the engine is nothing more than the material covered in Chapter 14 in the dynamic analysis of engines, except that the solution is made for many positions of the crank. Consequently, the work involved for a complete analysis is time-taking, although repetitious.

The first step is to determine a probable indicator card for the engine, based upon thermodynamic considerations of the events in a cycle of operation. If actual indicator cards of actual engines are available, the problem is simplified. The probable indicator card will vary considerably, depending upon the type of engine—whether the engine is a steam engine, spark ignition engine as used in automobiles, compression engine, as diesel engines, or whether a reciprocating type air compressor or ammonia compressor is being analyzed. Again, a two-cycle engine indicator card is quite different from a four-cycle engine indicator card. The theoretical card is probably relatively easy to predetermine, but the actual indicator card may be quite different owing to variations in valve setting, heat conduction from the engine, which affects the type of expansion and compression curves, carburetion, throttle setting, incomplete combustion, etc. For purposes of illustrating procedure of attack, an assumed indicator diagram is shown in Fig. 16.2a, together with the assumed crank positions for the events in the cycle.

The expansion curve and compression curve follow the relation

$$PV^n = \text{constant}$$

where n may vary from 1.2 to 1.4 for actual engines. Sharp corners have been rounded off from the theoretical curve.

The crank is rotating clockwise. The intercept p on the indicator card is the pressure above atmospheric pressure acting on the piston for the position shown. The total force applied to the piston for the position shown is $F = pA$, where A is the area of the piston.

Procedure in Determining Flywheel Requirements

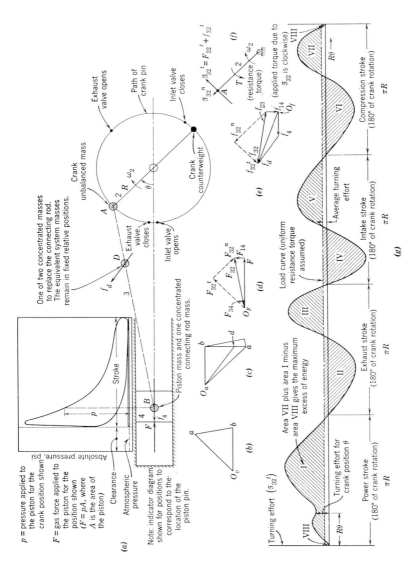

Fig. 16.2. (a) Indicator diagram. (b) Velocity polygon. (c) Acceleration polygon. (d) Static force polygon. (e) Inertia force polygon. (f) Forces on the crank at the crank pin. (g) Plot of the turning effort versus the distance traveled by the crank pin.

Note that the center of the piston pin serves as the reference point in determining the corresponding pressure acting on the piston.

Analysis. The analysis in the determination of the turning effort may be carried out in many ways. Some of the methods are:

(1) *Graphical Solution.* In the graphical solution, the acceleration is determined graphically, the inertia force of each member is determined, and the dynamic analysis, either with separate static and inertia force analyses, or by a combined force analysis, is made. The analysis is made for various crank positions, 15-degree intervals, for instance.

(2) *Kinetically Equivalent System, Graphical Acceleration and Force Analysis.* Simplification can be obtained by the use of a kinetically equivalent system wherein the mass of the connecting rod may be replaced by two concentrated masses, whose positions are fixed on the links. One concentrated mass may be located, for instance, at the piston pin and the other mass at a fixed point on the connecting rod; or one mass may be located at the crank pin and the other mass at a fixed position on the connecting rod. The accelerations can be determined graphically for the piston, and the mass at the piston, and also for the concentrated mass on the connecting rod. The inertia forces would be computed next. A graphical solution for the force analysis could be used to determine the turning effort applied to the crank. The advantage of the kinetically equivalent system is that the inertia force of the concentrated mass on the connecting rod would pass through the center of the concentrated mass, or the location of the resultant inertia force of the connecting rod would not have to be determined.

(3) *Kinetically Equivalent System—Analytical Determination of the Accelerations.* A third method is to use a kinetically equivalent system where the accelerations of the concentrated masses are determined analytically and the force analysis made graphically.

(4) *Approximate Kinetically Equivalent System.* An approximate kinetically equivalent system is sometimes used, with, however, the introduction of an error in the turning effort diagram. The approximation is to consider the mass of the connecting rod replaced by two concentrated masses, one of which is considered concentrated at the piston pin and the other at the crank pin (see pages 295-300).

(5) *Special Methods of Attack.* There are special methods available for the solution of individual parts of the analysis. However, since the methods are limited to slider-crank mechanisms, only the general

Procedure in Determining Flywheel Requirements

method of approach will be discussed, even though the problem selected is a slider-crank mechanism.

Solution. The method used in the solution of the given problem is that described in method 2 above, the kinetically equivalent system with a graphical acceleration and force analysis. The step-by-step procedure is as follows for the one position of the crank shown in Fig. 16.2a.

(1) The velocity polygon is drawn (Fig. 16.2b). The crank is assumed rotating at a constant speed.

(2) A kinetically equivalent system is determined, to give concentrated masses at the piston pin, B, and at D, as shown in Fig. 16.2a.

(3) The acceleration polygon is drawn, Fig. 16.2c, with the acceleration of the piston pin and point D of the connecting rod determined.

(4) The inertia force of the piston and connecting rod mass considered concentrated at the piston pin are determined; the inertia force of the concentrated mass at D is determined.

(5) The static force* analysis is made, as shown in Fig. 16.2d, with the total force on the piston used. The component of F_{32} perpendicular to the crank is determined. This component is the turning effort applied to the crank as a result of the static forces.

(6) The inertia force* analysis is made next, as shown in Fig. 16.2e. The component of f_{32} perpendicular to the crank is found. This is called the turning effort applied to the crank as a result of the inertia forces.

Note that the engine is assumed to be driven, for the inertia force analysis, by a couple applied to the crank, although, actually, for the position shown, the effort comes from the pressure in the cylinder. No difference results in the final value of torque on the crank.

(7) The resultant force perpendicular to the crank is found by the algebraic sum of $F_{32}{}^t$ and $f_{32}{}^t$. Note that for the position being analyzed, the resultant turning effort gives a torque in the same direction as the rotation of the crank; or there is power output from the engine (Fig. 16.2f). Note also that all the power coming from the work on the piston does not go into the output of the engine, but part of it goes into the flywheel, for control of the speed variation in the cycle.

(8) The plot of the turning effort versus the distance traveled by the crank pin is made (Fig. 16.2g).

---

* Separate static and inertia force analyses are made, using a force system as shown in Fig. 14.7b. The total force applied to the piston is considered a static force.

254 Flywheel Analysis

(9) The work done by the engine per cycle is the net area* of the curve. Areas above the axis are positive; areas below the curve are negative. If the engine is assumed to be under constant load, the load curve will be a straight line, parallel to the horizontal axis (Fig. 16.2g). The ordinate is determined by dividing the output work per cycle by $4\pi R$.

(10) Figure 16.2g shows the portions of the cycle when there is an excess of energy which the flywheel must absorb and also the portions of the cycle when there is a deficiency of energy which the flywheel must supply. For instance, areas I, III, V, and VII are portions of the cycle when excesses of energy are provided; areas II, IV, VI, and VIII are portions of the cycle when there is a deficiency of energy.

The maximum variation of speed will take place when there is a maximum excess or deficiency of energy. It is necessary, therefore, to determine the maximum area above or below the mean turning effort. For this case, area VII plus area I minus area VIII gives the maximum variation in energy, with a maximum variation in crank speed. Taking into account the scales used, we may determine the maximum variation of energy, E, in the proper units.

When the maximum variation of energy has been determined, the necessary weight of a flywheel rim may be determined from

$$W = \frac{Eg}{V^2 \delta}$$

if the weight is considered concentrated at the rim with about a 10% reduction to allow for the hub and spokes. Or, if the flywheel is a solid disk, the weight of the flywheel may be determined from

$$W = \frac{2Eg}{V^2 \delta}$$

Conclusions. It is to be noted that the analysis presented affords a method of obtaining the indicated horsepower of an engine. The method presented affords also the opportunity of determining the forces applied on the pins, the method being, basically, a continuation of the force analysis.

Several assumptions have been made that introduce very little error in the analysis. One of the assumptions is that the crank rotates

* Areas may be determined by the use of a planimeter, by the method of integration discussed in Chapter 21, or by approximate methods, such as Simpson's rule or Durand's rule. The reader is referred to handbooks for these two rules.

at a constant speed. The angular acceleration of the crank is relatively small and is neglected. Another assumption is that the assumed average speed of the engine is the same for all portions of the cycle. This is not true, inasmuch as the average speed for each portion of the cycle may vary. Again, the error introduced in the analysis for the flywheel weight is small.

PROBLEMS

16.1. A ½-in. hole is to be punched in a ⅝-in. plate which is made of SAE 1030 steel. Twenty holes per minute are to be punched. The actual punching takes place in one-fifth the interval between punches. The driving motor runs at 1200 rpm, with the necessary velocity reduction through a gear train to give the twenty punching operations per minute. The flywheel rotates at 300 rpm. The maximum force during the punching occurs around three-eighths of the depth of the hole. The resistance to shear may be taken as 52,000 psi.
  (a) Find the size of motor required, without the use of a flywheel.
  (b) Find the size of the motor to be used, assuming that a flywheel is used.
  (c) Assuming that the flywheel is of the rim-type construction, determine the weight of flywheel rim necessary, assuming that 90% of the flywheel effect comes from the rim. The maximum speed of the mean radius of the flywheel is 4000 ft/min. The permissible speed drop is 15%.
  (d) If the total weight of the flywheel is approximately 125% of the rim weight, determine the approximate total weight of the flywheel.
  (e) If the flywheel rim has a square cross section, determine the necessary dimensions. The rim is made of cast iron, which weighs 0.255 lb/cu in.
  (f) If the flywheel is made from a solid circular plate of cast iron, determine the necessary flywheel weight for a maximum peripheral velocity of 4000 ft/min. The permissible speed drop is 15%, as used above.

16.2. An engine develops 50 hp at 300 rpm. The maximum energy variation per revolution has been found to be 30% of the mean energy and the total speed variation to be 1%. Determine the rim weight and dimensions of a square section for a mean rim speed of 4000 ft/min, assuming that 90% of the needed flywheel effect is provided by the rim. Assume that cast iron is to be used, and that cast iron weighs 0.255 lb/cu in.

16.3. The indicated horsepower of a steam engine is 80 at 250 rpm. If the maximum energy variation per revolution is 15% of the mean energy and the allowable coefficient of speed fluctuation is 1/40, determine the cast iron rim weight and dimensions of a square section for a mean rim speed of 4000 ft/min, assuming that 90% of the needed flywheel effect is provided by the rim. The weight of cast iron may be taken as 0.255 lb/cu in.

16.4. A crusher drive rotates at 60 rpm and requires an average power input of 10 hp. If the maximum energy variation during the cycle is equal to the mean energy of the cycle and the speed must not drop more than 10% during the crushing operation, determine the required weight of a flywheel rim with a mean diameter of 80 in. The crushing operation occurs in each revolution of the drive shaft. The flywheel is mounted on the drive shaft.

16.5. List all assumptions and approximations made in the process of determining the weight of flywheel necessary to maintain the speed fluctuation within specified limits for a single cylinder, single-acting engine driving a generator.

16.6. A steam engine having a piston 20 in. in diameter runs at 100 rpm. From a tangential effort diagram, the maximum excess of energy is found to be 1.5 sq in. The scales used are: 1 in. = 10,000 lb for the turning effort (the force perpendicular to the crank at the crank pin); and 1 in. = 3 in. for the distance traveled by the crank pin. Assuming a mean diameter of 10 ft for the flywheel, determine the weight of the rim required if the engine is to drive a d-c generator with a coefficient of fluctuation of 0.0065. Assume that the weight of the arms and the hub of the flywheel is such that it gives 10% of the total flywheel effect.

16.7. Determine the size and weight of a cast iron rimmed flywheel for the following single-cylinder, spark ignition, four-cycle, internal-combustion engine to be used for driving an electric generator. The coefficient of speed fluctuation is to be 0.003.

Although the expected indicator card may be predicted rather closely, and should be used in the analysis, an ideal air standard Otto cycle will be assumed for simplification for this problem, without sacrifice of any of the basic points in the dynamic analysis.

Figure 16.7 shows an ideal Otto cycle where compression takes place from point 1 to point 2, ignition takes place from 2 to 3, expansion takes place from

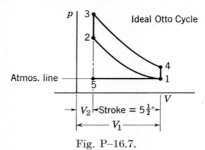

Fig. P–16.7.

3 to 4, exhaust takes place from 4 to 5, and intake takes place from point 5 to point 1. For an ideal cycle, the compression and expansion follow the relation $pV^k = C$, where $k = 1.4$ for air. It is assumed that there is no energy loss in the valves and ports in the exhaust and intake strokes. The following data are given:

Crank speed = 1800 rpm
p_1 = 15 psi, absolute
$\dfrac{p_3}{p_2} = 2.5$
Compression ratio = $V_1/V_2 = 6$
Bore = $4\tfrac{1}{2}$ in.
Stroke = $5\tfrac{1}{2}$ in.
Weight of piston, piston pin, rings = $2\tfrac{1}{4}$ lb
Weight of connecting rod = $3\tfrac{1}{2}$ lb
Weight of crank (considered concentrated at the crank pin) = 2.0 lb
Time for twenty complete oscillations of the connecting rod about the center of the bearing at the piston pin end = 18 sec
Connecting rod length (distance between the center of the piston pin and crank pin) = $9\tfrac{1}{2}$ in.
Center of gravity of the connecting rod is 4 in. from the center of the crank pin.

Counterbalance on the crank, diametrically opposite the crank pin, = 6 lb. The counterbalance is considered as a concentrated weight at the crank radius.

Analyze the forces acting on the engine in 15-degree intervals. Use a kinetically equivalent system in the graphical analysis. Plot the turning force as a function of the distance traveled by the crank pin for two revolutions of the crank. Determine the mean torque applied to the crank, and the maximum variation of energy from the mean energy. Note that the horsepower rating which may be found from the mean torque and speed may be checked by determining the work put into the engine as found from the indicator card.

Determine the weight of the rim of a flywheel for the given value of coefficient of speed fluctuation. Assume that the mean speed of the mean radius of the rim of the flywheel is 4000 ft/min.

Plot a polar diagram of the total force applied to the crank pin. Note that the crank pin is rotating as well as the total variable force applied to the crank pin. Plot, also, the polar diagram of the total force applied to the piston pin. The piston pin is fastened rigidly to the piston and therefore does not rotate.

Cast iron weighs 0.255 lb/cu in.

16.8. Determine the size and the weight of a solid cylindrical steel flywheel for a single-cylinder, four-cycle diesel engine, with the following specifications.

The indicator card which will be assumed for this problem is an ideal air standard diesel cycle, without sacrifice of any of the basic points in the dynamic analysis.

Figure 16.8 shows an ideal diesel cycle where compression takes place from point 1 to 2, fuel is admitted from points 2 to 3, and burns as a result of the temper-

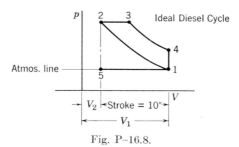

Fig. P–16.8.

ature reached from the compression, expansion takes place from points 3 to 4, exhaust takes place from points 4 to 5, and intake takes place from 5 to 1. In an ideal cycle, the compression and the expansion curves are assumed to follow the relation $pV^k = C$, where $k = 1.4$ for air. It is assumed that there is no energy loss in the valves and ports in the exhaust and intake strokes. The following data are given:

Crank speed = 2400 rpm
$p_1 = 15$ psi, absolute
Fuel injection occurs for 30% of the stroke
Compression ratio = 16 = V_1/V_2
Bore = $8\frac{1}{2}$ in.
Stroke = 10 in.
Weight of piston, piston pin, rings = 5 lb
Weight of connecting rod = 9 lb

258 Flywheel Analysis

Weight of the crank, considered concentrated at the crank pin, = 4 lb

Time for twenty-four complete oscillations of the connecting rod about the center of the bearing at the piston pin end = 40 sec

Connecting rod length (distance between the center of the piston pin and crank pin) = $17\frac{1}{2}$ in.

Center of gravity of the connecting rod is $7\frac{1}{2}$ in. from the center of the crank pin. The crank is counterbalanced by a counterweight of 4 lb. The counterbalance is considered as a concentrated weight at the crank radius.

Analyze the forces acting on the engine in 15-degree intervals. Use a kinetically equivalent system in the graphical analysis. Plot the turning force as a function of the distance traveled by the crank pin for two revolutions of the crank. Determine the mean torque applied to the crank, and the maximum variation of energy from the mean energy. Note that the horsepower rating which may be found from the mean torque and speed may be checked by determining the work put into the engine as found from the indicator card.

Determine the weight and proportions of a solid cylindrical steel flywheel for a coefficient of speed fluctuation of 0.05. Assume that the mean peripheral speed of the flywheel is 4000 ft/min. Steel weighs 0.283 lb/cu in.

Plot the polar diagram of the total force applied to the crank pin. Note that the crank pin is rotating as well as the variable force applied to the crank pin.

Plot, also, the polar diagram of the total force applied to the piston pin. The piston pin is fastened rigidly to the piston and therefore does not rotate.

CHAPTER 17

Balancing Rotating Masses

Considerable space has been devoted to the determination of inertia forces for various mechanisms. The effect of the inertia forces in setting up shaking forces on the structure has also been discussed. The question now is what can be done about the shaking forces.

It is possible to balance wholly, or in part, the inertia forces in a system by introducing additional inertia forces which will serve to counteract the effect of the original forces. This system of balancing is applied to two different types of problems, the first being that of rotating masses, as illustrated by an automobile crankshaft, and the second being that of reciprocating masses, as illustrated by a slider-crank mechanism. Balancing rotating masses are discussed in this chapter; discussion of the balancing of reciprocating masses is deferred to Chapters 19 and 20.

17.1 Single rotating mass

For an illustration of the principles involved, consider Fig. 17.1a, where a concentrated weight W_1 is rotating about a fixed axis at a constant speed. The inertia force is $(W_1/g)R\omega^2$. The inertia force would set up varying reactions on the bearings A and B if the weight were allowed to rotate with no other weight present. If a weight W_2 were located directly opposite W_1, as shown in Fig. 17.1b, and at a radius R_2 such that the inertia force of W_2 balanced that of W_1, a system of balance would be effected:

$$\frac{W_1}{g} R_1 \omega^2 = \frac{W_2}{g} R_2 \omega^2$$

or
$$W_1 R_1 = W_2 R_2$$

For a given W_1 and R_1, the value of W_2 or R_2 may be assumed, the value of the other quantity then being determined by the equation. Usually, a convenient radius for the balancing weight is selected, with the radius as large as the system permits to obtain the smallest weight, and the weight is then calculated.

If the shaft of Fig. 17.1a is assumed to be horizontal, we shall note that the shaft is not in equilibrium for all positions of the weight W_1. The weight will cause the shaft to rotate. However, with the addition of W_2, in Fig. 17.1b, the shaft will be in equilibrium as far as rotation of the shaft is concerned, as may be seen from the following equations.

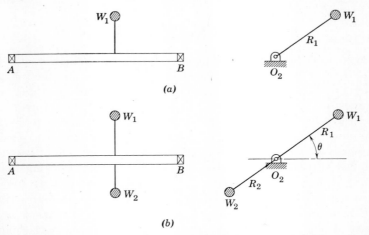

Fig. 17.1. Balancing a single rotating mass. Equation of balance same as equation for static equilibrium.

For equilibrium of the shaft as far as rotation of the shaft is concerned, the sum of the moments of the weights about O_2 must be zero:

$$W_1(R_1 \cos \theta) - W_2(R_2 \cos \theta) = 0$$

or
$$W_1 R_1 = W_2 R_2$$

So, for this simple case, the condition of static equilibrium will be satisfied if the inertia forces are balanced, since both are satisfied by the same equation.

It is well to point out before proceeding to a more general case that sometimes a single weight is balanced by two weights, as in a single-throw crankshaft, for instance.

17.2 Two rotating weights

Figure 17.2 shows two weights on a shaft. For simplification, W_2 is located diametrically opposite W_1, but displaced along the shaft. A condition of static balance is obtained, as in the previous case, by having $W_1 R_1 = W_2 R_2$ (which is also the condition for having the

Multi-weight System   261

inertia forces balance for the special case of two opposed weights). There is not complete balance of the weights, however, inasmuch as the centrifugal force of the two weights sets up a couple which will set up varying forces on the structure.

This special case serves to illustrate the fact that having static equilibrium is not sufficient to insure dynamic balance. It is necessary to introduce additional weights in the system to give both static and dynamic balance. For this particular case it is necessary to introduce two weights for balance since it is not possible to balance a couple with a single force. A weight may be located opposite W_1 to balance W_1, and a weight may be located opposite W_2 to balance W_2. It may not be possible in a particular machine to locate weights in such a fashion because of the construction. The problem now is to

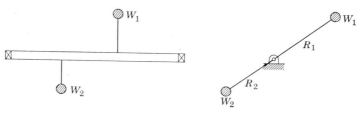

Fig. 17.2. A two-mass system giving a couple, even though static balance exists.

determine the magnitudes of the weights necessary for balance, the minimum number of balancing weights that can be used for complete balance, and the location of the weights for a general multi-weight system.

17.3 Multi-weight system*

Figure 17.3a shows a general system of weights located along a shaft which is rotating at a constant angular speed. $\theta_1, \theta_2, \theta_3$ are the angles which W_1, W_2, W_3, respectively, make with the x-axis. It is desired to show that the system can be balanced by the addition of two weights, one in plane A and the other in plane B. It is also desired to find the relations which must be satisfied.

Consider W_1 first, as shown isolated in Fig. 17.3b. If two equal and opposite imaginary forces, $(W_1 R_1 \omega^2/g)$, are added in plane A, as shown in Fig. 17.3b, no change is made in the system. The forces which are added are parallel to the inertia force of W_1. The resultant of the three forces is: (1) a force $(W_1 R_1 \omega^2/g)$ in plane A parallel to and in

---

* This method of balancing rotating weights is due to Professor W. E. Dalby as presented in *The Balancing of Engines*, by W. E. Dalby, published by Edward Arnold & Co., London, 1902, Fourth edition, 1929.

the same direction as the original force, and (2) a couple equal to $(W_1 R_1 \omega^2/g)(a_1)$, where a_1 is the distance from plane A to W_1.

Since a couple can be considered as two equal, opposite, and parallel forces, the couple $(W_1 R_1 \omega^2/g)(a_1)$ can be considered as also being

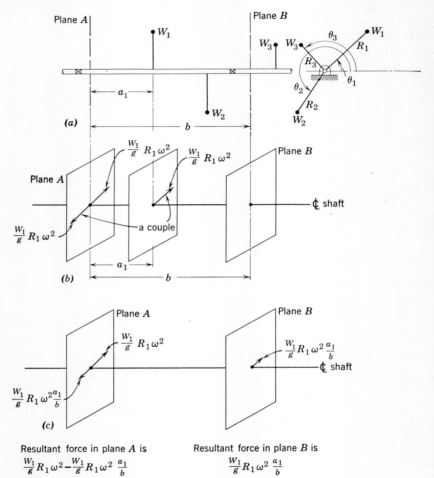

Fig. 17.3a, b, c. Inertia force replaced by forces in two reference planes.

obtained by locating a force $(W_1 R_1 \omega^2/g)(a_1/b)$ in plane A and the same force in plane B, as shown in Fig. 17.3c. The forces must be parallel to the original inertia force of W_1. The distance b is the distance between planes A and B. Figure 17.3c shows the forces so far that give the same effect as the original inertia force of W_1.

Multi-weight System

Summing up, we shall note the following conclusion. The inertia force of W_1 can be replaced by two components, the component in plane A being $(W_1 R_1 \omega^2/g) - (W_1 R_1 \omega^2/g)(a_1/b)$ and the component in plane B being $(W_1 R_1 \omega^2/g)(a_1/b)$. Thus, if the original inertia force

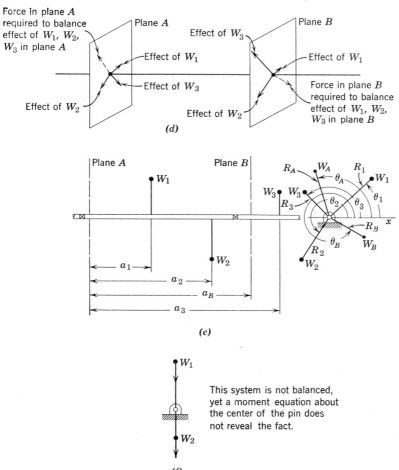

Fig. 17.3d, e, f. Since all inertia forces can be replaced by components in two reference planes, two weights, one in each reference plane, can be used to effect complete balance.

is to be replaced by components, at least two reference planes must be used.

In a similar fashion, the other inertia forces may be replaced by components in reference planes A and B, as indicated in Fig. 17.3d. The force components in plane A may be balanced by a single balancing force in plane A; the force components in plane B may be balanced by a single balancing force in plane B, as indicated in Fig. 17.3d. Consequently, the problem is reduced to finding four unknowns: direction and magnitude of the balancing force in plane A and the magnitude and direction of the balancing force in plane B.

The analysis given in the preceding sections is the basis of solution for the balancing of rotating masses. There are two procedures available for solution, analytical and graphical. Each method is based upon the application of the equations of equilibrium.

17.4 Analytical method

The equations applicable to an analytical solution of equilibrium of a system of rotating masses are developed as follows. We consider first the balance of horizontal and vertical force components and refer to Fig. 17.3e:

(a) For balance of the horizontal forces:

$$\frac{W_1}{g} R_1\omega^2 \cos\theta_1 + \frac{W_2}{g} R_2\omega^2 \cos\theta_2 + \cdots + \frac{W_A}{g} R_A\omega^2 \cos\theta_A$$
$$+ \frac{W_B}{g} R_B \omega^2 \cos\theta_B = 0$$

Divide through by ω^2/g:

$$\Sigma WR \cos\theta = 0 \qquad (1)$$

(b) For balance of the vertical forces:

$$\frac{W_1}{g} R_1\omega^2 \sin\theta_1 + \frac{W_2}{g} R_2\omega^2 \sin\theta_2 + \cdots + \frac{W_A}{g} R_A\omega^2 \sin\theta_A$$
$$+ \frac{W_B}{g} R_B\omega^2 \sin\theta_B = 0$$

or $$\Sigma WR \sin\theta = 0 \qquad (2)$$

(c) For balance of moments about plane A of the horizontal forces, where a represents the distance from each force to the reference plane A:

$$\frac{W_1}{g} R_1\omega^2 a_1 \cos\theta_1 + \frac{W_2}{g} R_2\omega^2 a_2 \cos\theta_2 + \cdots$$
$$+ \frac{W_B}{g} R_B\omega^2 a_B \cos\theta_B = 0$$

or $$\Sigma WRa \cos\theta = 0 \qquad (3)$$

Analytical Method

(d) For balance of moments about plane A of the vertical forces:

$$\frac{W_1}{g} R_1\omega^2 a_1 \sin\theta_1 + \frac{W_2}{g} R_2\omega^2 a_2 \sin\theta_2 + \cdots + \frac{W_B}{g} R_B\omega^2 a_B \sin\theta_B = 0$$

or $\qquad \Sigma WRa \sin\theta = 0 \qquad (4)$

The conclusion at this point is that for equilibrium of the inertia forces by themselves, the following four equations must be satisfied:

$$\mathrm{I}\begin{cases} \Sigma WR \cos\theta = 0 \text{ (horizontal forces balance)} \\ \Sigma WR \sin\theta = 0 \text{ (vertical forces balance)} \\ \Sigma WRa \cos\theta = 0 \text{ (moments balance in a horizontal plane about} \\ \qquad\qquad\qquad \text{plane } A) \\ \Sigma WRa \sin\theta = 0 \text{ (moments balance in a vertical plane about plane} \\ \qquad\qquad\qquad A) \end{cases}$$

The equations above are essentially two force equations and two moment equations. Instead of using two force equations, two moment equations may be substituted:

$$\mathrm{II}\begin{cases} \Sigma WRa \cos\theta = 0 \\ \Sigma WRa \sin\theta = 0 \\ \Sigma WRb \cos\theta = 0 \\ \Sigma WRb \sin\theta = 0 \end{cases}$$

where b is the distance of a given weight to reference plane B.

It is evident that two force equations and two moment equations, with moments taken about plane B, may be used for I above.

A second interpretation of the two force equations, $\Sigma WR \cos\theta = 0$ and $\Sigma WR \sin\theta = 0$, is helpful in visualizing why static balance is obtained if these two equations are satisfied. Figure 17.3e shows an end view of the general system. Each weight will cause a moment tending to turn the shaft. For the position shown, the condition necessary for equilibrium is:

$$W_1(R_1 \cos\theta_1) + W_2(R_2 \cos\theta_2) + W_3(R_3 \cos\theta_3) + W_4(R_4 \cos\theta_4) \\ + W_A R_A \cos\theta_A + W_B R_B \cos\theta_B = 0$$

or $\qquad \Sigma WR \cos\theta = 0 \qquad (5)$

To insure equilibrium of the shaft for all positions, and to avoid the possibility of a condition shown in Fig. 17.3f, where the weights may be located so as not to appear in the above equation and thus not have the shaft in equilibrium for all positions, consider rotating the shaft 90 degrees counterclockwise. The relation that results is:

$$W_1[R_1 \cos(\theta_1 + 90)] + W_2[R_2 \cos(\theta_2 + 90)] + W_3[R_3 \cos(\theta_3 + 90)]$$
$$+ W_4[R_4 \cos(\theta_4 + 90)] + W_A R_A \cos(\theta_A + 90°)$$
$$+ W_B R_B \cos(\theta_B + 90°) = 0$$

or
$$\Sigma WR \sin \theta = 0 \qquad (6)$$

Thus, for *static* equilibrium, equations 5 and 6 must be satisfied. These equations are the same as those that satisfy the balance of *inertia* forces.

It should be emphasized that the equilibrium of inertia forces does not take into account the bearing reactions. The bearing reactions will be present as a result of the necessity of supporting the shaft

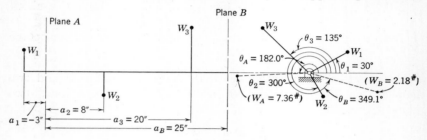

Fig. 17.4. Data for illustrative example.

$W_1 = 10$ lb	$R_1 = 6''$	$\theta_1 = 30°$	$W_A = ?$	$\theta_A = ?$
$W_2 = 16$ lb	$R_2 = 4''$	$\theta_2 = 300°$	$W_B = ?$	$\theta_B = ?$
$W_3 = 5$ lb	$R_3 = 9''$	$\theta_3 = 135°$		
	$(R_A = 10'')$			
	$(R_B = 10'')$			

and weights. However, the bearings will not be subjected to any other forces than the force of the weights of the parts if the inertia forces are balanced in themselves.

Example. The following example illustrates the procedure of solution of a problem by the analytical method to determine the necessary balancing weights of a system of rotating weights.

Three weights, W_1, W_2, and W_3, which rotate in transverse planes 1, 2, and 3 are to be balanced by the addition of two rotating weights, W_A in the plane A, and W_B in plane B. $W_1 = 10$ lb, $W_2 = 16$ lb, and $W_3 = 5$ lb. The location to the center of gravity of each weight is given by $R_1 = 6$ in., $R_2 = 4$ in., and $R_3 = 9$ in. Also, $\theta_1 = 30$ degrees, $\theta_2 = 300$ degrees, and $\theta_3 = 135$ degrees. The distance of each weight from plane A is given by $a_1 = -3$ in., $a_2 = 8$ in., $a_3 = 20$ in., and $a_B = 25$ in. Figure 17.4 shows a sketch of the arrangement.

To start with, there are six unknowns, W_A, W_B, R_A, R_B, θ_A, and θ_B.

Analytical Method

However, there are only four equations that can be applied. It is therefore necessary to assume two of the unknowns. The quantities which cannot be assumed are the angles θ_A and θ_B, which are fixed by the known inertia forces. Since the product of W and R is involved in each equation, it is evident that the balancing weight may be made as small as desired by having the radius as large as possible. Practically, the value of R is limited by the machine. It will be assumed that for this particular case the value of R_A and R_B may be taken as 10 in. Thus there will be four unknowns: W_A, W_B, θ_A, and θ_B.

Two force equations and two moment equations will be used for the solution of this problem. For simplicity, a tabular form will be used. The table below gives the known quantities.

W	R	θ	a	$\cos\theta$	$\sin\theta$	$WR\cos\theta$	$WR\sin\theta$	$WRa\cos\theta$	$WRa\sin\theta$
$W_1 = 10$	6″	30°	−3″	+0.866	+0.500	+52.0	+30.0	−156	−90
$W_2 = 16$	4″	300°	+8″	+0.500	−0.866	+32.0	−55.4	+256	−443
$W_3 = 5$	9″	135°	+20″	−0.707	+0.707	−31.8	+31.8	−636	+636
W_A	10″		0					0	0
W_B	10″		+25″						
						$\Sigma = 0$	$\Sigma = 0$	$\Sigma = 0$	$\Sigma = 0$
						(Static balance)		(Dynamic balance)	

The value of $W_B R_B a_B \cos\theta_B$ and $W_B R_B a_B \sin\theta_B$ may be found immediately, since these values are the only unknowns in the corresponding columns, and the summation of the column must be zero.

$$W_B R_B a_B \cos\theta = +536$$
$$W_B R_B a_B \sin\theta = -103$$

Divide each value by $a_B = +25$.

$$W_B R_B \cos\theta = +21.4$$
$$W_B R_B \sin\theta = -4.1$$

With the above values known, there is only one unknown in each column for $WR\cos\theta$ and $WR\sin\theta$: $W_A R_A \cos\theta_A$ and $W_A R_A \sin\theta_A$. They are found by noting that the sum of the quantities in each column is zero:

$$W_A R_A \cos\theta_A = -73.6$$
$$W_A R_A \sin\theta_A = -2.3$$

268   Balancing Rotating Masses

W_B may be found by noting that

$$\sqrt{(W_B R_B \cos \theta_B)^2 + (W_B R_B \sin \theta_B)^2} = W_B R_B$$

or
$$\sqrt{(21.4)^2 + (4.1)^2} = W_B(10)$$

From which, $W_B = 2.18$ lb.

In a similar fashion, W_A is found to be 7.36 lb.

θ_B may be found from the following:

$$\frac{W_B R_B \sin \theta_B}{W_B R_B \cos \theta_B} = \tan \theta_B = \frac{-4.1}{+21.4} = -0.193$$

Or $\theta_B = 349.1°$. The angle is in the fourth quadrant inasmuch as the sine is negative and the cosine is positive.*

Similarly, $\quad \tan \theta_A = \dfrac{-2.3}{-73.6} = +0.031$

Or $\theta_A = 182.0°$. This angle is in the third quadrant inasmuch as the sine and cosine are both negative.

The complete table is reproduced below to show the final form.

W	R	θ	a	$\cos \theta$	$\sin \theta$	$WR \cos \theta$	$Wr \sin \theta$	$WRa \cos \theta$	$WRa \sin \theta$
$W_1 = 10$	6″	30°	−3″	+0.866	+0.500	+52.0	+30.0	−156	−90
$W_2 = 16$	4″	300°	+8″	+0.500	−0.866	+32.0	−55.4	+256	−443
$W_3 = 5$	9″	135°	+20″	−0.707	+0.707	−31.8	+31.8	−636	+636
$W_A = 7.36$	10″	182.0°	0			−73.6	−2.3	0	0
$W_B = 2.18$	10″	349.1°	+25″			+21.4	−4.1	+536	−103
						$\Sigma = 0$	$\Sigma = 0$	$\Sigma = 0$	$\Sigma = 0$

Note that W_A could have been found independently of W_B by the use of a moment equation about plane B.

Attention is called to the units. Apparently, $WR \cos \theta$ has units of pound-inches, the units of a moment. However, the quantity represents a force, since ω^2/g was canceled out of the original equation. Similarly with $WRa \cos \theta$. The latter quantity represents a moment,

* Use of $\dfrac{W_B R_B \sin \theta_B a_B}{W_B R_B \cos \theta_B a_B} = \tan \theta_B$ to determine the quadrant where the balancing weight is located might, in some cases, cause the student to locate the direction improperly, if attention is not paid to the sign of the distance a_B, that is, if a_B is negative.

Moments as Vectors 269

even though the apparent units are pound-inches^2. Again, this is due to the canceling of ω^2/g in the original derivation of the equation.

17.5 Graphical analysis

The second method for solution of problems of balancing of rotating masses is the purely graphical method.

The equations that must be satisfied are the same, but are expressed a little differently, that is, are expressed as vector equations:

$$\sum \vec{WR} = 0$$

$$\sum \vec{WRa} = 0$$

The first equation is the representation of all the vector inertia forces adding up to a resultant equal to zero. The second equation is the representation of all the vector moments about plane A adding up to a resultant equal to zero.

As in the analytical procedure, in the graphical solution a force and a moment equation can be used or two moment equations can be used. The equations that apply if moments are taken are:

$$\sum \vec{WRa} = 0$$

$$\sum \vec{WRb} = 0$$

Moments may be treated as vectors, and handled accordingly, as will be shown in the next section.

17.6 Moments as vectors

Figure 17.5a shows a couple M_1 acting in plane 1 and a couple M_2 acting in plane 2. It is desired to obtain the vector sum of the two couples.

Figure 17.5b shows the same system expressed differently. The couple M_1 is replaced by two equal and opposite forces P at a distance p such that $Pp = M_1$. Also, M_2 is replaced by two equal and opposite forces P but at a distance q such that $Pq = M_2$. One force P of each couple is located in the line of intersection of the two planes, and are placed opposite to each other so that they cancel each other.

270 Balancing Rotating Masses

The other two forces form a couple equal to Pr, where r is the distance between the two remaining forces. The moment of the couple can be expressed in terms of the original distances by expressing r as a function of p, q, and ϕ by the law of cosines, where ϕ is the angle between the two planes:

$$r^2 = p^2 + q^2 - 2pq \cos \phi$$

Or the resultant couple is

$$M_R = P \sqrt{(p^2 + q^2 - 2pq \cos \phi)} \qquad (1)$$

It is now necessary to show that if the couples M_1 and M_2 are expressed as vectors the same result will be obtained as is given by Eq.

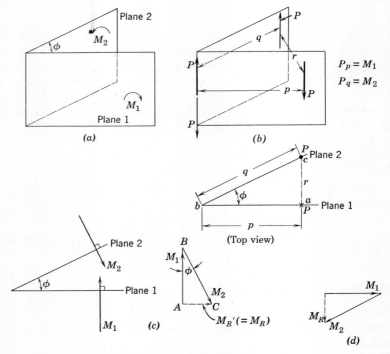

Fig. 17.5. Vector addition of couples.

1 above. The right-hand rule is used to give the vector of a couple. If the fingers of the right hand are curved in the direction of the couple, the thumb will represent the direction of the vector.

Figure 17.5c shows the vectors of the couples by the above rule.

Also shown in the same figure is the vector sum of M_1 and M_2: M_R'. The angle between M_1 and M_2 is ϕ. The resultant is

$$M_R' = M_1 + M_2$$

By the law of cosines:

$$M_R'^2 = M_1^2 + M_2^2 - 2M_1M_2 \cos \phi$$

But $M_1 = Pp$ and $M_2 = Pq$. Therefore,

$$M_R'^2 = (Pp)^2 + (Pq)^2 - 2(Pp)(Pq) \cos \phi$$
$$= P^2(p^2 + q^2 - 2pq \cos \phi)$$

or
$$M_R' = P \sqrt{(p^2 + q^2 - 2pq \cos \phi)} \tag{2}$$

Comparison of Eqs. 1 and 2 shows that the same magnitude of couple results by either method.

To complete the proof that M_R is the vector that represents the vector sum of M_1 and M_2, we must show that M_R is perpendicular to the plane in which the couple Pr is located. Since M_1, M_2, and M_R are respectively Pp, Pq, and Pr, the corresponding moment is proportional to p, q, and r, respectively. If the vector diagram of Fig. 17.5c is drawn to a scale so that M_1 and M_2 are proportional to p and q, respectively, M_R will be proportional to r. Or triangle abc of Fig. 17.5b will be similar to triangle ABC of Fig. 17.5c. And since M_1 is perpendicular to plane 1 and M_2 is perpendicular to plane 2, M_R will be perpendicular to the plane of couple Pr. The direction of the vector also satisfies the right-hand rule used to determine M_1 and M_2.

Sometimes the vector diagram of Fig. 17.5c is rotated 90°, as shown in Fig. 17.5d, to give a quicker solution. Such will be the case in the graphical solution of the illustrative example on balancing of rotating weights.

Note that the moments due to couples in parallel planes may be added algebraically.

Example. The following example illustrates the procedure of solution in balancing rotating masses, using the graphical method.

Figure 17.6a shows Fig. 17.4 reproduced. The same problem worked analytically will be worked graphically.

Since there are four unknowns, W_A, W_B, θ_A, and θ_B (R_A and R_B are each assumed to be $10''$), it is not possible to start with a vector force equation. Moments will have to be taken, as in the analytical solution.

272 Balancing Rotating Masses

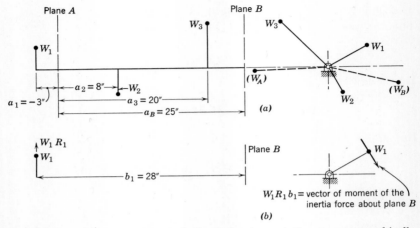

Fig. 17.6a, b. Illustrative example for balancing rotating masses graphically.

$W_1 = 10$ lb	$R_1 = 6''$	$\theta_1 = 30°$	$W_A = ?$	$\theta_A = ?$
$W_2 = 16$ lb	$R_2 = 4''$	$\theta_2 = 300°$	$W_B = ?$	$\theta_B = ?$
$W_3 = 5$ lb	$R_3 = 9''$	$\theta_3 = 135°$		
	$(R_A = 10'')$			
	$(R_B = 10'')$			

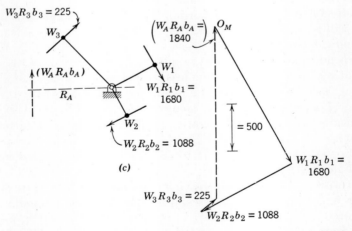

Fig. 17.6c. Moments about the reference plane B treated as vectors by the right-hand rule to find W_A.

Moments as Vectors 273

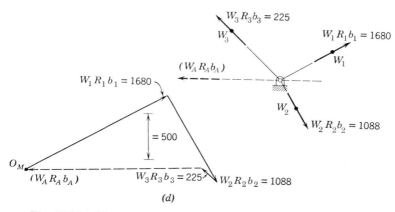

Fig. 17.6d. Moment vectors rotated 90 degrees counterclockwise.

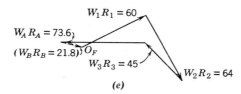

Fig. 17.6e. Force equation used to find W_B.

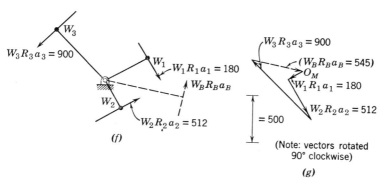

Fig. 17.6f, g. Moment equation about plane A used to find W_B.

274 Balancing Rotating Masses

A tabular form of the known data will simplify the problem procedure.

W	R	WR	a	WRa	b	WRb
$W_1 = 10$	6	60	-3	-180	$+28$	$+1680$
$W_2 = 16$	4	64	$+8$	$+512$	$+17$	$+1088$
$W_3 = 5$	9	45	$+20$	$+900$	$+5$	$+225$
$W_A = ?$	10	?	0	0	$+25$	?
$W_B = ?$	10	?	$+25$	?	0	0
		$\vec{\sum} = 0$		$\vec{\sum} = 0$		$\vec{\sum} = 0$

To find W_A:

Consider taking moments about *plane B* first, and consider the moment of W_1. Figure 17.6b shows W_1, with the inertia force of W_1. The moment of the inertia force may be considered to be $W_1 R_1 b_1$, since ω^2/g is common to every term and may be considered to be canceled, as shown in the analytical equations derived. By the right-hand rule, the moment of the centrifugal force may be represented by a vector, as shown in Fig. 17.6b. So it is, also, with all the moments in that they may be represented by a vector perpendicular to the radius, as shown in Fig. 17.6c. The moment to give a resultant equal to zero is the closing line of the polygon: $W_A R_A b_A$. This vector is perpendicular to the radius, R_A, thus determining the angular position of R_A.

The analysis may be simplified if each vector is turned through 90 degrees counterclockwise so that each moment vector then corresponds in direction to the radius of each weight, as shown in Fig. 17.6d. The closing line of the polygon, $W_A R_A b_A$, gives the proper direction of W_A.

The magnitude of W_A can be found now by scaling of $W_A R_A b_A$ and dividing the product $R_A b_A$ into it:

$$\frac{W_A R_A b_A}{R_A b_A} = W_A = \frac{1840}{(10)(25)} = 7.36 \text{ lb}$$

The angle θ_A reads 182 degrees.

To find W_B:

There are two methods available for finding W_B:

(1) *Using a force equation.* The vector sum of the forces must be zero, or $\vec{\sum} WR = 0$. Since W_A has been found, W_B may be found

Problems   275

from the force polygon shown in Fig. 17.6e. The direction of W_B can be determined also.

$W_B R_B$ scales off to 21.8. Thus,

$$\frac{W_B R_B}{R_B} = W_B = \frac{21.8}{10} = 2.18 \text{ lb}$$

θ_B reads 349 degrees.

(2) *Applying an independent moment equation about plane B.* In this way, W_A may be found independently of W_B. This method will be discussed to indicate the procedure of attack where weights are on both sides of the reference plane.

Figure 17.6f shows the direction of the moment vector of $W_1 R_1 a_1$, $W_2 R_2 a_2$, and $W_3 R_3 a_3$. Note that if the vectors are rotated 90 degrees clockwise, the vectors are as shown in Fig. 17.6g. *Those vectors on the positive side of the plane A are directed outward along the radius while the vector on the negative side of plane A is directed towards the center of the shaft.*

Consequently, noting that the moment vectors on the positive side of plane A are directed outward from the center of the shaft and those on the negative side are directed towards the center of the shaft, we may obtain Fig. 17.6g. $W_B R_B a_B$ scales off to 545. Thus $W_B = 2.18$, where $R_B = 10$ and $a_B = 25$. $\theta_B = 349°$.

The complete tabular form is:

W	R	WR	a	WRa	b	WRb
$W_1 = 10$	6	60	-3	-180	$+28$	$+1680$
$W_2 = 16$	4	64	$+8$	$+512$	$+17$	$+1088$
$W_3 = 5$	9	45	$+20$	$+900$	$+5$	$+225$
$W_A = [7.36]$	10	[73.6]	0	0	$+25$	$+[1840]$
$W_B = [2.18]$	10	[21.8]	$+25$	$+[545]$	0	0
		$\overset{\rightarrow}{\sum} = 0$		$\overset{\rightarrow}{\sum} = 0$		$\overset{\rightarrow}{\sum} = 0$

It is important to note again that the same procedure of attack applies to the analytical as well as to the graphical solution. One method may be used to check the results obtained by the other method.

PROBLEMS

17.1 Figure 17.1 shows a system of two weights, W_A and W_B, which have been found to be necessary to balance a system of weights, not shown. However,

276   Balancing Rotating Masses

because of space limitations, it is decided to replace W_B by two weights, W_C and W_D, in planes C and D, where planes C and D are prescribed in position as shown.

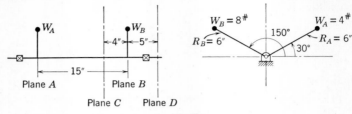

Fig. P–17.1.

What are the magnitude and the angular location of W_C and W_D, assuming that the radius to the center of gravity of each weight is 6 in.?

17.2 Figure 17.2 shows a system of two weights, W_A and W_B, which have been found to be necessary to balance a system of weights, not shown. However, it is

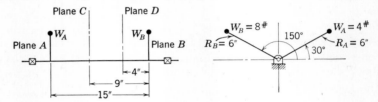

Fig. P–17.2.

decided to replace W_B by two weights, W_C and W_D, in planes C and D, where planes C and D are prescribed in position as shown.

What are the magnitude and the angular location of W_C and W_D, assuming that the radius to the center of gravity of each weight is 6 in.?

17.3 Figure 17.3 shows a system of four weights rotating in a plane. Determine the magnitude and the angular location of the single weight necessary to

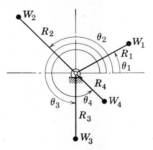

Fig. P–17.3.

$W_1 = 8$ lb $R_1 = 9''$ $\theta_1 = 30°$
$W_2 = 4$ lb $R_2 = 12''$ $\theta_2 = 135°$
$W_3 = 6$ lb $R_3 = 10''$ $\theta_3 = 270°$
$W_4 = 5$ lb $R_4 = 6''$ $\theta_4 = 315°$

balance the rotating weights. Assume that the radius to the center of gravity of the balancing weight is 8 in. The shaft is rotating at 1200 rpm.
   (a) Solve analytically.
   (b) Solve graphically.

17.4. Refer to Fig. 17.3. It is desired to balance the system of weights rotating in a plane by the use of two balance weights, where one balance weight is twice the other. Determine, graphically, the angular location of each balance weight. If the radius to the center of gravity of each balance weight is 8 in., determine the magnitude of each balance weight.

17.5 Figure 17.5 shows a single rotating weight, $W = 8$ lb, at a radius $R = 4$ in. from the axis of rotation. If two equal counterweights are to be used to balance

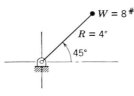

Fig. P–17.5.

the single rotating weight, and each counterweight is to have the same radius $R = 4$ in., determine the magnitude and the angular location of the two counterweights with the condition that neither counterweight is to exceed 5 lb.

17.6. (a) The data for a rotating unbalanced system are given with Fig. 17.6. If no balancing weights are used, what will be the total bearing reactions on the

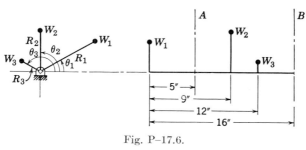

Fig. P–17.6.

$W_1 = 20$ lb	$R_1 = 3''$	$\theta_1 = 30°$
$W_2 = 30$ lb	$R_2 = 2''$	$\theta_2 = 90°$
$W_3 = 40$ lb	$R_3 = 1''$	$\theta_3 = 150°$
$W_A = ?$	$R_A = 3''$	$\theta_A = ?$
$W_B = ?$	$R_B = 2''$	$\theta_B = ?$

bearings in planes A and B? Assume that the shaft is rotating at 800 rpm. Note that, as the shaft rotates, the direction of each bearing force changes.

(b) If bearings are located elsewhere on the shaft other than in planes A and B, what weights will be required in planes A and B at the given radii in order to eliminate any bearing reactions due to the inertia forces? And what must their angular location be in order to eliminate any bearing reactions due to the inertia forces?

17.7. Refer to Fig. 17.7. Three rotating weights, W_1, W_2, and W_3, are to be balanced by two balancing weights W_A and W_B in planes A and B. Plane B

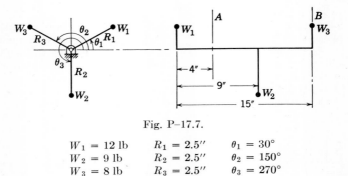

Fig. P–17.7.

$W_1 = 12$ lb	$R_1 = 2.5''$	$\theta_1 = 30°$
$W_2 = 9$ lb	$R_2 = 2.5''$	$\theta_2 = 150°$
$W_3 = 8$ lb	$R_3 = 2.5''$	$\theta_3 = 270°$
$W_A = ?$	$R_A = 2.5''$	$\theta_A = ?$
$W_B = ?$	$R_B = 2.5''$	$\theta_B = ?$

coincides with the plane of rotation of W_3. Determine the magnitude and the angular location of the balancing weights in planes A and B.

17.8 Figure 17.8 shows a system of four rotating weights. Determine the magnitude and angular location of the balancing weights in planes A and B necessary to balance the system.

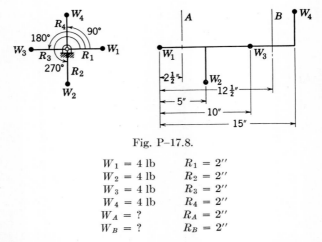

Fig. P–17.8.

$W_1 = 4$ lb	$R_1 = 2''$
$W_2 = 4$ lb	$R_2 = 2''$
$W_3 = 4$ lb	$R_3 = 2''$
$W_4 = 4$ lb	$R_4 = 2''$
$W_A = ?$	$R_A = 2''$
$W_B = ?$	$R_B = 2''$

17.9. Investigate the system of rotating weights shown in Fig. 17.9 for unbalance. Consider the planes of weights 1 and 6 as the reference planes. If unbalance occurs, what balancing weights, and angular location of the balancing weights, are necessary in planes 1 and 6.

Note that the arrangement shown corresponds to the arrangement used in a six-cylinder, in-line type of automobile engine.

Problems

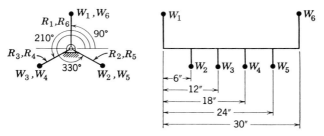

Fig. P-17.9.

$$W_1 = W_2 = W_3 = W_4 = W_5 = W_6 = 5 \text{ lb}$$
$$R_1 = R_2 = R_3 = R_4 = R_5 = R_6 = 2''$$

17.10 Figure 17.10 shows the schematic arrangement of the crankshaft of a two-cylinder engine. Two flywheels are used, one in plane A and the other in plane

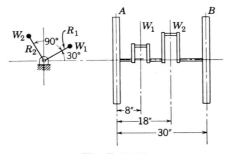

Fig. P-17.10.

$W_1 = 14$ lb	$R_1 = 5''$
$W_2 = 14$ lb	$R_2 = 5''$
$W_A = ?$	$R_A = 15''$
$W_B = ;$	$R_B = 15''$

B. The crankshaft is to be balanced by two counterweights, one in plane A and the other in plane B, the weights having a radial distance of 15 in. to the center of gravity from the axis of rotation.

Determine the magnitude and the angular location of the two counterweights. Express the angular location of each weight with respect to the position of crank W_1.

17.11 (Fig. 17.11). Three rotating weights are in balance. Known values are W_3, R_1, R_2, R_3, θ_1. Determine the unknown values W_1, W_2, θ_2, and θ_3.

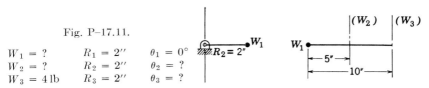

Fig. P-17.11.

$W_1 = ?$	$R_1 = 2''$	$\theta_1 = 0°$
$W_2 = ?$	$R_2 = 2''$	$\theta_2 = ?$
$W_3 = 4$ lb	$R_3 = 2''$	$\theta_3 = ?$

17.12 (Fig. 17.12). Four rotating weights are in balance. Known values are W_2, W_3, W_4, R_1, R_2, R_3, R_4, and θ_1. Unknown values are W_1, θ_2, θ_3, θ_4. Determine the unknown values. (Note that there are two solutions for the angular

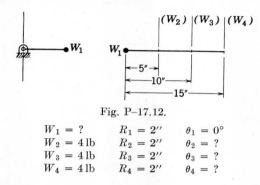

Fig. P–17.12.

$W_1 = ?$	$R_1 = 2''$	$\theta_1 = 0°$
$W_2 = 4$ lb	$R_2 = 2''$	$\theta_2 = ?$
$W_3 = 4$ lb	$R_3 = 2''$	$\theta_3 = ?$
$W_4 = 4$ lb	$R_4 = 2''$	$\theta_4 = ?$

position of W_2, W_3, and W_4, but only one value for each of the weights W_2, W_3, and W_4.)

17.13 Figure 17.13 shows a shaft with four known rotating weights. It is desired to balance the system with balancing weights in the three planes A, B, and

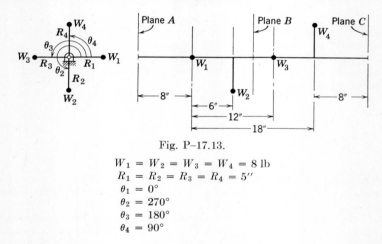

Fig. P–17.13.

$W_1 = W_2 = W_3 = W_4 = 8$ lb
$R_1 = R_2 = R_3 = R_4 = 5''$
$\theta_1 = 0°$
$\theta_2 = 270°$
$\theta_3 = 180°$
$\theta_4 = 90°$

C. Space limitations in plane B limit the amount of counterweight that can be used in that plane. It is arbitrarily decided to make the counterweights in planes A and C equal, and the counterweight in plane B one half that of the counterweight in plane A or that in plane C. What are the magnitude and the angular location of each counterweight?

17.14. The shaft shown in Fig. 17.14 carries two eccentric disks at (1) and (3). The disks weigh 20 lb each, and the center of gravity of each is 5 in. from the center of the shaft. The weight of the crank is 14 lb, and its center of gravity is 3 in. from the center of the shaft. Determine the magnitude and the angular location of the two counterbalancing weights to be placed in planes A and B. The radius to the center of gravity of each counterbalancing weight is 4 in.

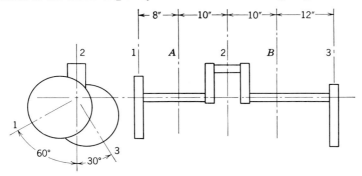

Fig. P–17.14.

17.15. A certain rotating machine part 25 in. long was tested in a balancing machine. The unbalanced "moment" $[\Sigma(dW)Ra]$ about the left end due to the unbalance in the machine part was found to be 7.32 lb-in.^{2} at an angle of 30 degrees with a reference line on the part. The unbalanced "moment" $[\Sigma(dW)Rb]$ about the right end was found to be 15.76 lb-in.^{2} at an angle of 60 degrees.

Weight may be added or removed to eliminate the unbalance, but, owing to the construction of the part, this can be done only at angles of 90 degrees and 180 degrees from the reference line used above, and at radii of 6.0 in.

Specify the magnitude of the two weights that must be added or removed, and the location of the weights with respect to the left end of the machine part.

Suggestion. Take moments of the weights about the left end and about the right end of the machine part.

CHAPTER 18

Balancing Machines

It was pointed out in Chapter 17 that any system of rotating weights could be balanced completely, theoretically, by the addition of two balancing weights, one in each of any two reference planes. However, in spite of all the care that may be taken in the mathematical analysis and in the shop in the manufacture of the part, whether it is a completely machined piece, casting, forging, or assembled piece as the rotor of an electric motor, it is an uncommon occurrence for the part to run smoothly; it is more uncommon if the operating speed is high.

Variations in machining, small as they may be, non-homogeneity of materials, eccentricity of bearing surfaces, methods of assembly, all contribute to the offsetting of the center of gravity from the axis of rotation. It is apparent that unmachined castings or forgings could not be held to as close dimensions as machined parts, and greater unbalance could be expected from such parts. In any event, it is impossible to design against the possible variations which may occur. Balancing machines are used to determine the magnitude and location of balancing weights for a rotating part after the part has been manufactured, regardless of the type of part and method of production.

The effect of the unbalance should be considered. Figure 18.1 shows a curve of the magnitude of centrifugal force for one ounce-inch unbalance for various speeds. (An ounce-inch is defined as one ounce of weight at one inch from the axis of rotation.*) The centrifugal force of one ounce-inch at 100 rpm is 0.0177 lb. At 10,000 rpm the centrifugal force of one ounce-inch is 177 lb, or ten thousand times as large; or the centrifugal force increases as the *square* of the speed. It is evident that the unbalanced centrifugal forces for large rotating weights can reach high values, even if the center of gravity of the

---

* Another interpretation of an ounce-inch is that it is the equivalent amount of unbalance in a part which can be corrected by adding or removing one ounce of material at a radius of one inch from the center of rotation.

rotating weight is displaced a small amount from the axis of rotation, and consequently set up large shaking forces on a given structure. As a further example of the magnitude of unbalance which may be expected, consider a two-ton rotor rotating at 3600 rpm, with the

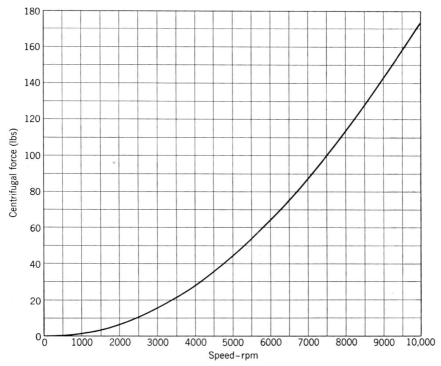

Fig. 18.1. Plot of centrifugal force unbalance for one ounce-inch unbalance for various speeds.

center of gravity of the rotor 0.001 in. from the axis of rotation. The unbalanced centrifugal force is

$$\frac{W}{g} R\omega^2 = \left(\frac{4000}{32.2}\right)\left(\frac{0.001}{12}\right)\left[\frac{(3600)(2\pi)}{60}\right]^2 = 1470 \text{ lb}$$

Such a force can cause considerable damage to the machine. If the unbalance should be a dynamic couple, the magnitude of the couple may reach an extremely high value, also.

Thus, while balancing may not be necessary for slow speeds, it is a necessity for high speeds. Noise, vibration, stresses due to vibration, wear, and loosened joints due to vibration and wear are results of an unbalanced system. It is a fallacy to think that if the machine is rigidly fastened to the foundation it is not necessary to balance the rotating parts. Internal damage to the machine may result if the unbalance is high.

18.1 Types of balancing

There are two types of balancing operations which are used in balancing a rotating part, the type depending upon various conditions: (1) Static balancing. (2) Dynamic balancing.

In *static balancing*, the equations

$$\Sigma WR \cos \theta = 0$$
$$\Sigma WR \sin \theta = 0$$

are satisfied. The process is one, basically, of satisfying the condition that no turning of the part because of unbalance will occur when the part is supported on level ways.

In *dynamic balancing*, the equations

$$\Sigma WR \cos \theta = 0 \qquad \Sigma WRa \cos \theta = 0$$
$$\Sigma WR \sin \theta = 0 \qquad \Sigma WRa \sin \theta = 0$$

are satisfied. These equations, when satisfied, will give a completely balanced system both for forces and couples due to inertia forces. Dynamic balancing automatically includes a static balance.

Types of balancing required. The type of balancing which is required for a particular piece is dictated by the size, speed, operating conditions, and economics involved. The following table, furnished by the Westinghouse Electric Corporation, indicates the type of balance used with horizontal machines and vertical exciters, and vertical machines, in their Generator Division. Although the table is the specification of the type of balance used with particular machines, it does, nevertheless, serve to illustrate the fact that the type of machine, size, and speed control the extent to which balancing is carried out, consistent with satisfactory operation and economics.

In general, static balancing is sufficient for (1) slow speeds, where the dynamic effect is small and thus negligible; and (2) narrow parts at moderate speed, as automobile wheels, narrow fans, airplane propellers. Static balancing must be used properly, or misleading results may be obtained. Static balance of a long rotor, for instance, might give more harmful results if the balancing weight is located so as to give a dynamic couple of worse effect than the unbalanced force.

Horizontal Machines and Vertical Exciters

Diameter of Rotor, inches	Speed, rpm	Type Balance
Up to and including 30	Below 400	None
,, ,, ,, ,, ,,	400 to 700, inclusive	Static
,, ,, ,, ,, ,,	Above 700	Dynamic
30 to 55, inclusive	Below 250	None
,, ,, ,, ,,	250 to 550, inclusive	Static
,, ,, ,, ,,	Above 550	Dynamic
55 to 85, inclusive	Below 250	None
,, ,, ,, ,,	250 to 400, inclusive	Static
,, ,, ,, ,,	Above 400	Dynamic
Above 85	Below 400	Static
,, ,,	400 and above	Dynamic

Vertical Machines (Except Vertical Exciters)

Diameter of Rotor, inches	Speed, rpm	Type Balance
Up to and including 85	Below 200	Static
,, ,, ,, ,, ,,	Above 200	Dynamic
Above 85	Below 400	Static
,, ,,	400 and above	Dynamic

Courtesy Westinghouse Electric Corp.

18.2 Methods of balancing

While it is impossible to discuss all types of balancing machines available, it is appropriate to describe some of the methods and machines for balancing.

(1) **Trial and error.** The trial and error method may be applied to either static or dynamic balancing.

In static balancing, the part is set on level ways, and the operator fastens temporary weights on the part until the part remains in equilibrium in any position on the level ways. After the amount and location of the unbalance are determined, permanent weights are fastened to the part. (Or the proper amount of material may be removed from the diametrically opposite side to achieve the same effect.)

In dynamic balancing, a pure trial and error method may be used, with, perhaps, considerable difficulty in locating the proper balancing weight in two reference planes. The method may be simplified and modified by the use of a specially designed test jig, where the rotor is supported in two bearings mounted on I-beams which are flexible in a lateral direction and can be clamped rigid. If one I-beam is clamped and the rotor is run at speed, the unbalance will cause the unclamped I-beam to vibrate. Trial and error can be used to locate a weight in one reference plane until the vibration at the unclamped support is a

286   Balancing Machines

minimum. By clamping the free support and unclamping the support previously clamped, the procedure may be repeated until a balance weight in a second reference plane is found to give the minimum vibration at the unclamped support. The process is repeated until smooth running of the rotor results with both supports unclamped.*

Fig. 18.2a. Static balancing machine used to balance clutch assemblies. (Courtesy of the Taylor Dynamometer and Machine Co.)

(2) **Taylor static balancing machine.** A machine developed specially for static balancing is the Taylor balancing machine.† Figure 18.2a shows the machine used for balancing clutch assemblies, and Fig. 18.2b shows the machine used for balancing automobile wheels.

* See *Mechanical Vibrations*, by Den Hartog, 3rd Edition, McGraw-Hill, pages 296–317, for a method to eliminate trial and error.
† Taylor Dynamometer and Machine Co., Milwaukee, Wisconsin.

Methods of Balancing

Fig. 18.2b. Static balancing machine used to balance automobile wheels. (Courtesy of the Taylor Dynamometer and Machine Co.)

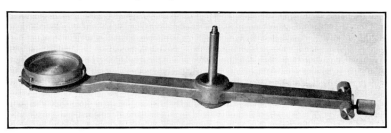

Fig. 18.2c. Spirit level used to show amount and location of unbalance in the Taylor static balancing machine. (Courtesy of the Taylor Dynamometer and Machine Co.)

The machine is so constructed that the raising of the workpiece by the operator causes the workpiece to be supported by a steel ball on a flat surface. The center of gravity of the supported system is located below the point of suspension, and any unbalance will cause the workpiece to rotate to bring the center of gravity directly below the point of suspension. The amount, as well as the position, of the unbalance is indicated on a spirit level, shown in Fig. 18.2c. The workpiece is mounted on the spirit level and the two operate as a unit in the balancing. The use of a ball for support minimizes the effect of friction.

(3) **Dynamic balancing machines.** There are a variety of dynamic balancing machines available. Some are based upon mechanical means of determining the magnitude and location of the unbalanced weights; others use electrical means. The electrical machines may be of the compensating type or of the indicating type. Since space limitations prevent the description of the various types in detail, only the Olsen type EAA, an electrical indicating machine, manufactured by the Tinius Olsen Testing Machine Company, Philadelphia, Pa., will be described.

The Tinius Olsen dynamic balancing machine is designed to give the magnitude and the location of the unbalance in two reference planes, so that a complete balance of a rotating part can be obtained. Refer to Fig. 18.3 for the parts described.

A rotor is supported on rollers, D. The brackets, I, support the rollers and are clamped to the vibratory cradle, C. The vibratory cradle, C, is supported on four flexible rods, two at each end of the cradle. The flexible rods permit the lateral motion of the cradle. There are two pivot brackets, J, which are clamped to the cradle rails, with the corresponding pivots and pivot locking brackets. These pivoting assemblies are adjusted by the operator so that one pivot is located in each of two planes of correction, the so-called reference planes, which have been selected for dynamic balancing. The knurled knob, K, operates both pivots, locking one pivot and simultaneously unlocking the other pivot.

When the pivots are located in two reference planes, unbalance readings taken in either plane, both as to amount and angular location, are entirely independent of readings taken or corrections made in the other reference plane. This is due to the fact that any unbalance or corrections made in the plane of the locked pivot have no moment with respect to the plane of the unlocked pivot, hence cannot affect the unbalance condition in the plane of the unlocked pivot.

A magnetic pickup B is located at each end of the cradle. The vibration of the rotor because of the unbalance moves a coil through a magnetic field set up by a permanent magnet, and the alternating current so developed is proportional to the amplitude of vibration and hence proportional to the unbalanced to be measured. An indicating ammeter shows the maximum current developed, the readings on the ammeter being calibrated for a particular part. Such calibration can be quickly made by putting known weights in the correction planes of a rotor and noting the change of meter reading.

Fig. 18.3. Type EAA Electrodyne balancing machine. (Courtesy of the Tinius Olsen Testing Machine Co.)

290   Balancing Machines

The unbalance is defined both as to amplitude and phase by electronic circuits. The Tinius Olsen Elecᵩdyne system provides a system whereby the readings are obtained on two meters, as seen in Fig. 18.3.

The Elecᵩdyne amplifier is a two-channel system. The first channel indicates the rectified a-c voltage of the pickup on the right-hand meter on the panel. The second channel uses pulse techniques, and its first portion forms a single pulse of the a-c voltage from the pickup. Its second portion receives a reference pulse from a pulse generator on the machine spindle. The phase relationship between these two pulses is determined electronically, and the angle of unbalance is read on the angle meter, which is the other meter on the panel.

It is to be noted that the basis equations of analysis are applied, with the machine performing the arithmetic in the application of two moment equations, and with the moments being taken about two reference planes, the planes of correction.

(4) **Electrically operated static balancing machine.** A machine developed by the Tinius Olsen Company for the static balancing of parts which are relatively narrow with respect to their diameter, their type EAA vertical machine, is shown in Fig. 18.4. The vibratory frame, however, is different from the cradle construction described in the previous section in that it is so arranged that it is restrained to vibrate in a horizontal plane about a vertical axis. The vibratory frame carrying the spindle connects to a magnetic pickup, which generates an alternating current proportional to the vibration of the vibratory frame, and hence proportional to the unbalance to be measured. The phase relation between pickup output and machine rotation is found electronically, as in the Tinius Olsen dynamic balancing machine.

18.3 Portable balancing equipment

It is quite often necessary to balance a rotor in the field, or to balance a rotor assembled in its own bearings. Portable balancing equipment is available for such purposes. Magnetic pickups are suspended by springs from fixtures which are attached to two points on the machine, such as the bearing pedestals, for instance, where it is desired to analyze the vibrations. The movable core of the magnetic pickups are connected by a push rod to the pedestal so that vibrations in a direction parallel to the push rods are transmitted to the core. The core carries a coil located in the air gap of a magnetic circuit. The body of the magnetic pickup, and hence the magnetic field, remains practically stationary in space while the coil is forced to vibrate with the machine pedestal. The relative motion of the coil and magnetic field generates an alternating current proportional to the vibration. Additional equipment is used to determine the phase relation between pickup output and machine rotation. Three runs at any constant

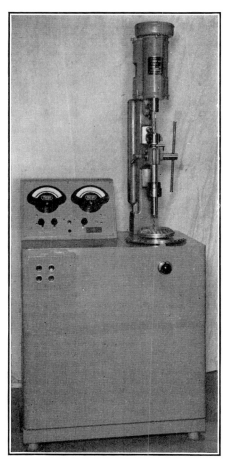

Fig. 18.4. Type EAA vertical Electrodyne balancing machine. (Courtesy Tinius Olsen Testing Machine Co.)

speed are necessary to obtain the data required to determine the location and amounts of correction. The first run is made under normal conditions without any corrective weights on the rotor. The second run is made with a known unbalance or trial weight, known in both amount and angle, inserted in one of the planes selected for correction.

The third run is made in a similar manner, but with a known unbalance in the second plane selected for correction, the trial weight having been removed from the first plane. The necessary balance weights are then calculated, or determined from a graphical layout of the proper vectors.

Field balancing cannot be expected to give the same accuracy of results as can be achieved in a balancing machine, and a second correction may be necessary to refine the unbalance.

18.4   Balancing rotors by electrical networks

A balancing machine which allows the simultaneous determination of the necessary balance weights by means of manipulation of electrical networks is the Westinghouse-Gisholt balancing machine, the principles of which are discussed completely in an article "Balancing Rotors by Means of Electrical Networks," by J. G. Baker and F. C. Rushing, published in the *Journal of the Franklin Institute*, in August, 1936. The vibration of the rotor due to unbalance is measured by electrical pickup units. However, an electrical network is used to screen out the effect of the unbalance in one plane as affecting the unbalance in a second plane. Thus the rotor is not clamped during the balancing operation, and the total motions at the two supports are used, with proper adjustment of the electrical circuits, to give electrical readings which are proportional to the unbalance in one plane only.

The reader is referred to the article for a complete discussion and mathematical analysis.

18.5   Balancing flexible rotors

In spite of balancing a rotating member in a balancing machine, it may still be necessary to resort to further balancing after the parts are assembled, if the speed at which the rotating member revolves is different from the speed at which the balance is made. The discussion to follow applies to flexible shafts, for instance, but does not apply to members which are rigid, as fans. A flexible rotor balanced at one speed will not necessarily be balanced at all speeds. This is illustrated in Fig. 18.5a, where an unbalanced weight is balanced by weights in two balancing planes, one in plane A and the other in plane B. Suppose that the shaft was run at a speed in the balancing machine which caused the shaft to deflect as shown in Fig. 18.5b. The insertion of balancing weights in planes A and B would be appropriate. Suppose, however, that the speed at which the shaft was run in its own bearings caused the shaft to assume the position shown in

Balancing Flexible Rotors

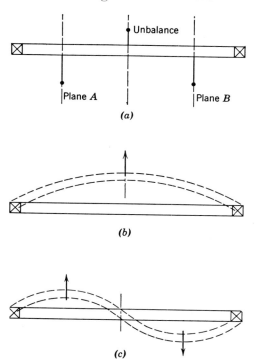

Fig. 18.5. Flexible rotors balanced at one speed are not necessarily balanced at all speeds because of deflection characteristics.

Fig. 18.5c. The extra balancing weights would cause now an unbalance which could not be detected in the balancing machine. Thus, "touching up" a flexible member after it is assembled in its own bearings and run at its proper speed might be necessary.

CHAPTER 19

Balancing Masses Reciprocating in a Plane

The transfer of power in a reciprocating engine involves the transfer of forces. In general, the forces are not constant. The gas pressure in the cylinder, as well as the inertia forces, vary during a cycle. The net result of the periodic nature of the forces in the engine is the transmission of periodic forces to the structure on which the engine is mounted, with consequent vibration, noise, and stresses.

As was pointed out in Chapter 14, the engine experiences resultant forces, the forces being due to the effect of the gas pressure in the cylinder and to the inertia effects of the moving parts. Also, separate static and inertia force analyses could be made, the resultant of the two analyses giving the same results as a combined force analysis. It was pointed out that different combinations of static and inertia force analyses could be made, with no change in the final results. (See Fig. 14.7a, b, and c.) Surprisingly enough, the effect of the inertia forces on the resultant force applied to the structure is the same for each case; it is the vector sum of the inertia forces. The location of the shaking force varies, however, and depends upon how the engine is driven. If the force causing the motion is applied to the piston, the location of the shaking force is different from the location if the engine is assumed to be driven by the flywheel. The shaking force, whatever its location, can be considered replaced by a force at the crank bearing and a couple. The effect of the static force on the structure is a couple, the couple being caused by the resultant action of the force on the crank bearings, cylinder wall, and cylinder head. Thus the effect of the forces can be considered as causing a couple acting on the frame of the engine and a shaking force, with the shaking force taken as acting through the center of the crank bearings.

Now the question is what can be done with the variable couple and shaking force so as to give a smooth-running engine. The variable couple applied to the frame of the machine can be offset by the use of a flywheel, by the use of a number of cylinders, or by a combination of the two. A flywheel cannot be used, however, to offset the effect of a shaking force, because a couple effect cannot be used to balance a force. A multicylinder engine can be designed in such a way that the inertia forces balance each other, but many engines are available where the inertia forces do not balance each other, and auxiliary means are used to give balance.

The purpose of this chapter, therefore, is to investigate the magnitude and variation of shaking forces, as well as the means by which the shaking forces may be balanced, in engines with masses reciprocating in a plane.

19.1 The connecting rod replaced by a kinetically equivalent system

In the slider-crank mechanism of Fig. 19.1a with link centers of gravity G_2, G_3, and G_4, the inertia forces may be found by analytical methods readily enough for links 2 and 4 and with somewhat greater difficulty for link 3. The procedure that is used to simplify the analysis is to replace the mass of the connecting rod concentrated at the center of gravity with two masses, one mass at the crank pin and the other mass at the piston pin. The weight of the connecting rod considered as concentrated at the crank pin, W_c', and the weight of the connecting rod considered as concentrated at the piston pin, W_p', may be found as shown in Fig. 19.1b, where the connecting rod is set on two scales, and the weight at the crank pin and the weight at the piston pin are determined. Analytically, the weights can be expressed by

$$W_c' = W(h_p/L)$$
$$W_p' = W(h_c/L)$$

where W is the total weight of the connecting rod,
$\quad h_p$ is the distance from the center of gravity of the connecting rod to the center of the piston pin,
$\quad h_c$ is the distance from the center of gravity of the connecting rod to the center of the crank pin,
$\quad L$ is the length of the connecting rod.

Figure 19.1c shows the system at this point, where the connecting rod is considered weightless.

The above method of replacing the connecting rod by two concentrated weights would seem, at first glance, to give an approximation in the magnitude of the shaking force, and in the direction of the shaking force. This is not so, however, as may be seen from an examination of the equations for a kinetically equivalent system:

Balancing Masses Reciprocating in a Plane

$$m_1 + m_2 = M \tag{1}$$

$$m_1 h_1 = m_2 h_2 \tag{2}$$

$$m_1 h_1{}^2 + m_2 h_2{}^2 = M k^2 \tag{3}$$

and the derived relation $h_1 h_2 = M k^2$.

Equations 1 and 2 are satisfied, but not Eq. 3, with the conclusion that the resultant inertia force of the two concentrated weights will

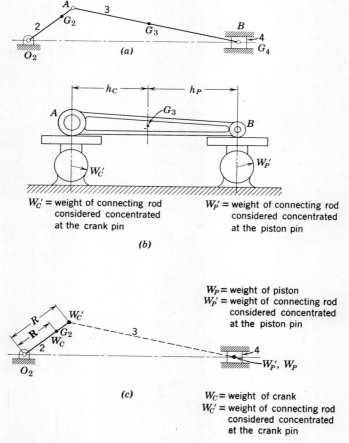

Fig. 19.1. Weight of connecting rod replaced by two concentrated weights, one at the piston pin and the other at the crank pin.

have the correct magnitude and direction, but the resultant inertia force will be parallel to and offset from the correct value. The offset stems from the fact that the moment of inertia of the original connecting rod is different from the system using two concentrated weights, giving different values of the distance from the center of gravity to the inertia force, as may be seen from $h = I\alpha/MA_g$.

Inasmuch as we are concerned with the vector sum of the inertia forces, the same vector sum will result for the original system as the equivalent one. The equivalent system will not give the same line of action of the shaking force, but that is not to be considered in the analysis.

19.2 Alternate analysis of replacement of weight of connecting rod by two concentrated weights

So that the student may appreciate fully the step which has been taken, an analytical analysis is presented. The magnitude of the

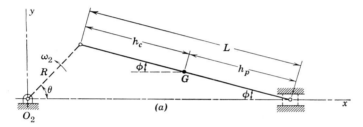

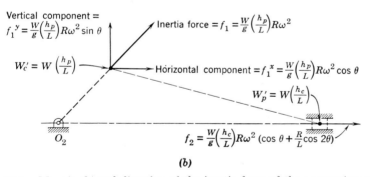

Fig. 19.2. Magnitude and direction of the inertia force of the connecting rod as found by considering the acceleration of the center of gravity is the same as that found by considering the connecting rod replaced by two concentrated weights, one at the piston pin and the other at the crank pin.

inertia force of the original system will be determined and compared to the inertia force of the equivalent system.

Inertia force of connecting rod. The x- and y-coordinates of the center of gravity, G, of the connecting rod, shown in Fig. 19.2a, are

Balancing Masses Reciprocating in a Plane

$$x = R \cos \theta + h_c \cos \phi$$

$$y = h_p \sin \phi$$

The vertical projection of the crank is

$$R \sin \theta = L \sin \phi$$

or

$$\sin \phi = \frac{R}{L} \sin \theta$$

and

$$\cos \phi = (1 - \sin^2 \phi)^{1/2} = \left[1 - \left(\frac{R}{L} \sin \theta\right)^2\right]^{1/2} \approx 1 - \frac{1}{2}\left(\frac{R}{L} \sin \theta\right)^{2*}$$

Therefore the x- and y-coordinates may be written

$$x = R \cos \theta + h_c \left[1 - \frac{1}{2}\left(\frac{R}{L} \sin \theta\right)^2\right]$$

$$y = h_p \left(\frac{R}{L} \sin \theta\right)$$

Differentiation of the above equations with respect to time gives the components of velocity in the x- and y-directions:

$$\frac{dx}{dt} = -R\omega \sin \theta - \frac{1}{2} h_c \left(\frac{R}{L}\right)^2 \omega \sin 2\theta$$

$$\frac{dy}{dt} = h_p \left(\frac{R}{L}\right) \omega \cos \theta$$

Differentiation again with respect to time gives the components of acceleration, with ω considered constant:

$$A_G{}^x = \frac{d^2 x}{dt^2} = -R\omega^2 \cos \theta - h_c \left(\frac{R}{L}\right)^2 \omega^2 \cos 2\theta$$

$$A_G{}^y = \frac{d^2 y}{dy^2} = -h_p \left(\frac{R}{L}\right) \omega^2 \sin \theta$$

Multiply the acceleration components by the mass to obtain the components of inertia force of the connecting rod, changing signs because the inertia force is in the opposite direction to the acceleration:

$$f_x = \frac{W}{g} R\omega^2 \cos \theta + \frac{W}{g} h_c \left(\frac{R}{L}\right)^2 \omega^2 \cos 2\theta \tag{1}$$

$$f_y = \frac{W}{g} h_p \frac{R}{L} \omega^2 \sin \theta \tag{2}$$

* See page 240 for expansion with binomial theorem.

It is necessary now to show that the same components of inertia force will result by replacing the mass of the connecting rod with two concentrated masses, one at the piston pin and the other at the crank pin.

The inertia force of the mass considered concentrated at the crank pin is

$$f_1 = \frac{W}{g}\left(\frac{h_p}{L}\right) R\omega^2$$

The horizontal and vertical components are

$$f_1{}^x = \frac{W}{g}\left(\frac{h_p}{L}\right) R\omega^2 \cos\theta$$

$$f_1{}^y = \frac{W}{g}\left(\frac{h_p}{L}\right) R\omega^2 \sin\theta$$

as shown in Fig. 19.2b.

The inertia force of the mass considered concentrated at the piston pin is

$$f_2 = \frac{W}{g}\frac{h_c}{L} R\omega^2 \left[\cos\theta + \frac{R}{L}\cos 2\theta\right]$$

as shown in Fig. 19.2b.

f_2 is horizontal. (See Chapter 15 for the derivation of the acceleration of the piston.)

The total horizontal inertia force of the connecting rod is

$$f_x = f_1{}^x + f_2 = \frac{W}{g}\left(\frac{h_p}{L}\right) R\omega^2 \cos\theta + \frac{W}{g}\left(\frac{h_c}{L}\right) R\omega^2 \left[\cos\theta + \frac{R}{L}\cos 2\theta\right]$$

$$= \frac{W}{g}\frac{(h_p + h_c)}{L} R\omega^2 \cos\theta + \frac{W}{g} h_c \left(\frac{R}{L}\right)^2 \omega^2 \cos 2\theta$$

Since $h_p + h_c = L$,

$$f_x = \frac{W}{g} R\omega^2 \cos\theta + \frac{W}{g} h_c \left(\frac{R}{L}\right)^2 \omega^2 \cos 2\theta \qquad (3)$$

The total vertical component of the inertia force of the connecting rod is

$$f_y = f_1{}^y = \frac{W}{g}\left(\frac{h_p}{L}\right) R\omega^2 \sin\theta \qquad (4)$$

Equations 1 and 2 are identical to Eqs. 3 and 4, respectively. Thus the use of two concentrated masses, one at the crank pin and the other at the piston pin, will give the same total inertia force of the connecting rod as when the mass of the connecting rod is considered to be at the center of gravity of the rod. Since the components by either analysis are the same, the same direction of force will result. The two resultant forces will not pass through the same points, however, but will be parallel. Since the point of application of the shaking force on the engine will not be considered, as mentioned before, no error will be introduced in the analysis of the determination of the total shaking force.

19.3 No counterbalancing weight

For the slider crank mechanism of Fig. 19.1c, therefore, there are four inertia forces to be considered in obtaining the total shaking force:

(1) The inertia force due to the weight W_p of the piston.

(2) The inertia force due to the weight W_p' of the connecting rod considered concentrated at the piston.

(3) The inertia force due to the weight W_c of the crank at the center of gravity $\mathbf{R}$ from the crank bearing.

(4) The inertia force due to the weight W_c' of the connecting rod considered concentrated at the crank pin.

Forces 1 and 2 may be combined into a single force, along the line of motion of the piston:

$$f_p = \frac{(W_p + W_p')}{g} R\omega^2 \left(\cos\theta + \frac{R}{L} \cos 2\theta \right) \qquad (1)$$

Forces 3 and 4 may be combined into a single force, directed outwards along the crank:

$$f_c = \frac{W_c \mathbf{R}\omega^2}{g} + \frac{W_c' R\omega^2}{g}$$

It is desirable to simplify the equations above by replacing W_c with an equivalent weight at the crank pin, W_c'' so that $W_c \mathbf{R} = W_c'' R$, and so that the weight of the crank can be considered at the crank pin:

$$f_c = \frac{(W_c'' + W_c')}{g} R\omega^2$$

Figure 19.3a shows the inertia forces of the piston and crank in position. As shown in Chapter 14, under the heading "shaking forces," (page 229), the total force exerted on the structure due to the inertia forces is the vector sum of the inertia forces of the individual

No Counterbalancing Weight

members. The inertia forces are shown added vectorially in Fig. 19.3b.

The picture of variation of shaking force for a complete cycle may be obtained quickly by drawing a polar diagram. In Fig. 19.3c, three circles are drawn with

$$r_1 = \frac{(W_c'' + W_c')}{g} R\omega^2$$

$$r_2 = \frac{(W_p + W_p')}{g} R\omega^2$$

$$r_3 = \frac{(W_p + W_p')}{g} R\omega^2 \left(\frac{R}{L}\right)$$

The inertia forces of the crank and piston may be obtained directly on the figure for any angle θ of the crank. For instance, for the position of the crank shown in Fig. 19.3c, the inertia force of the crank,

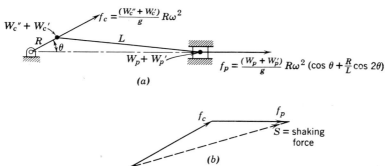

(a)

(b)

Fig. 19.3a, b. Inertia forces are shown with the connecting rod replaced by two concentrated weights. The shaking force is the vector sum of the inertia forces.

f_c, is given by OM directly, and is shown in the correct position, along the crank. The inertia force of the piston is obtained next in two steps.

Write the inertia force of the piston, Eq. 1 as:

$$f_p = \frac{(W_p + W_p')}{g} R\omega^2 \cos\theta + \frac{(W_p + W_p')}{g} R\omega^2 \left(\frac{R}{L}\right) \cos 2\theta$$

The first quantity, called the primary force, is represented by the projection of MN, that is, MP. The second quantity, called the secondary force, is represented by the projection of QR, where OQR is at an angle of 2θ with the horizontal.

Thus, $f_c \nrightarrow f_p$ is the same as $OM \nrightarrow MP \nrightarrow QS$. The resultant is shown in Fig. 19.3c, which is the same as shown in Fig. 19.3b.

Balancing Masses Reciprocating in a Plane

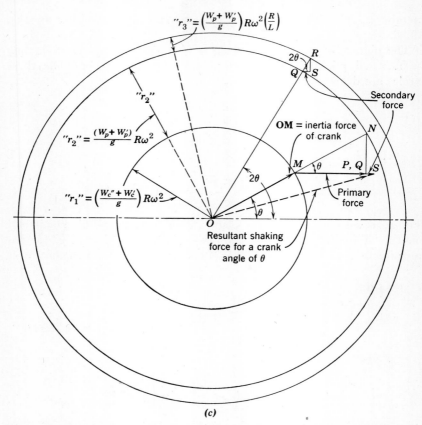

Fig. 19.3c. Construction for determination of the shaking force of a slider-crank mechanism for one position of the crank.

Example. The following data are given for a single-cylinder horizontal diesel engine:

Speed of crank	1200 rpm
Stroke	6 in. = $2R$
Length of connecting rod	15 in. = L
Distance from the center of gravity of the connecting rod to crank pin	5 in. = h_c
Equivalent unbalanced weight of the crank at 3-in. radius	4 lb = W_c''
Weight of piston	8 lb = W_p
Weight of connecting rod	18 lb

It is required to find the polar diagram of shaking forces for one complete revolution of the crank.

The weight of the connecting rod considered concentrated at the crank pin is $W_c' = (\tfrac{10}{15})(18) = 12$ lb. The weight of the connecting

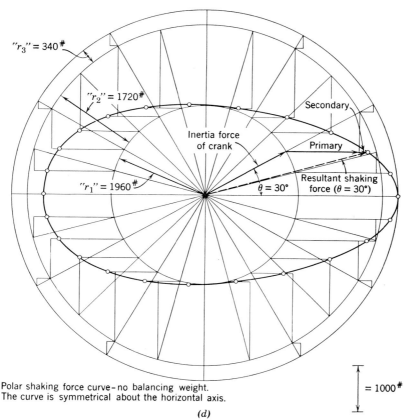

Fig. 19.3d. Polar shaking force curve—no counterbalancing.

rod considered concentrated at the piston pin is $W_p' = (\frac{5}{15})(18) = 6$ lb. Thus,

$$W_p + W_p' = 8 + 6 = 14 \text{ lb}$$

$$W_c'' + W_c' = 4 + 12 = 16 \text{ lb}$$

$$r_1 = \frac{(W_c'' + W_c')}{g} R\omega^2 = \frac{(4+12)}{32.2} \frac{3}{12} \left[\frac{(1200)(2\pi)}{60}\right]^2$$
$$= 1960 \text{ lb}$$

$$r_2 = \frac{(W_p + W_p')}{g} R\omega^2 = \frac{(8+6)}{32.2} \frac{3}{12} \left[\frac{(1200)(2\pi)}{60}\right]^2$$
$$= 1720 \text{ lb}$$

$$r_3 = \frac{(W_p + W_p')}{g} R\omega^2 \frac{R}{L} = \frac{(8+6)}{32.2} \frac{3}{12} \left[\frac{(1200)(2\pi)}{60}\right]^2 \frac{3/12}{15/12}$$
$$= 340 \text{ lb}$$

Figure 19.3d shows the polar force diagram plotted for 15-degree intervals of the crank. A smooth curve drawn through the ends of each shaking force vector gives a graphic picture of the variation of shaking force through a complete cycle. Note that the shaking force vector does not lie along the crank for a general position of the crank. The shaking force does lie along the crank for two positions, when $\theta = 0°$ and $\theta = 180°$.

The diagram shows rather large unbalances, and also shows that the maximum horizontal shaking force is larger than the maximum vertical shaking force for the horizontal engine.

The question now is: What can be done to reduce the shaking forces or to eliminate the shaking forces?

19.4 Adding counterweights

First, it is not possible to balance the shaking forces of the engine considered so far, for all positions of the crank, with a single counterbalancing weight. Inspection of Fig. 19.4a will show the reason. A

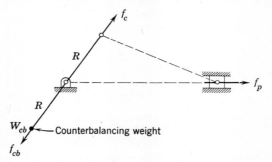

Fig. 19.4a. Counterweight used to give partial balancing.

counterbalance located as shown on the crank may be selected to balance the crank completely, but the piston will then be unbalanced. Figure 19.4b shows an actual steam engine with a counterweight. If more counterbalancing weight is used to introduce a horizontal component to attempt to balance the piston, a vertical component of unbalance will be introduced. Or, expressed differently, it is not possible to balance by a rotating weight a force which is horizontal (the inertia force of the piston) for all positions of the crank. It is possible, however, to alleviate certain conditions by selecting a proper counterbalancing weight. For instance, it is possible to minimize the horizontal unbalance, introducing, perhaps, greater vertical unbalance. Or it is possible to obtain smaller maximum shaking forces. Let us examine the effect of using different counterbalancing weights.

Counterbalancing weight equal to the weight of the crank.
A counterbalancing weight equal to the weight of the crank and connecting rod considered concentrated at the crank pin will balance the crank completely. The unbalance will be that of the piston, and the shaking force will be horizontal for the horizontal engine, for all positions of the crank.

Fig. 19.4b. Horizontal "Universal Unaflow" steam engine built by the Skinner Engine Co. The crank disk is a solid steel casting into which a forged steel crank pin is pressed and then riveted over on the back side. The disk itself is shrunk and keyed onto the main crankshaft. The counterbalance web is hollow, and lead is poured through suitable holes into this hollow web to produce the desired amount of counterbalance. (Courtesy Skinner Engine Co.)

Counterbalancing weight equal to the weight of the piston and the weight of the crank. The other extreme to no counterbalancing weight is a counterweight equal to the weight of the crank and piston. That is,

$$W_{cb} = (W_c'' + W_c') + (W_p + W_p')$$

The center of gravity of the counterweight will be located at R.

The inertia force of the counterweight is

$$f_{cb} = \frac{(W_c'' + W_c' + W_p + W_p')}{g} R\omega^2$$

$$= \frac{(4 + 12 + 8 + 6)}{32.2} \left(\frac{3}{12}\right) \left[\frac{1200(2\pi)}{60}\right]^2$$

$$= 3680 \text{ lb}$$

The inertia force of the counterbalancing weight is shown added to the inertia force of the crank and piston, in Fig. 19.5a, to give the polar diagram of shaking forces.

Counterbalancing weight equal to the weight of the crank and one half the weight of the piston. An intermediate value of

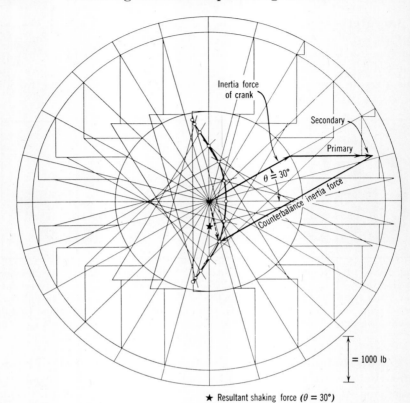

Fig. 19.5a. Polar shaking force curve. Counterbalancing weight is equal to the sum of the equivalent weight of the crank and the equivalent weight of the piston. (The curve is symmetrical about the horizontal axis.)

counterbalancing weight may be taken as the weight of crank and one half the weight of the piston:

$$W_{cb} = (W_c'' + W_c') + \tfrac{1}{2}(W_p + W_p')$$

The inertia force of the counterbalancing weight is

$$\begin{aligned}
f_{cb} &= \left[\frac{W_c'' + W_c' + \tfrac{1}{2}(W_p + W_p')}{g}\right] R\omega^2 \\
&= \left[\frac{4 + 12 + \tfrac{1}{2}(8 + 6)}{32.2}\right]\left(\frac{3}{12}\right)\left[\frac{(1200)(2\pi)}{60}\right]^2 \\
&= 2820 \text{ lb}
\end{aligned}$$

The inertia force of the counterweight is added vectorially to the inertia force of the crank and piston to give the shaking force diagram shown in Fig. 19.5b.

It is evident that the shaking force diagram may be changed considerably by the addition of various counterweights. Practically, the

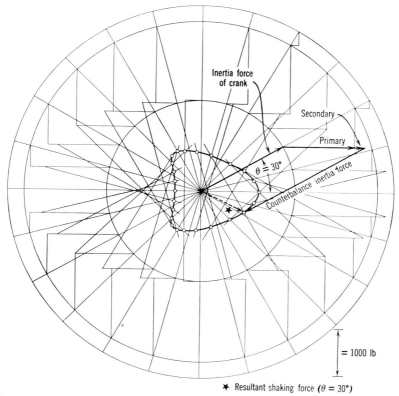

Fig. 19.5b. Polar shaking force curve. Counterbalancing weight is equal to the sum of the equivalent weight of the crank and *one half* the equivalent weight of the piston. (The curve is symmetrical about the horizontal axis.)

magnitude of counterweight is determined by the operating conditions, the location of the machine, and the type of unbalance which would be more objectionable, that is, whether the vertical unbalance would be more objectionable than the horizontal, or vice versa. If the problem is to obtain the smallest overall shaking force, the counterweight that would be used at the crank radius would be the sum of the weights of the crank considered concentrated at the crank pin and of the connecting rod considered concentrated at the crank pin, plus about one half to two thirds the sum of the weights of the piston (and reciprocating parts) and of the connecting rod considered concentrated at the piston pin. That is,

$$W_{cb} = (W_c'' + W_c') + (1/2 \text{ to } 2/3)(W_p + W_p')$$

where W_{cb} is located at R inches from the center of rotation.

19.5 Theoretical complete balance of a single-cylinder engine

It was stated that a single-cylinder engine could not be balanced completely by a single rotating weight, which is true if practical proportions are maintained in the design. A theoretical means of obtaining complete balance is shown in Fig. 19.6, where the connecting

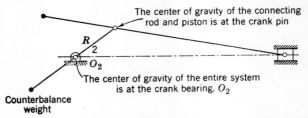

Fig. 19.6. Arrangement to balance completely a single-cylinder engine with one rotating counterweight.

rod is extended so that the center of gravity of the connecting rod and piston is at the crank pin. The weight at the crank pin could be balanced by a single counterweight on the crank. Note that the center of gravity of the moving parts would be located at the center of the crank bearing.

19.6 Other methods for balancing wholly or in part the inertia forces of a single-cylinder engine

(a) Dummy pistons. The rotating parts of a single-cylinder engine may be balanced by a counterweight on the crank. It is not possible to balance the inertia force of the reciprocating mass by a rotating mass. It is possible to balance the inertia force of a reciprocating mass by another reciprocating mass, so located that its acceleration is always equal to and opposite to that of the mass to be balanced, and thus the inertia forces would always be equal and opposite to each other.

Figure 19.7 shows the arrangement of the parts. Essentially, the system is an engine of the opposed type, discussed in detail in the next section. The dummy pistons are merely reciprocating masses, and no steam or gas pressure is applied to the dummy pistons. Note the use of two dummy pistons so as not to introduce any offset which would create a couple, even though the inertia forces are balanced.

It should be mentioned that the main piston connecting rod need not be the same length as the dummy piston connecting rods, as long as the ratio of crank to connecting rod is the same for the main piston and the dummy pistons. The inertia forces of the main piston and the dummy pistons may be made the same by adjusting the mass of the dummy pistons.

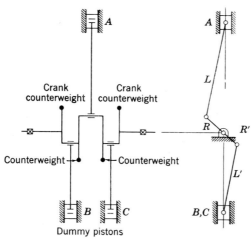

Fig. 19.7. Dummy pistons used to balance a single-cylinder engine.

(b) Gobron-Brillé engine.*, † Figure 19.8 shows the general scheme of arrangement of parts of the Gobron-Brillé engine. The two pistons, A and B, work in the same cylinder with the combustion space between them. The inertia force of piston A is partly balanced by the inertia force of piston B. If it is assumed that the weights are so proportioned that the crank for piston A balances the crank for piston B, and if it is assumed that the weight of piston A is the same as the weight of piston B, the primary forces are balanced, but the secondary forces are not balanced. The total unbalance is twice

---

* See *Internal Combustion Engines*, by J. A. Polson, 2nd edition, Wiley, 1942, pages 348–352, for a description of a Sun-Doxford opposed piston type of two-cycle diesel engine.

† The early Junker engine has the same basic design. Another Junker aircraft type of engine has two separate crankshafts with a gear train connecting the two crankshafts. Power is taken off one shaft. See Fig. 19.13.

310   Balancing Masses Reciprocating in a Plane

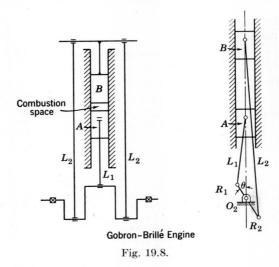

Fig. 19.8.

the secondary force of one piston, for the conditions assumed. The analysis is discussed in detail in the next section.

19.7 Opposed types of engines

Type I

(a) *Coaxial Type.* Figure 19.9 shows the first type of opposed engine to be considered. It is arranged so that the pistons move along

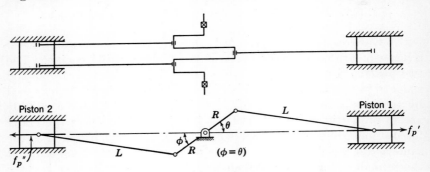

Fig. 19.9. Coaxial type of opposed engine.

the same axis. The weight of each connecting rod is considered replaced by concentrated weights at the corresponding piston pin and crank pin. The crankshaft is assumed to be balanced. It will be assumed, also, that the equivalent weight of each piston is the same.

The inertia force of piston 1 is:

$$f_p' = \frac{(W_p + W_p')}{g} R\omega^2 \left(\cos\theta + \frac{R}{L}\cos 2\theta\right)*$$

If θ is considered a small angle, f_p' is positive, or is directed to the *right*. The inertia force of piston 2 is

$$f_p'' = \frac{(W_p + W_p')}{g} R\omega^2 \left(\cos\phi + \frac{R}{L}\cos 2\phi\right)$$

If ϕ is small, f_p'' is positive, and is directed to the *left*. Since $\theta = \phi$, f_p' is equal and opposite to f_p'', or the inertia forces of the pistons balance each other for the position shown and for all values of θ.

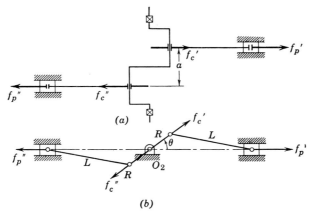

Fig. 19.10. Offset type of opposed engine.

Thus, this type of engine is balanced completely as regards the inertia forces.

(b) *Offset Type.* The second type of opposed engine in this classification is shown in Figs. 19.10a and 19.10b. The offset a between the two planes of motion of the piston is shown. The same analysis as for the coaxial type may be carried out to show that the inertia forces are balanced completely. However, the offset of the pistons will set up a couple of magnitude

$$\left[\frac{(W_p + W_p')}{g} R\omega^2 \left(\cos\theta + \frac{R}{L}\cos 2\theta\right)\right][a]$$

* The angle θ in $R\omega^2 \left(\cos\theta + \frac{R}{L}\cos 2\theta\right)$ is measured from the line of motion of the piston to the crank. See Chapter 15.

312 Balancing Masses Reciprocating in a Plane

The offset of the *cranks* will set up a couple of the inertia forces of the *crank*. This couple may be balanced completely by introducing counterweights on the cranks.

The unbalanced couple due to the inertia forces of the pistons may be kept small by keeping the distance a as small as possible. The offset rather than the coaxial type is the usual arrangement.

Type II. Figure 19.11 shows a second type of opposed engine. The crank is assumed to be balanced. To determine whether the

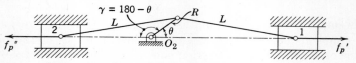

Fig. 19.11. Opposed engine with a common crank.

pistons balance each other, set up the expressions for the inertia force of each piston.

The inertia force of piston 1 is

$$f_p' = \frac{(W_p + W_p'')}{g} R\omega^2 \left(\cos\theta + \frac{R}{L}\cos 2\theta\right)$$

Consider θ as a small angle, with f_p' thus being positive, or to the *right*.

The inertia force of piston 2 is

$$f_p'' = \frac{(W_p + W_p'')}{g} R\omega^2 \left(\cos\gamma + \frac{R}{L}\cos 2\gamma\right)$$

If γ is considered to be a small angle, f_p'' is positive and to the *left*.

The net force to the right is the difference of f_p' and f_p'':

$$f_p' \nrightarrow f_p'' = \frac{(W_p + W_p')}{g} R\omega^2 \left(\cos\theta + \frac{R}{L}\cos 2\theta\right) - \frac{(W_p + W_p'')}{g} R\omega^2 \left(\cos\gamma + \frac{R}{L}\cos 2\gamma\right)$$

Since $\gamma = (180 - \theta)$,

$$f_p' \nrightarrow f_p'' = 2\frac{(W_p + W_p')}{g} R\omega^2 \cos\theta$$

Thus the secondary components cancel each other, whereas the primary components add.

Type III. The third type of opposed engine is shown in Fig. 19.12. The Gobron-Brillé engine discussed in the previous section comes in this classification. It will be assumed that the ratio of crank

Opposed Types of Engines 313

to connecting rod is the same for each part. That is, $(R_1/L_1) = (R_2/L_2)$. It will be assumed also that the crank, and the weight of the connecting rods considered concentrated at the crank pin, is balanced by a counterweight on the crank. It will be assumed further

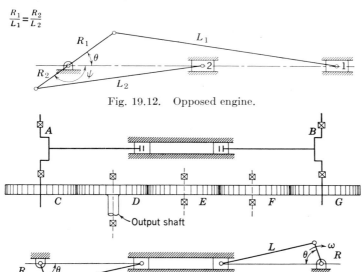

Fig. 19.12. Opposed engine.

Fig. 19.13. Opposed engine with the cranks coupled together by gears. Junker-type high-compression aircraft engine. A and B are the crankshafts. Gears C and G are on the crankshafts A and B. Gears D, E, F form a gear train with power taken off gear D.

that the weight of the pistons is such that $(W_p + W_p')R$ for piston 1 is the same as for piston 2.

The inertia force of piston 1 is

$$f_p' = \frac{(W_p + W_p')_1}{g} R_1 \omega^2 \left(\cos \theta + \frac{R_1}{L_1} \cos 2\theta \right) \quad (1)$$

f_p' is positive for a small angle θ, and is directed to the right for the position shown.

The inertia force of piston 2 is

$$f_p'' = \frac{(W_p + W_p')_2}{g} R_2 \omega^2 \left(\cos \psi + \frac{R_2}{L_2} \cos 2\psi \right) \quad (2)$$

f_p'' is positive to the right for a small angle ψ.

Thus the sum of the inertia forces if $f_p' + f_p''$. Add Eqs. 1 and 2, substituting $\psi = (180 - \theta)$:

314 Balancing Masses Reciprocating in a Plane

$$f_p' + f_p'' = 2 \frac{(W_p + W_p')_1}{g} R_1 \omega^2 \frac{R_1}{L_1} \cos 2\theta$$

The equation above reveals that the primary forces are balanced and the secondary forces are not balanced. Or this type of engine is not balanced.

Type IV. Figure 19.13 shows a type of opposed engine where the cranks are coupled together by gears. The cranks are balanced with counterweights. The inertia forces of the pistons balance each other.

Type V. Figure 19.14 shows an arrangement which is similar to that shown in Fig. 19.13 except for the relation of cranks. The cranks

Fig. 19.14. Opposed engine.

are balanced with counterweights. The inertia forces of the pistons balance each other.

19.8 Auxiliary gear arrangements for balancing

Various gear arrangements are available for balancing the primary and/or the secondary forces in a single-cylinder engine.

(a) **Balance of the primary force only.** Figure 19.15a shows a series of gears, A, B, C, D, and E. Gears A and B are fastened to the crankshaft and rotate at the same speed as the crankshaft, ω_2. Gear B drives gear C, and at the same speed ω_2, with the diameter of C the same as that of B. D is an idler gear between gears A and E. Gears A, D, and E are of the same diameter. Thus gears C and E rotate at the same speed but in opposite directions.

It will be assumed that the crank, and the portion of the connecting rod assumed concentrated at the crank pin, are balanced by a counterweight.

Counterweights F and G are fastened to gears C and E, respectively. They are located so that the line through the center of gravity of the counterweight and gear bearing makes an angle of θ, the crank angle, with the vertical as shown in the figure. If W_{CB} is the weight of the counterweight on gears C and D, and if R_c is the distance from the center of rotation to the center of gravity of the counterweight, the inertia force of the counterweight is

$$f_c = \frac{W_{CB}}{g} R_c \omega_2^2$$

Auxiliary Gear Arrangements for Balancing 315

The component of each force along the line of motion of the piston is $\frac{W_{CB}}{g} R_c \omega_2^2 \cos \theta$. The components of the inertia forces of the counterweights balance each other in a direction perpendicular to the line of motion of the piston; or the resultant of the inertia forces of the

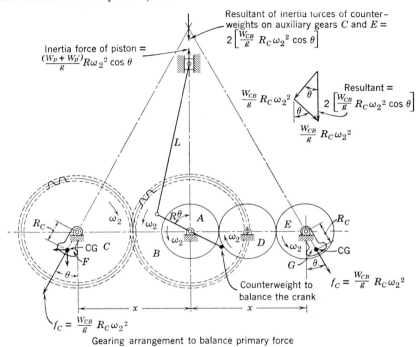

Gearing arrangement to balance primary force

(a)

Fig. 19.15a. From *Technische Dynamik*, by Biezeno-Grammel. Lithoprinted by Edwards Bros., Inc., Ann Arbor, Michigan.

counterweights is $2 \frac{W_{CB}}{g} R_c \omega_2^2 \cos \theta$. The resultant force is directed along the line of motion of the piston, and is opposite in direction to the primary inertia force of the piston. The magnitude of W_{CB} may be found from

$$2 \frac{W_{CB}}{g} R_c \omega_2^2 \cos \theta = \frac{(W_p + W_p{}')}{g} R \omega_2^2 \cos \theta$$

or

$$W_{CB} = \frac{(W_p + W_p{}')}{2} \frac{R}{R_c}$$

(b) Balance of the secondary force only. Figure 19.15b shows an arrangement of gears and counterweights to balance the secondary force of a single-cylinder engine. Gear A rotates at the same speed as the crank. Gears A and B have the same diameters. Gears C

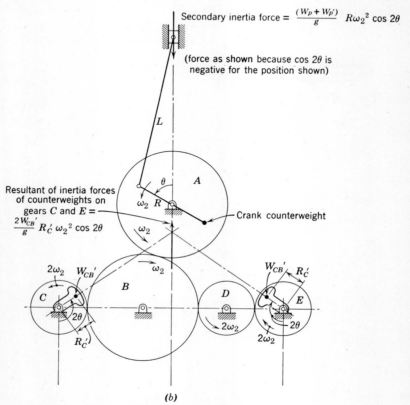

(b)

Gearing arrangement to balance secondary force

Fig. 19.15b. From *Technische Dynamik*, by Biezeno-Grammel. Lithoprinted by Edwards Bros., Inc.

and E are one half the diameter of gears A and B. D is an intermediate gear of such a diameter that gears C and E are symmetrically placed about the centerline. Thus gears C and E rotate at twice the speed of gears A and B.

The resultant inertia force of the counterweights on gears C and E is $2\dfrac{W_{CB}'}{g} R_c'(2\omega_2)^2 \cos 2\theta$ and acts along the line of motion of the piston in the opposite direction to that of the secondary force of the piston.

The weight of a counterweight may be found by equating the inertia force of the counterweights to the secondary force:

$$2\frac{W_{CB}'}{g} R_c'(2\omega_2)^2 \cos 2\theta = \frac{(W_p + W_p')}{g} R\omega_2^2 \frac{R}{L} \cos 2\theta$$

or
$$W_{CB}' = \frac{(W_p + W_p')}{8} \frac{R^2}{R_c' L}$$

(c) Balance of the primary and secondary forces. The primary and secondary forces may be balanced completely by a combined gear system, not shown.

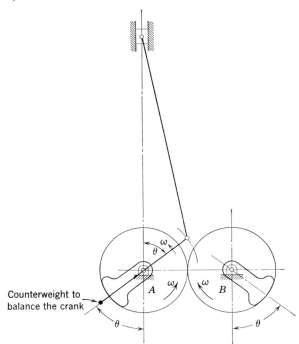

Fig. 19.16. Brush arrangement for balancing of the primary force.

(d) Brush arrangement for balancing of the primary force. Figure 19.16 shows an arrangement used in one of the early single-cylinder automobiles to balance the primary force. Gears A and B are the same size, and run at crankshaft speed. The counterweight on gear A may be considered to be on the crankshaft instead of on gear A inasmuch as the gear A and the crankshaft rotate at the same speed.

For the arrangement shown, a couple will be set up. The couple

318   Balancing Masses Reciprocating in a Plane

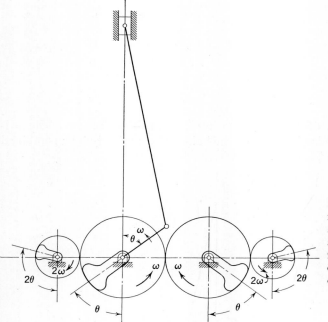

Fig. 19.17. Arrangement to balance the primary and secondary forces, with, however, the introduction of a couple.

may be eliminated by the addition of other gears with counterweights, not shown.

(e) Balance of the primary and secondary forces. Figure 19.17 shows an arrangement to balance the primary and secondary forces of a single-cylinder engine. Couples will be set up, which can be balanced by use of other gears with counterweights, not shown.

19.9 Two-cylinder V-engine

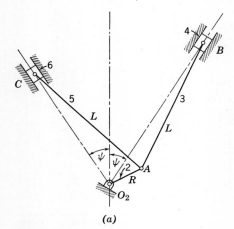

(a)

Fig. 19.18a. Two-cylinder V-engine, with the connecting rods connected to the crank at a common point.

An engine which can be considered a modification of the opposed type of engine is the V-type. There are two types* of construction used in V-engines,

---
* A third type, discussed in Chapter 20, has the connecting rods offset. See Fig. 20.10.

distinguished by the manner of connection of the connecting rod to the crank. Figure 19.18a shows the connecting rods pinned to the crank pin A. Figures 19.18b and 19.18c, furnished by General Motors, Electro-Motive Division, show the arrangement of connecting rods in their model 567B, two-cycle diesel engines, which are of the V-type.

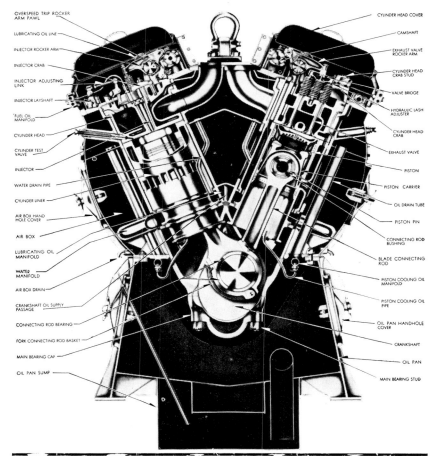

Fig. 19.18b. Courtesy of General Motors Corp., Electro-Motive Division.

The connecting rods are of the interlocking, blade and fork construction. The blade rod oscillates on the back of the upper bearing shell and is held in place by a counterbore in the fork rod.

One side of the blade rod bearing surface is longer than the other, and is known as the "long toe." The blade rods are installed in the *right* bank of the engine, with the long toe towards the center of the engine.

The fork rods are installed in the left bank of the engine. Serrations on

320   Balancing Masses Reciprocating in a Plane

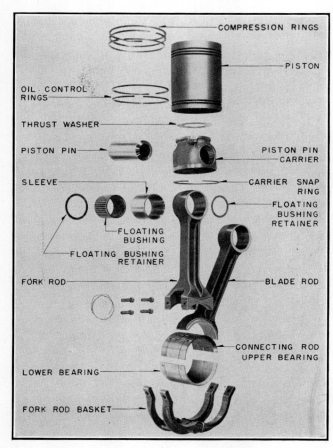

Fig. 19.18c. Courtesy of General Motors Corp., Electro-Motive Division.

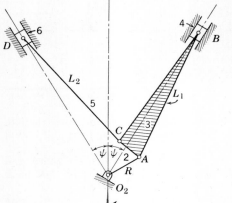

Fig. 19.18d. Two-cylinder V-engine with a master rod and articulated rod.

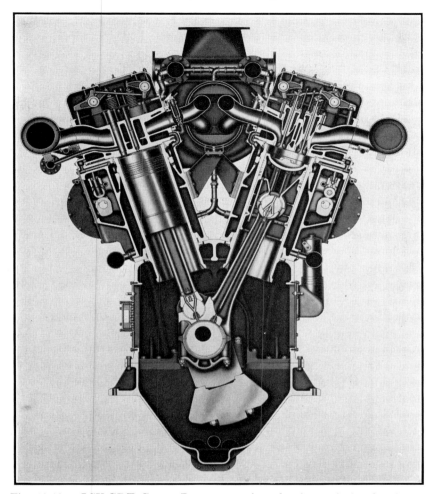

Fig. 19.18e. LSV-GDT Cooper-Bessemer engine, showing articulated rod construction. The 16-cylinder LSV-16 Cooper-Bessemer gas-diesel engine is one recently installed in the Yuma Plant of the Arizona Edison Co. (Courtesy Cooper-Bessemer Corp., Mount Vernon, Ohio.)

the sides of the rod at the bottom match the serrations on the two-piece hinged bearing basket. Rods and bearing baskets are machine fitted in sets and numbered to match.

The novel arrangement of parts is such that the pistons and connecting rods are in the same plane, as shown schematically in Fig. 19.18a. The pistons have the same stroke.

A second arrangement is shown schematically in Fig. 19.18d, where

322   Balancing Masses Reciprocating in a Plane

the connecting rod L_1 serves as a master rod to which the other connecting rod L_2 is pinned at B. Connecting rod L_2 is called an articulated rod. Figure 19.18e shows a photograph, furnished by the Cooper-Bessemer Corporation, which shows the actual construction.

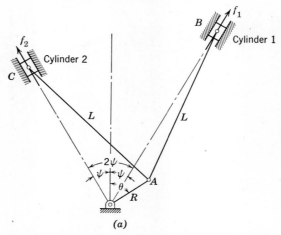

(a)

Fig. 19.19a.   V-engine analyzed for unbalanced inertia forces.

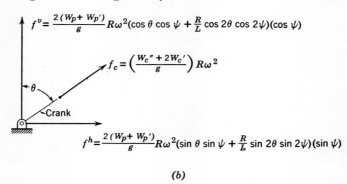

$$f^v = \frac{2(W_p + W_p')}{g} R\omega^2 (\cos\theta \cos\psi + \frac{R}{L}\cos 2\theta \cos 2\psi)(\cos\psi)$$

$$f_c = \left(\frac{W_c'' + 2W_c'}{g}\right) R\omega^2$$

$$f^h = \frac{2(W_p + W_p')}{g} R\omega^2 (\sin\theta \sin\psi + \frac{R}{L}\sin 2\theta \sin 2\psi)(\sin\psi)$$

(b)

Fig. 19.19b.   Vertical and horizontal components of the inertia forces, and the inertia force of the crank.

The usual simplifying assumption made in the analysis of the type shown in Fig. 19.18d is to treat it as shown in Fig. 19.18a.

The V-type engine shown in Fig. 19.19a will be analyzed for the unbalanced inertia forces. The cylinders are taken in the same plane. The connecting rods are pinned to the crank at A. It will be assumed that the reciprocating parts are identical and that the connecting rods are identical. The angle between the two cylinders is 2ψ, and

Two-Cylinder V-engine

the angle between the central plane and the crank is θ. The inertia force of each piston is

$$f_1 = \frac{(W_p + W_p')}{g} R\omega^2 \left[\cos(\theta - \psi) + \frac{R}{L} \cos 2(\theta - \psi) \right]$$

$$f_2 = \frac{(W_p + W_p')}{g} R\omega^2 \left[\cos(\theta + \psi) + \frac{R}{L} \cos 2(\theta + \psi) \right]$$

If f_1 and f_2 are considered positive, the inertia forces will be directed away from the crank bearing, as shown in Fig. 19.19a.

The vertical components of f_1 and f_2 are both upwards and are

$$f_1{}^v = \frac{(W_p + W_p')}{g} R\omega^2 \left[\cos(\theta - \psi) + \frac{R}{L} \cos 2(\theta - \psi) \right] \cos \psi$$

$$f_2{}^v = \frac{(W_p + W_p')}{g} R\omega^2 \left[\cos(\theta + \psi) + \frac{R}{L} \cos 2(\theta + \psi) \right] \cos \psi$$

Noting that $\cos(x + y)$ may be written $\cos x \cos y - \sin x \sin y$, we may expand the equation above to read

$$f_1{}^v = \frac{(W_p + W_p')}{g} R\omega^2 \left[\cos \theta \cos^2 \psi + \sin \theta \sin \psi \cos \psi \right.$$
$$\left. + \frac{R}{L} \cos 2\theta \cos 2\psi \cos \psi + \frac{R}{L} \sin 2\theta \sin 2\psi \cos \psi \right]$$

$$f_2{}^v = \frac{(W_p + W_p')}{g} R\omega^2 \left[\cos \theta \cos^2 \psi - \sin \theta \sin \psi \cos \psi \right.$$
$$\left. + \frac{R}{L} \cos 2\theta \cos 2\psi \cos \psi - \frac{R}{L} \sin 2\theta \sin 2\psi \cos \psi \right]$$

The total vertical inertia force of the reciprocating weights is $f_1{}^v + f_2{}^v = f^v$:

$$f^v = f_1{}^v + f_2{}^v = \frac{(W_p + W_p')}{g} R\omega^2 \left[2 \cos \theta \cos^2 \psi \right.$$
$$\left. + \frac{2R}{L} \cos 2\theta \cos 2\psi \cos \psi \right]$$

$$= 2 \frac{(W_p + W_p')}{g} R\omega^2 \left(\cos \theta \cos \psi + \frac{R}{L} \cos 2\theta \cos 2\psi \right)(\cos \psi) \quad (1)$$

The horizontal components of the inertia forces of the reciprocating weights are

$$f_1^h = \frac{(W_p + W_p')}{g} R\omega^2 \left[\cos(\theta - \psi) + \frac{R}{L}\cos 2(\theta - \psi)\right]\sin\psi$$

$$f_2^h = \frac{(W_p + W_p')}{g} R\omega^2 \left[\cos(\theta + \psi) + \frac{R}{L}\cos 2(\theta + \psi)\right]\sin\psi$$

f_1^h is to the right; f_2^h is to the left. The *net* force to the *right* is $f_1^h - f_2^h = f^h$. Expanding, and collecting terms, we obtain f^h:

$$f^h = 2\frac{(W_p + W_p')}{g} R\omega^2 \left(\sin\theta\sin\psi + \frac{R}{L}\sin 2\theta\sin 2\psi\right)(\sin\psi) \quad (2)$$

If f^h is positive, it is the net force to the right; if f^h is negative, it is the net force to the left.

The result of the analysis so far shows that the inertia forces of the pistons may be added to give the components of equations 1 and 2. At the same time, the inertia force of the crank is present, and may be expressed by

$$f_c = \frac{(W_c'' + 2W_c')}{g} R\omega^2$$

where W_c'' is the weight of the crank considered concentrated at R, and $2W_c'$ is the portion of the weight of the two connecting rods which may be considered as concentrated at the crank pin. The inertia force of the crank is directed along the crank.

The summary of the resultant forces in the horizontal and vertical directions is shown in Fig. 19.19b, as well as the inertia force of the crank.

Example. This numerical example shows the effect of various counterbalances used in a two-cylinder V-engine:

Speed of crank	1200 rpm
Crank length	3 in. $= R$
Connecting rod length	15 in. $= L$
2ψ	60°
Weight of a piston	8 lb $= W_p$
Weight of crank at 3 in.	4 lb $= W_c''$
Weight of a connecting rod at the crank pin	12 lb $= W_c'$
Weight of a connecting rod at the piston pin	6 lb $= W_p'$

Two-Cylinder V-engine 325

The calculations for the data given are

$$f^v = 2\frac{(W_p + W_p')}{g} R\omega^2 \left[\cos\theta \cos\psi + \frac{R}{L}\cos 2\theta \cos 2\psi\right](\cos\psi)$$

$$= 2\frac{(8+6)}{32.2}\frac{3}{12}\left[\frac{(1200)(2\pi)}{60}\right]^2$$

$$\left[\cos\theta\cos 30 + \frac{3/12}{15/12}(\cos 2\theta)(\cos 60)\right](\cos 30)$$

$$= 2580\cos\theta + 300\cos 2\theta \text{ lb}$$

$$f^h = 2\frac{(W_p + W_p')}{g} R\omega^2 \left[\sin\theta \sin\psi + \frac{R}{L}\sin 2\theta \sin 2\psi\right](\sin\psi)$$

$$= 2\frac{(8+6)}{32.2}\frac{3}{12}\left[\frac{(1200)(2\pi)}{60}\right]^2$$

$$\left[\sin\theta\sin 30 + \frac{3/12}{15/12}\sin 2\theta \sin 60\right](\sin 30)$$

$$= 860\sin\theta + 300\sin 2\theta \text{ lb}$$

$$f_c = \frac{(W_c'' + 2W_c')}{32.2} R\omega^2$$

$$= \left[\frac{4 + 2(12)}{32.2}\right]\frac{3}{12}\left[\frac{(1200)(2\pi)}{60}\right]^2$$

$$= 3420 \text{ lb}$$

A polar diagram, plotted in a fashion similar to that for a single-cylinder engine, is given in Fig. 19.19c. There are four curves shown for different conditions of counterbalance:

Curve I

No counterbalance is used for curve I. The curve shows the variation of the shaking force due to $f^v \leftrightarrow f^h \leftrightarrow f_c$.

Curve II

The variation of shaking force with a counterbalance to balance the crank only is shown in curve II.

$$f_{cb} = \frac{(W_c'' + 2W_c')}{32.2} R\omega^2$$

$$= \left[\frac{4 + 2(12)}{32.2}\right]\frac{3}{12}\left[\frac{(1200)(2\pi)}{60}\right]^2$$

$$= 3420 \text{ lb}$$

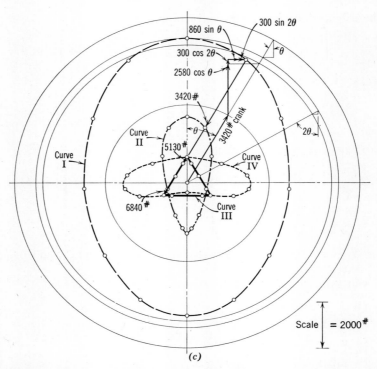

Fig. 19.19c. Polar shaking force curves. Curve I. No counterbalance. Curve II. Crank only counterbalanced. Curve III. Counterbalance equal to the sum of the equivalent weight of the crank and the equivalent weight of one piston. Curve IV. Counterbalance equal to the sum of the equivalent weight of the crank and twice the equivalent weight of one piston.

Curve III

The variation of shaking force with a counterbalance equal to the equivalent weight of the crank plus the equivalent weight of one piston is shown in curve III.

$$f_{cb} = \left[\frac{(W_c'' + 2W_c') + (W_p + W_p')}{32.2}\right] R\omega^2$$

$$= \left[\frac{4 + 2(12) + (8 + 6)}{32.2}\right] \frac{3}{12} \left[\frac{(1200)(2\pi)}{60}\right]^2$$

$$= 5130 \text{ lb}$$

Curve IV

The variation of shaking force with a counterbalance equal to the equivalent weight of the crank plus twice the equivalent weight of one piston is shown in curve IV.

$$f_{cb} = \left[\frac{(W_c'' + 2W_c') + 2(W_p + W_p')}{g}\right] R\omega^2$$

$$= \left[\frac{4 + 2(12) + 2(8 + 6)}{32.2}\right] \frac{3}{12} \left[\frac{(1200)(2\pi)}{60}\right]^2$$

$$= 6840 \text{ lb}$$

19.10 Ninety-degree V-engine

The equations for the vertical and horizontal unbalanced inertia forces of the reciprocating weights of a V-engine, with no counterbalancing weight, are repeated here:

$$f^v = 2\frac{(W_p + W_p')}{g} R\omega^2 \left(\cos\theta \cos\psi + \frac{R}{L} \cos 2\theta \cos 2\psi\right)(\cos\psi)$$

$$f^h = 2\frac{(W_p + W_p')}{g} R\omega^2 \left(\sin\theta \sin\psi + \frac{R}{L} \sin 2\theta \sin 2\psi\right)(\sin\psi)$$

For the special case of a 90-degree V-engine, $2\psi = 90°$, or $\psi = 45°$. The components of the primary inertia forces for this case are

$$f^v = 2\frac{(W_p + W_p')}{g} R\omega^2 \frac{\cos\theta}{2}$$

$$f^h = 2\frac{(W_p + W_p')}{g} R\omega^2 \left[\frac{\sin\theta}{2} + \left(\frac{R}{L} \sin 2\theta\right)\left(\frac{\sqrt{2}}{2}\right)\right]$$

The resultant of the primary force components is found by

$$\sqrt{\left[2\frac{(W_p + W_p')}{g} R\omega^2 \frac{\cos\theta}{2}\right]^2 + \left[2\frac{(W_p + W_p')}{g} R\omega^2 \frac{\sin\theta}{2}\right]^2}$$

which is equal to $\frac{(W_p + W_p')}{g} R\omega^2$. This resultant force is directed outward along the crank, and its magnitude is independent of the position of the crank. Consequently, the primary forces may be balanced by a rotating weight on the crank. If the balancing weight is located at a distance R from the crank bearing, the magnitude will be

328   Balancing Masses Reciprocating in a Plane

$(W_p + W_p')$. Or the magnitude of counterbalancing weight to balance out the primary forces is the sum of the weight of one piston plus the portion of the weight of one connecting rod considered concentrated at the piston pin, located at the crank radius.

There is left the secondary force component in the horizontal direction,

$$\sqrt{2}\,\frac{(W_p + W_p')}{g}\,R\omega^2 \left(\frac{R}{L}\right)\sin 2\theta$$

which is still unbalanced. It is not possible to balance this force by a single rotating counterbalancing weight.

In conclusion, then, the primary force components for a 90-degree V-engine may be balanced by a single counterbalancing weight, but the secondary force component cannot be balanced by a single counterbalancing weight. (The crank and the weight of the connecting rods assumed concentrated at the crank pin can be balanced by a rotating weight.)

19.11   W-engine

Figure 19.20 shows a W-type engine. The engine is essentially a V-engine with another piston added in the central plane. The inertia

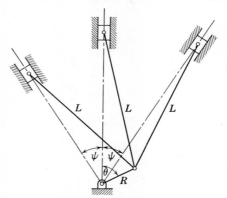

Fig. 19.20.   W-engine.

force components may be found by adding to the inertia force components of a two-cylinder V-engine the inertia force of the piston in the central plane. It will be assumed that the pistons are alike, the connecting rods are alike also, and that the crank and that portion of the connecting rod considered concentrated at the crank pin are balanced. θ is the angle between the crank and central plane; 2ψ is the total V-angle.

The total vertical force is

$$f^v = 2\frac{(W_p + W_p')}{g} R\omega^2 \left(\cos\theta \cos\psi + \frac{R}{L}\cos 2\theta \cos 2\psi\right)(\cos\psi)$$

$$+ \frac{(W_p + W_p')}{g} R\omega^2 \left(\cos\theta + \frac{R}{L}\cos 2\theta\right)$$

or $f^v = \frac{(W_p + W_p')}{g} R\omega^2$

$$\left[\cos\theta(2\cos^2\psi + 1) + \frac{R}{L}\cos 2\theta(2\cos 2\psi + 1)\right]$$

The horizontal force is

$$f^h = 2\frac{(W_p + W_p')}{g} R\omega^2 \left(\sin\theta \sin\psi + \frac{R}{L}\sin 2\theta \sin 2\psi\right)(\sin\psi)$$

It is not possible to balance either the primary force or secondary force components with a single rotating counterweight. Modification of the total unbalance may be made with a rotating counterweight, but such an analysis will not be given for the W-engine in this book.

19.12 X-type engine

Figure 19.21a shows four cylinders arranged in the form of an X to give a so-called X-type engine. The crank is shown at an angle θ from the vertical. The cylinders 1 and 2 are set at an angle of 2ψ apart, or each cylinder forms an angle of ψ with the vertical. Also, the angle which cylinders 3 and 4 makes with the vertical is ψ, as shown.

It will be assumed that the crank and the weight of the connecting rod which is assumed as concentrated at the crank pin are balanced by a rotating weight. With the assumption that the connecting rods are identical, it is desired to find the amount of unbalance of the inertia forces due to the reciprocating parts.

The inertia forces of pistons 1, 2, 3, and 4 along the line of motion of the respective pistons are

$$f_1 = \frac{(W_p + W_p')}{g} R\omega^2 \left[\cos(\theta - \psi) + \frac{R}{L}\cos 2(\theta - \psi)\right]$$

$$f_2 = \frac{(W_p + W_p')}{g} R\omega^2 \left[\cos(\theta + \psi) + \frac{R}{L}\cos 2(\theta + \psi)\right]$$

Balancing Masses Reciprocating in a Plane

$$f_3 = \frac{(W_p + W_p')}{g} R\omega^2 \left[\cos[180 - (\theta - \psi)] + \frac{R}{L} \cos 2[180 - (\theta - \psi)] \right]$$

$$f_4 = \frac{(W_p + W_p')}{g} R\omega^2 \left[\cos[180 - (\theta + \psi)] + \frac{R}{L} \cos 2[180 - (\theta + \psi)] \right]$$

The vector addition of the inertia forces will be simplified by considering the inertia force resultant of f_1 and f_3 first, and then considering the inertia force resultant of f_2 and f_4 next. The two resultants will then be combined into a single resultant.

Examination of pistons 1 and 3 reveals that they form an opposed engine which has already been discussed on page 312. It was found that the resultant inertia force, along the line of motion of the pistons, was the sum of the primary forces. Here the resultant would be

$$f_1 + f_3 = 2 \frac{(W_p + W_p')}{g} R\omega^2 \cos(\theta - \psi)$$

If the resultant is positive, it will be directed upwards to the right. The equation above may be considered to be derived by adding f_1 and f_3, noting that the positive direction of inertia force, f_1, is upwards to the right, and the positive direction of inertia force, f_2, is downwards to the left.

In a similar manner, the sum of the inertia forces of pistons 2 and 4 is

$$f_2 + f_4 = 2 \frac{(W_p + W_p')}{g} R\omega^2 \cos(\theta + \psi)$$

If the resultant is positive, it will be directed upwards to the left.

Figure 19.21b shows the resultant forces so far, with the resultant force broken up into the components in the horizontal and vertical directions also shown.

The vertical components add to

$$f^v = 2 \frac{(W_p + W_p')}{g} R\omega^2 \cos(\theta - \psi) \cos \psi$$

$$+ 2 \frac{(W_p + W_p')}{g} R\omega^2 \cos(\theta + \psi) \cos \psi$$

$$= 4 \frac{(W_p + W_p')}{g} R\omega^2 \cos \theta \cos^2 \psi$$

X-type Engine

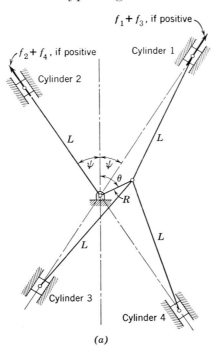

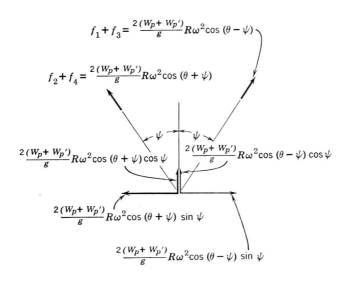

Fig. 19.21. X-engine, with the components of the inertia forces.

The net difference to the right of the horizontal components is

$$f^h = 2\frac{(W_p + W_p')}{g} R\omega^2 \cos(\theta - \psi) \sin \psi$$

$$- 2\frac{(W_p + W_p')}{g} R\omega^2 \cos(\theta + \psi) \sin \psi$$

$$= 4\frac{(W_p + W_p')}{g} R\omega^2 \sin\theta \sin^2 \psi$$

The total resultant inertia force R is found from $[(f^v)^2 + (f^h)^2]^{1/2}$, which is, for this case, equal to

$$R = 4\frac{(W_p + W_p')}{g} R\omega^2 [(\cos\theta \cos^2\psi)^2 + (\sin\theta \sin^2\psi)^2]^{1/2}$$

For a general value of ψ, the resultant inertia force cannot be balanced by a single rotating weight. However, if $\psi = 45°$, the expression reduces to

$$R = 2\frac{(W_p + W_p')}{g} R\omega^2$$

which can be balanced by a single balancing counterweight of $2(W_p + W_p')$ located at a radius of R. Or, expressing it differently, an X-engine, with an angle of 90 degrees between cylinders, can be balanced completely by a single rotating weight at a radius equal to the crank radius, the weight being twice the reciprocating weight of one piston, including the weight of the connecting rod considered concentrated at the piston pin. The weight to balance the crank and the weight of the connecting rod considered concentrated at the crank pin is not included.

It is interesting to note that an X-engine with 90 degrees between cylinders is a special case of the radial engine to be discussed in the next section.

19.13 Radial engine

Figure 19.22a shows a schematic arrangement of parts of a five-cylinder radial-type engine. Articulated rods are connected to the master rod. The usual simplification made in the analysis of radial engines is to consider the system replaced by that shown in Fig. 19.22b, where it is assumed that all the connecting rods are connected to the same crank pin. It will also be assumed that all the connecting rods are identical, and all the pistons are identical. The assumption

Radial Engine

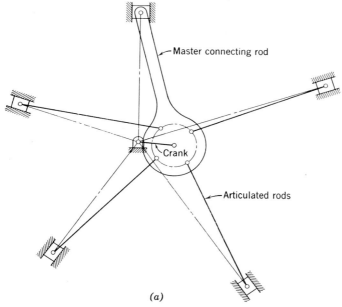

Fig. 19.22a. Radial engine with the master rod and articulated rods.

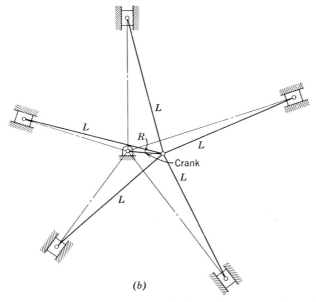

Fig. 19.22b. Assumed arrangement for analysis, with all connecting rods assumed connected to the same crank pin.

334   Balancing Masses Reciprocating in a Plane

that the connecting rods are identical and connected to the same crank pin means that the strokes of the pistons are assumed the same, which is not the case in the actual engine. However, the simplification in analysis is considerable, with good practical results.

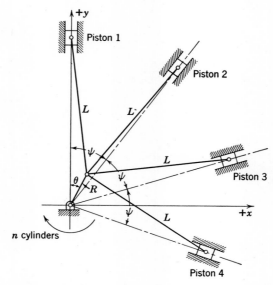

Fig. 19.23. General case of a radial engine with n cylinders.

Consider the general case of a radial engine of n cylinders, Fig. 19.23, with the angle between cylinders equal to ψ. The angle ψ is equal to $360°/n$. For instance, for a five-cylinder engine, $\psi = 360°/5 = 72°$. The inertia force of each piston is

$$\text{Piston 1:} \quad f_1 = \frac{(W_p + W_p')}{g} R\omega^2 \left(\cos \theta + \frac{R}{L} \cos 2\theta \right)$$

which may be written, without any change, as

$$f_1 = \frac{(W_p + W_p')}{g} R\omega^2 \left[\cos (\theta - 0) + \frac{R}{L} \cos 2(\theta - 0) \right]$$

The reason for the manner of expression above will be apparent when the inertia forces of the other pistons are determined and are found to be in the same form.

Radial Engine

Piston 2: $f_2 = \dfrac{(W_p + W_p')}{g} R\omega^2 \left[\cos(\theta - \psi) + \dfrac{R}{L} \cos 2(\theta - \psi) \right]$

Piston 3: $f_3 = \dfrac{(W_p + W_p')}{g} R\omega^2 \left[\cos(\theta - 2\psi) + \dfrac{R}{L} \cos 2(\theta - 2\psi) \right]$

Piston 4: $f_4 = \dfrac{(W_p + W_p')}{g} R\omega^2 \left[\cos(\theta - 3\psi) + \dfrac{R}{L} \cos 2(\theta - 3\psi) \right]$

. . .

Piston n: $f_n = \dfrac{(W_p + W_p')}{g} R\omega^2 \Big[\cos(\theta - (n-1)\psi)$

$+ \dfrac{R}{L} \cos 2(\theta - (n-1)\psi) \Big]$

If the inertia force for a particular piston is positive, it is directed away from the crank bearing outward along the line of motion; if it is negative, it is directed towards the crank bearing.

The resultant inertia force may be found in the quickest way by taking components and adding components. Thus, using the x- and y-coordinates shown in Fig. 19.23, the vertical components of the inertia forces of the pistons are

$$f_1^v = f_1 \cos 0$$
$$f_2^v = f_2 \cos \psi$$
$$f_3^v = f_3 \cos 2\psi$$
$$f_4^v = f_4 \cos 3\psi$$

. . .

$$f_n^v = f_n \cos(n-1)\psi$$

Similarly, the horizontal components are

$$f_1^h = f_1 \sin 0$$
$$f_2^h = f_2 \sin \psi$$
$$f_3^h = f_3 \sin 2\psi$$
$$f_4^h = f_4 \sin 3\psi$$

. . .

$$f_n^h = f_n \sin(n-1)\psi$$

If horizontal and vertical components of forces are taken, we can show the following results:

The vertical components of the primary forces add up to

$$f^v_{\text{primary}} = \frac{n}{2} \frac{(W_p + W_p')}{g} R\omega^2 \cos \theta$$

The horizontal components of the primary forces add up to

$$f^h_{\text{primary}} = \frac{n}{2} \frac{(W_p + W_p')}{g} R\omega^2 \sin \theta$$

The vertical components of the secondary forces add up to zero; the horizontal components of the secondary forces also add up to zero.

The total shaking force is $\frac{n}{2} \frac{(W_p + W_p')}{g} R\omega^2$, which is directed outward along the crank. The shaking force can be balanced by a single counterweight equal to $\frac{n}{2}(W_p + W_p')$, or one half the total weight of all the pistons and the connecting rods considered concentrated at the piston pins, located at the crank radius.

PROBLEMS

19.1. The following data are given for a single-cylinder diesel engine, with the engine mounted horizontally:

Speed = 1200 rpm.
Stroke = 6 in.
Length of connecting rod = 12 in.
Distance from the center of gravity of the connecting rod to the crank pin = 4 in.
Equivalent unbalanced weight of the crank at a 3 in. radius = 6 lb
Weight of the piston = 7 lb
Weight of the connecting rod = 15 lb

Determine the magnitude of the shaking force for an angle of 150 degrees between the crank and the line of motion of the piston if a counterbalancing weight equal to the sum of the rotating weight concentrated at the crank pin and the reciprocating weight concentrated at the piston pin is used. Determine the value analytically, and check the result by a graphical solution of velocity, acceleration, and force analysis. What is the direction of the shaking force?

Consider the counterbalancing weight as located at the crank radius.

19.2. For the data of the single-cylinder engine given in problem 19.1, draw the polar shaking force diagram, in 15-degree intervals, for the following cases of counterbalancing weights:

(a) No counterbalancing weights.

(b) A counterbalancing weight equal to the sum of the weight of the crank at the crank radius, the weight of the connecting rod considered concentrated at the crank pin, the weight of the piston, and the weight of the connecting rod considered concentrated at the piston pin.

Problems

(c) A counterbalancing weight equal to the sum of the weight of the crank at the crank radius, the weight of the connecting rod considered concentrated at the crank pin, and one half the weight of the piston and connecting rod considered concentrated at the piston pin.

19.3. For the data of the single-cylinder engine given in problem 19.1, determine the necessary counterweight to give:
  (a) The smallest horizontal shaking force.
  (b) The smallest vertical shaking force.
Consider the counterbalancing weight as located at the crank radius.

19.4. For the data of the single-cylinder engine given in problem 19.1, draw the polar shaking force diagram, in 15-degree intervals, the diagram to be made considering only the primary force of the piston and the inertia force of the crank. Draw the polar shaking force diagram for the three cases of counterweights:
  (a) A counterweight is used such that the crank and portion of the connecting rod considered concentrated at the crank pin are balanced.
  (b) A counterweight is used such that the weight of the counterweight at the crank radius is equal to the weight of the crank plus the weight of the connecting rod considered concentrated at the crank pin plus the equivalent piston weight.
  (c) A counterweight is used such that the weight of the counterweight at the crank radius is equal to the weight of the crank plus the weight of the connecting rod considered concentrated at the crank pin plus one half the equivalent weight of the piston.

Show that in part c the resultant polar shaking force curve is a circle with the shaking force of constant magnitude. Show also that the shaking force vector rotates at the same speed as the crankshaft, but in the opposite direction. Show that the arrangement given in Fig. 19.16 of the text will balance the crank and primary force of the piston inasmuch as the counterweight on gear B rotates at the same speed but in the opposite direction to that of the crankshaft.

19.5. Determine an arrangement, using only two gears rotating at twice the crankshaft speed, that will balance the secondary force of a single-cylinder engine. How must the counterweights compare with the equivalent weight of the piston to effect balance of the secondary force?

19.6. Refer to Fig. 19.7 of the text. The equivalent weight of piston $A = 12$ lb, $R = 5$ in., $L = 17\frac{1}{2}$ in., $R' = 3$ in. The crank as well as the weight of the connecting rods considered concentrated at the crank pins are balanced. Determine the value of L' and determine the necessary weight of each dummy piston for balance of the inertia forces.

19.7. Refer to Fig. 19.8 of the text. Determine the magnitude of the shaking force of the Gobron-Brillé engine for one position of the crank: $\theta = 90°$, for a crank speed of 1800 rpm.

The equivalent weight of each piston, including the effect of the connecting rod, is 8 lb. Assume that $R_1 = 1\frac{1}{2}$ in., $R_2 = 2$ in., $L_1 = 6\frac{1}{2}$ in., and $L_2 = 17$ in.

The equivalent unbalanced weight of each crank, including the effect of the connecting rods at the crank pins, is balanced by counterweights on the crankshaft.

What is the equivalent stroke for the two pistons?

19.8. Refer to Fig. 19.10a and b of the text. If the equivalent weight of each piston is $5\frac{1}{2}$ lb, and the unbalanced equivalent weight of each crank, with the effect of one connecting rod, is 4 lb at the crank radius, determine the magnitude of the total shaking force for one position of the crank, $\theta = 30°$, for a crank speed of 1800 rpm. $R = 2\frac{1}{2}$ in., and $L = 9$ in.

Determine also the various couples due to the shaking forces. Specify the planes in which the couples are applied.

19.9. Refer to Fig. 19.11a and b of the text. The equivalent weight of each piston is $5\frac{1}{2}$ lb, including the effect of a connecting rod. The crank, with the effect of the connecting rods, is balanced. $R = 2\frac{1}{2}$ in. $L = 9$ in. Crank speed $= 1800$ rpm.

What is the total unbalance for (a) $\theta = 0°$? For (b) $\theta = 60°$? For (c) $\theta = 270°$?

19.10. A two-cylinder V-engine is operated with a crank speed of 3000 rpm. The stroke of each piston is 6 in. Each connecting rod is 12 in. The weight of each piston is 6 lb. The V-angle is 60 degrees. Center of gravity of each connecting rod is 4 in. from the crank pin. Weight of each rod is 9 lb. Weight of the crank at the crank radius is 4 lb.

Draw the polar shaking force curves, for 15-degree intervals of the crank, for the following cases of counterbalancing weights:

(a) No counterbalance.

(b) A counterbalance with only the crank and portion of the connecting rods considered concentrated at the crank pin balanced.

(c) A counterbalance at the crank radius equal to the sum of the equivalent unbalanced weight of the crank at the crank radius and the equivalent weight of one piston.

(d) A counterbalance at the crank radius equal to the equivalent unbalanced weight of the crank and twice the equivalent weight of one piston.

19.11. Determine the resultant shaking force in a two-cylinder V-type motorcycle engine (the included angle between the cylinders is 90 degrees), given the following data:

Reciprocating weight at each piston pin $= 4$ lb.

Total equivalent rotating weight at crankpin $= 6$ lb.

Rpm of engine $= 3000$.

$R = 3$ in. $L = 12$ in.

No counterbalance weight is used.

The crank is 30 degrees from cylinder number 1 (the right-hand cylinder) and 60 degrees from cylinder number 2.

19.12. A two-cylinder opposed piston motorcycle engine is running at 4200 rpm. The total reciprocating weight in each cylinder is 3 lb, and the two connecting rods (each 7 in. long) are joined to a common crank having a 2-in. throw.

Determine the magnitude and the direction of the total resultant inertia force of the system when the crank angle of one of the cylinders is 30 degrees. Show a sketch of the engine upon which the direction of this force is shown. The crank is counterbalanced.

19.13. A two-cylinder opposed piston motor boat engine is running at 4200 rpm. The total reciprocating weight in each cylinder is 3 lb, and the two connecting rods (each 7 in. long) are attached to 2-in. crank throws which are 180 degrees apart.

Determine the magnitude and direction of the total resultant unbalanced force when the crank angle of one of the cylinders is 30 degrees. Show a sketch of the engine upon which the direction of this force is shown.

The crank is balanced.

19.14 (Fig. 19.14). The two cranks are of length R, and the connecting rods of length L. Analyze this combination to determine the magnitude and direction of the resultant primary forces in terms of W, R, L, ω (speed of shaft), and θ (angle of shaft rotation measured as indicated). State clearly what your results mean in regard to the manner in which the force varies during the rotation of the shaft.

Show how the resultant primary force may be balanced.

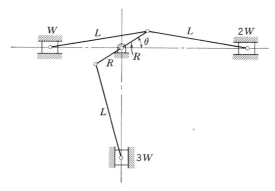

Fig. P-19.14.

19.15. It is proposed to construct an engine air compressor unit having two power cylinders and one compressor cylinder. The two power cylinders are to be 180 degrees apart and will share a common crank. The compressor cylinder axis is to be at 90 degrees with the axis of the power cylinders, and the compressor crank at 180 degrees with the power crank. The reciprocating weight for each power cylinder is to be W and for the compressor cylinder $2W$. The two cranks are of length R; the connecting rods are of length L.

Analyze this combination to determine the magnitude and the direction of the resultant of the primary forces in terms of W, R, L, ω (speed of shaft), and θ (angle of shaft rotation measured from any convenient reference). State clearly in words what your results mean in regard to the manner in which the force varies during the rotation of the shaft.

Show how the resultant primary force may be balanced.

19.16 (Fig. 19.16). A four-cylinder radial engine with cylinders spaced at 90 degrees is shown in the sketch. All connecting rods are the same length, L, and the reciprocating weight, W, is the same for each cylinder.

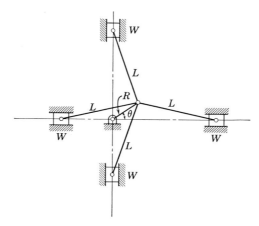

Fig. P-19.16.

340 Balancing Masses Reciprocating in a Plane

By analysis of the engine, determine the magnitude and the direction of the resultant of the primary forces in terms of W, R, L, ω (shaft speed), and θ (angle of shaft rotation measured as indicated.) State clearly what your results mean in regard to the manner in which the force varies during rotation of the shaft.

19.17. Determine the resultant shaking force in a three-cylinder radial engine (without counterbalancing), given the following data:

Reciprocating weight at each piston pin = 3 lb.
Total equivalent rotating weight at crank pin = 8 lb.
Rpm of engine = 2800.
R = 2 in.
L = 8 in.

Cylinder number one is vertical, and the piston in this cylinder is at the head end dead center position.

19.18. The following data apply to a five-cylinder radial engine:

Speed = 1250 rpm.
Stroke = 6.5 in.
Effective connecting rod length = 16.5 in.
Distance from center of gravity of connecting rod to crank pin = 4.8 in.
Crank weight = 52 lb.
Radius to center of gravity of crank = 1.5 in.
Weight of piston = 8.3 lb.
Weight of connecting rod = 14.7 lb.

Would a counterweight on the crankshaft improve the balance of the engine? What size counterweight would you suggest using, and where should it be placed? Explain.

CHAPTER 20

Balancing Masses Reciprocating in Several Planes

The engines analyzed in Chapter 19 were restricted, in general, to those wherein the pistons and connecting rods move in the same plane. This chapter will be devoted to the determination of the kind and the amount of unbalance in multicylinder engines where the masses reciprocate in several planes. The possible means of balancing engines by auxiliary arrangements which can be and are used will be discussed. Firing order and uniformity of turning effort will be discussed for two-cycle and four-cycle engines.

20.1 Equations for the total shaking force in a multicylinder in-line engine

Figure 20.1a shows the schematic arrangement of an engine with n cylinders. The total shaking force for the engine is to be found. The angle which crank 1 makes with the vertical is θ_1. The angle between cranks is fixed, since the crankshaft is rigid. The angle between crank 2 and crank 1 is expressed by ϕ_2, that between cranks 3 and 1 is ϕ_3, and so on for all the cranks. The angle between crank n and crank 1 is ϕ_n. Figure 20.1b is an isometric sketch of the crankshaft.

It will be assumed in the analysis of all the engines in this chapter that the crankshaft and the weight of the connecting rods considered as concentrated at the crank pins are balanced by the methods given for balancing rotating weights. Thus the balance of the reciprocating weights only will be considered.

The inertia force of each piston is given by

$$f_1 = \frac{(W_p + W_p')}{g} R\omega^2 \left[\cos(\theta_1 + \phi_1) + \frac{R}{L}\cos 2(\theta_1 + \phi_1) \right]$$

$$f_2 = \frac{(W_p + W_p')}{g} R\omega^2 \left[\cos(\theta_1 + \phi_2) + \frac{R}{L}\cos 2(\theta_1 + \phi_2) \right]$$

342 Balancing Masses Reciprocating in Several Planes

$$f_3 = \frac{(W_p + W_p')}{g} R\omega^2 \left[\cos(\theta_1 + \phi_3) + \frac{R}{L} \cos 2(\theta_1 + \phi_3) \right]$$

$$\vdots \qquad \vdots \qquad \vdots \qquad \vdots$$

$$f_n = \frac{(W_p + W_p')}{g} R\omega^2 \left[\cos(\theta_1 + \phi_n) + \frac{R}{L} \cos 2(\theta_1 + \phi_n) \right]$$

where W_p, the reciprocating weight of the parts of each cylinder, and W_p', the portion of the connecting rod considered concentrated at the

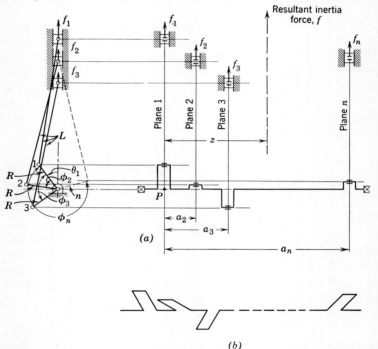

Fig. 20.1. General arrangement of a multicylinder in-line engine showing inertia forces.

piston pin, are assumed to be the same for each cylinder. Note that $\phi_1 = 0$, but is included for symmetry of the equations. The sum of the inertia forces is obtained by algebraic addition and can be expressed by

$$f = \frac{(W_p + W_p')}{g} R\omega^2 \left[\sum_{n=1}^{n=n} \cos(\theta_1 + \phi_n) + \frac{R}{L} \sum_{n=1}^{n=n} \cos 2(\theta_1 + \phi_n) \right]$$

Conditions for Balance of the Inertia Forces

The expression above may be simplified by expansion of the trigonometric functions:

$$\cos(\theta_1 + \phi_1) = \cos\theta_1 \cos\phi_1 - \sin\theta_1 \sin\phi_1$$
$$\cos(\theta_1 + \phi_2) = \cos\theta_1 \cos\phi_2 - \sin\theta_1 \sin\phi_2$$
$$\cos(\theta_1 + \phi_3) = \cos\theta_1 \cos\phi_3 - \sin\theta_1 \sin\phi_3$$
$$\vdots$$
$$\cos(\theta_1 + \phi_n) = \cos\theta_1 \cos\phi_n - \sin\theta_1 \sin\phi_n$$

or

$$\sum_{n=1}^{n=n} \cos(\theta_1 + \phi_n) = [\cos\theta_1(\cos\phi_1 + \cos\phi_2 + \cos\phi_3 + \cdots + \cos\phi_n) - \sin\theta_1(\sin\phi_1 + \sin\phi_2 + \sin\phi_3 + \cdots + \sin\phi_n)]$$

$$= \cos\theta_1 \sum_{n=1}^{n=n} \cos\phi_n - \sin\theta_1 \sum_{n=1}^{n=n} \sin\phi_n$$

In a similar fashion, it can be shown that

$$\sum_{n=1}^{n=n} \cos 2(\theta_1 + \phi_n) = \cos 2\theta_1 \sum_{n=1}^{n=n} \cos 2\phi_n - \sin 2\theta_1 \sum_{n=1}^{n=n} \sin 2\phi_n$$

Summing up, we can express the total unbalanced inertia force as

$$f = \frac{(W_p + W_p')}{g} R\omega^2 \left[\cos\theta_1 \sum_{n=1}^{n=n} \cos\phi_n - \sin\theta_1 \sum_{n=1}^{n=n} \sin\phi_n \right.$$
$$\left. + \frac{R}{L} \cos 2\theta_1 \sum_{n=1}^{n=n} \cos 2\phi_n - \frac{R}{L} \sin 2\theta_1 \sum_{n=1}^{n=n} \sin 2\phi_n \right] \quad (1)$$

20.2 Conditions for balance of the inertia forces

If the inertia forces are to be balanced not only for one position of the crank but for all positions of the crank, that is, if the total inertia force is not to be a function of the angle θ_1, it is necessary that the following relations hold:

I. $\left. \begin{array}{l} \sum_{n=1}^{n=n} \cos\phi_n = 0 \\ \sum_{n=1}^{n=n} \sin\phi_n = 0 \end{array} \right\}$ primary forces balanced

II. $\left. \begin{array}{l} \sum_{n=1}^{n=n} \cos 2\phi_n = 0 \\ \sum_{n=1}^{n=n} \sin 2\phi_n = 0 \end{array} \right\}$ secondary forces balanced

Balancing Masses Reciprocating in Several Planes

The above quantities are labeled for primary and secondary forces, inasmuch as each is separated to facilitate the analysis where unbalance exists and auxiliary balancing arrangements are used. Both equations in I above must be satisfied for balance of the primary forces, and both equations in II must be satisfied for balance of the secondary forces.

20.3 Location of an unbalanced shaking force

The location of the resultant inertia force can be found from the fact that the moment of the resultant force about any point is the same as the moment of the components about the same point. The usual practice is to take moments about a point in the plane of crank 1, as point P in Fig. 20.1a. Therefore, the moment of the components, that is, the moment of f_1, f_2, f_3, etc. about plane 1 is

$$M = f_1 a_1 + f_2 a_2 + f_3 a_3 + \cdots + f_n a_n$$

or
$$\begin{aligned}M = \frac{(W_p + W_p')}{g} R\omega^2 \Bigg(& a_1 \left[\cos(\theta_1 + \phi_1) + \frac{R}{L}\cos 2(\theta_1 + \phi_1)\right] \\ + & a_2 \left[\cos(\theta_1 + \phi_2) + \frac{R}{L}\cos 2(\theta_1 + \phi_2)\right] \\ + & a_3 \left[\cos(\theta_1 + \phi_3) + \frac{R}{L}\cos 2(\theta_1 + \phi_3)\right] + \cdots \\ + & a_n \left[\cos(\theta_1 + \phi_n) + \frac{R}{L}\cos 2(\theta_1 + \phi_n)\right] \Bigg)\end{aligned}$$

The equation above can be further simplified to give

$$\begin{aligned}M = \frac{(W_p + W_p')}{g} R\omega^2 \\ \bigg[\cos\theta_1(a_1\cos\phi_1 + a_2\cos\phi_2 + a_3\cos\phi_3 + \cdots + a_n\cos\phi_n) \\ - \sin\theta_1(a_1\sin\phi_1 + a_2\sin\phi_2 + a_3\sin\phi_3 + \cdots + a_n\sin\phi_n) \\ + \frac{R}{L}\cos 2\theta_1(a_1\cos 2\phi_1 + a_2\cos 2\phi_2 + a_3\cos 2\phi_3 + \cdots + a_n\cos 2\phi_n) \\ - \frac{R}{L}\sin 2\theta_1(a_1\sin 2\phi_1 + a_2\sin 2\phi_2 + a_3\sin 2\theta_3 + \cdots + a_n\sin 2\phi_n) \bigg]\end{aligned}$$

Resultant Effect

or $\quad M = \dfrac{(W_p + W_p')}{g} R\omega^2 \left[\cos\theta_1 \displaystyle\sum_{n=1}^{n=n} a_n \cos\phi_n \right.$

$\qquad\qquad - \sin\theta_1 \displaystyle\sum_{n=1}^{n=n} a_n \sin\phi_n + \dfrac{R}{L}\cos 2\theta_1 \displaystyle\sum_{n=1}^{n=n} a_n \cos 2\phi_n$

$\qquad\qquad\qquad\qquad \left. - \dfrac{R}{L}\sin 2\theta_1 \displaystyle\sum_{n=1}^{n=n} a_n \sin 2\phi_n \right] \quad (1)$

The location of the resultant force can now be found from $fz = M$, or $z = M/f$, where M is found from Eq. 1 above, and f is found from Eq. 1, Section 20.1.

Conditions for moments about plane 1 to be zero. If the moment of the inertia forces about plane 1 is to be zero for every position of the crank, or the moment is not to be a function of the crank angle θ_1, the following must hold:

$\left. \begin{array}{l} \displaystyle\sum_{n=1}^{n=n} a_n \cos\phi_n = 0 \\[1em] \displaystyle\sum_{n=1}^{n=n} a_n \sin\phi_n = 0 \end{array} \right\}$ moment of the primary forces about plane 1 $= 0$

$\left. \begin{array}{l} \displaystyle\sum_{n=1}^{n=n} a_n \cos 2\phi_n = 0 \\[1em] \displaystyle\sum_{n=1}^{n=n} a_n \sin 2\phi_n = 0 \end{array} \right\}$ moment of the secondary forces about plane 1 $= 0$

20.4 Resultant effect

There are four conditions which might arise in the analysis of the resultant force and moment about plane 1:

(1) The resultant force is zero and the moment about plane 1 is zero for any crank angle θ_1. This indicates complete balance.

(2) The resultant inertia force is zero, but the moment about plane 1 is not zero for any crank angle θ_1. This condition indicates unbalance, the unbalance being due to a couple. When such a condition exists in the engines to be analyzed, the letter C will be used instead of M to indicate a couple.

(3) The resultant inertia force is not zero and the moment about plane 1 is not zero for any crank angle θ_1. This indicates that the unbalance is due to a single resultant force, whose location can be found from $z = M/f$.

(4) The resultant force is not zero, but the moment of the inertia forces about plane 1 is zero, for any crank angle θ_1. This indicates that the resultant effect of unbalance is a single force in the reference plane.

Let us examine various engines for the unbalance which may be present.

20.5 Two-cylinder engine—cranks at 90 degrees

Figure 20.2 shows a two-cylinder engine with cranks at 90 degrees, the engine being the so-called quarter-crank engine. The arrangement

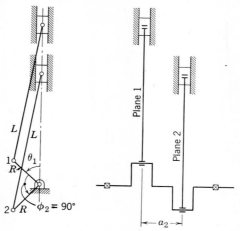

Fig. 20.2. Two-cylinders engine with cranks at 90 degrees.

of cylinders is that found on the common steam locomotive. The problem is to find the kind of unbalance and to determine if a simple balancing system can be used. The tabular form below will simplify the analysis:

ϕ	$\cos \phi$	$\sin \phi$	2ϕ	$\cos 2\phi$	$\sin 2\phi$	a	$a \cos \phi$	$a \sin \phi$	$a \cos 2\phi$	$a \sin 2\phi$
$\phi_1 = 0°$	1	0	0°	1	0	0	0	0	0	0
$\phi_2 = 90°$	0	1	180°	-1	0	a_2	0	a_2	$-a_2$	0
Sum	1	1	....	0	0	...	0	a_2	$-a_2$	0
	Primary forces unbalanced			Secondary forces balanced			Primary moments unbalanced		Secondary moments unbalanced	

The secondary forces only are balanced in this engine. The resultant effect of the primary forces and primary moments about plane 1

Two-Cylinder Engine—Cranks at 180 Degrees 347

is a single resultant force. The secondary forces cause a couple, since the secondary forces are balanced and there is a moment about plane 1 from the secondary forces. The resultant *total* unbalance is a single force. The force and moment about plane 1 are, from Eq. 1, Section 20.1 and Eq. 1, Section 20.3.

$$f = \frac{(W_p + W_p')}{g} R\omega^2 [\cos\theta_1 - \sin\theta_1]$$

$$M = \frac{(W_p + W_p')}{g} R\omega^2 \left[-\sin\theta_1(a_2) + \frac{R}{L}\cos 2\theta_1(-a_2) \right]$$

If the location of the resultant force is found from $z = M/f$, it will be found to be a function of θ_1. Or the resultant force is a variable quantity, and its location is a variable quantity. It is not possible to balance the quarter-crank engine by a single counterweight. It would be possible to balance the secondary couple by a system of rotating counterweights, and it would be possible to balance the primary force, but the system would be too involved to warrant attention.

20.6 Two-cylinder engine—cranks at 180 degrees

A second type of two-cylinder engine is one with cranks at 180 degrees, as shown in Fig. 20.3a. The following table shows that only the primary forces are balanced.

ϕ	$\cos\phi$	$\sin\phi$	2ϕ	$\cos 2\phi$	$\sin 2\phi$	a	$a\cos\phi$	$a\sin\phi$	$a\cos 2\phi$	$a\sin 2\phi$
$\phi_1 = 0°$	1	0	0°	1	0	0	0	0	0	0
$\phi_2 = 180°$	-1	0	360°	1	0	a_2	$-a_2$	0	a_2	0
Sum	0	0	....	2	0	...	$-a_2$	0	a_2	0
	Primary forces balanced			Secondary forces unbalanced			Primary moments unbalanced		Secondary moments unbalanced	

In this engine, the primary forces only are balanced. The total unbalanced force is, from Eq. 1, Section 20.1:

$$f = \frac{(W_p + W_p')}{g} R\omega^2 \left[\frac{R}{L}\cos 2\theta_1(2) \right]$$

The moment about plane 1 is, from Eq. 1, section 20.3:

$$M = \frac{(W_p + W_p')}{g} R\omega^2 \left[\cos\theta_1(-a_2) + \frac{R}{L}\cos 2\theta_1(a_2) \right]$$

348 Balancing Masses Reciprocating in Several Planes

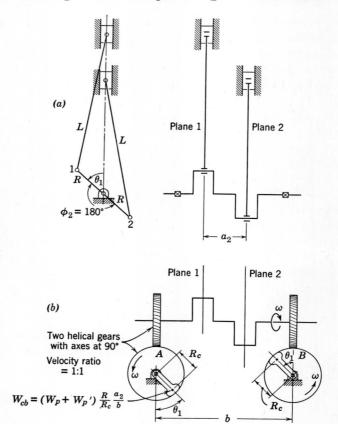

Arrangement to balance the primary couple only

Fig. 20.3a, b. Two-cylinder engine with cranks at 180 degrees, and an arrangement to balance the primary couple.

Therefore, the total unbalance in this engine is a single resultant force. However, let us consider separately the effect of the primary and secondary forces. The primary forces cause a couple of magnitude

$$C = \frac{(W_p + W_p')}{g} R\omega^2 [\cos \theta_1 (-a_2)]*$$

The resultant of the secondary forces is a single force of magnitude

$$f = \frac{(W_p + W_p')}{g} R\omega^2 \left[\frac{R}{L} \cos 2\theta_1 (2) \right]$$

* The minus sign indicates a clockwise couple.

Two-Cylinder Engine—Cranks at 180 Degrees

and the location is found from $z = M/f$, which can be expressed by
$$z = \frac{\cos 2\theta_1(a_2)}{\cos 2\theta_1(2)} = \frac{a}{2}.$$
Thus the secondary force acts through a fixed point in the engine, a point midway between the two cylinders.

Balancing the primary couple. Figure 20.3b illustrates how the primary couple can be balanced by means of a gearing system. A pair of helical gears, having a velocity ratio of 1:1, are set with axes at 90 degrees on each end of the crankshaft. Gears A and B rotate in the same direction, and with the angular speed of the crankshaft. Counterweights mounted integrally on the gears set up a couple to oppose the couple caused by the primary forces. Since the primary couple is clockwise, as seen from the negative sign in the equation for the couple, with $\cos \theta_1$ positive, the balancing couple should be counterclockwise for the corresponding angle θ_1.

The expression for the couple created by the balancing weights is
$$\frac{W_{cb} R_c}{g} \omega^2 b \cos \theta_1$$

where W_{cb} is the weight of each counterweight, R_c is the distance from the center of gravity of each counterweight to the center of rotation, and b is the distance between the centers of rotation of the counterweights. The necessary weight of each counterweight can be found from
$$\frac{(W_p + W_p')}{g} R\omega^2 \cos \theta_1(a_2) = \frac{W_{cb}}{g} R_c \omega^2 b \cos \theta_1$$

or
$$W_{cb} = (W_p + W_p')\left(\frac{R}{R_c}\right)\left(\frac{a_2}{b}\right)$$

It is left as an exercise for the student to sketch the arrangement of counterweights if both gears A and B rotate clockwise.

Balancing of the secondary forces. Figure 20.3c shows an arrangement whereby the secondary forces may be balanced. The resultant of the inertia forces of the counterweights, which rotate at twice crankshaft speed in opposite directions, is such that it acts through the center of the two cranks and opposes the secondary forces. The weight of counterweight required can be found from
$$\frac{(W_p + W_p')}{g} R\omega^2 \frac{R}{L} \cos 2\theta_1(2) = \frac{2 W_{cb}}{g} R_c \omega^2 \cos 2\theta_1$$

or
$$W_{cb} = (W_p + W_p') \frac{R^2}{L R_c}$$

350 Balancing Masses Reciprocating in Several Planes

An alternate arrangement in Fig. 20.3d permits the balancing of the secondary forces, with a simpler system, at the expense of introducing a rocking couple. This arrangement, called the Lanchester balancer, has been used in an early four-cylinder automobile engine where the

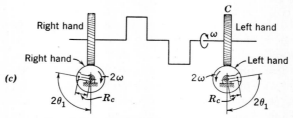

An arrangement to balance the secondary force only.

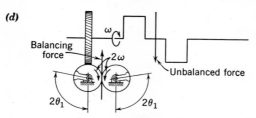

Alternate arrangement to balance the secondary force, introducing, however, a secondary couple.
(Lanchester balancer)

Fig. 20.3c, d. Arrangements to balance the secondary couple in a two-cylinder engine with cranks at 180 degrees.

unbalanced secondary force acts in the same way as in the two-cylinder engine with cranks at 180 degrees.*, †

20.7 Two-cylinder engine—cranks at zero degrees

The two-cylinder engine with cranks at zero degrees can be considered to be two identical single-cylinder engines, with the unbalance twice that of one engine, and with the resultant force acting midway between cranks.

20.8 Comparison of four types of two-cylinder, four-cycle engines

Comparison of the unbalanced force equations for the four types of two-cylinder engines, cranks at zero degrees, 90 degrees, 180 degrees,

---

* For the arrangement devised by Lanchester, see page 754 of *Automobile and Aircraft Engines*, by A. W. Judge, Pitman, 2nd edition, 1931.

† See *Balancing of Oil Engines*, by W. Ker Wilson, Lippincott, 1929, page 140, for the arrangement used to balance a four-cylinder four-cycle oil engine installed on the passenger ship *Empress of Australia*.

and the opposed two-cylinder engine with cranks at 180 degrees, reveals quite a variation in amount of unbalance. The selection of an engine for a particular application is not based necessarily upon the minimum unbalance. A very important consideration is the smoothness of the torque applied to the crankshaft, which in turn affects the speed variation and flywheel size; and the smoothness of the torque is affected by the firing order. Let us examine the firing order in a four-cycle engine.

The firing order is composed of four different events in the cycle: (1) the power stroke, *P*, which begins when a piston is approximately at its outer limit of travel and extends for 180 degrees rotation of the crank; (2) the exhaust stroke, *E*, for the next 180 degrees rotation of the crank; (3) the intake or suction stroke, *I*, for the next 180 degrees; and (4) the compression stroke, *C*, for the next 180 degrees. Thus a power stroke takes place for one half a revolution of the crank every two revolutions, or power is being supplied only one fourth the time. It is therefore necessary to distribute the power strokes as evenly as possible in order to distribute the power evenly. Table A shows the various possibilities of firing order.

TABLE A

Comparison of Firing Order of Various Two-Cylinder Engines (Four-Cycle)

Possibility I		Possibility II		Possibility I		Possibility II		Possibility I		Possibility II		Possibility I		Possibility II		Crank Angle θ_1
Cyl. 1	Cyl. 2	Cyl. 1	Cyl. 2	Cyl. 1	Cyl. 2	Cyl. 1	Cyl. 2	Cyl. 1	Cyl. 2	Cyl. 1	Cyl. 2	Cyl. 1	Cyl. 2	Cyl. 1	Cyl. 2	
					P		*I*									0°
P	*P*	*P*	*I*	*P*		*P*		*P*	*C*	*P*	*E*	*P*	*P*	*P*	*I*	
					E		*C*									180°
E	*E*	*E*	*C*	*E*		*E*		*E*	*P*	*E*	*I*	*E*	*E*	*E*	*C*	
					I		*P*									360°
I	*I*	*I*	*P*	*I*		*I*		*I*	*E*	*I*	*C*	*I*	*I*	*I*	*P*	
					C		*E*									540°
C	*C*	*C*	*E*	*C*		*C*		*C*	*I*	*C*	*P*	*C*	*C*	*C*	*E*	
					P		*I*									720°
Cranks at 0 degrees				Cranks at 90 degrees				Cranks at 180 degrees				Opposed, cranks at 180 degrees				

The following table summarizes the balance and evenness of power distribution, relative evaluation of the four types of engines being used:

352 Balancing Masses Reciprocating in Several Planes

	Balance	Evenness of Power Strokes
Cranks at 0°	poor	good (possibility II, Table A)
Cranks at 90°	good	fair (possibility II, Table A)
Cranks at 180°	fair	poor (same for I and II)
Opposed, cranks at 180°	excellent	good (possibility II, Table A)

Two-cycle engines. The discussion of firing order has been limited to four-cycle engines. The same procedure of analysis can be made with two-cycle engines, except that a power stroke can be considered as occurring every 90 degrees, or a complete cycle takes place every revolution of the crank. Whether the engine is a two- or four-cycle engine, the inertia forces remain unchanged.

20.9 Three-cylinder engine

Figure 20.4a shows a crankshaft used on a three-cylinder engine, with counterbalances for the crank only, whereas Fig. 20.4b shows the schematic arrangement of a three-cylinder engine where the cranks are assumed to be equally spaced. The table below indicates the unbalance.

ϕ	$\cos\phi$	$\sin\phi$	2ϕ	$\cos 2\phi$	$\sin 2\phi$	a	$a\cos\phi$	$a\sin\phi$	$a\cos 2\phi$	$a\sin 2\phi$
$\phi_1 = 0°$	1	0	0°	1	0	0	0	0	0	0
$\phi_2 = 240°$	$-\tfrac{1}{2}$	$-\tfrac{1}{2}\sqrt{3}$	480°	$-\tfrac{1}{2}$	$\tfrac{1}{2}\sqrt{3}$	a	$-\tfrac{1}{2}a$	$-\tfrac{1}{2}a\sqrt{3}$	$-\tfrac{1}{2}a$	$\tfrac{1}{2}a\sqrt{3}$
$\phi_3 = 120°$	$-\tfrac{1}{2}$	$\tfrac{1}{2}\sqrt{3}$	240°	$-\tfrac{1}{2}$	$-\tfrac{1}{2}\sqrt{3}$	$2a$	-1	$a\sqrt{3}$	$-a$	$-a\sqrt{3}$
Sum	0	0	...	0	0	...	$-\tfrac{3}{2}a$	$\tfrac{1}{2}a\sqrt{3}$	$-\tfrac{3}{2}a$	$-\tfrac{1}{2}a\sqrt{3}$

Since the forces are balanced, but the moments about plane 1 are not balanced, the unbalance in the engine is a couple. From Eq. 1, Section 20.3, the couple is

$$C = \frac{(W_p + W_p')}{g} R\omega^2 \left[\cos\theta_1 \left(-\frac{3}{2}a\right) - \sin\theta_1 \left(\frac{1}{2}a\sqrt{3}\right) \right.$$
$$\left. + \frac{R}{L}\cos 2\theta_1 \left(-\frac{3}{2}a\right) - \frac{R}{L}\sin 2\theta_1 \left(-\frac{1}{2}a\sqrt{3}\right) \right]$$
$$= \frac{(W_p + W_p')}{2g} R\omega^2 a \left[-3\cos\theta_1 - \sqrt{3}\sin\theta_1 - 3\frac{R}{L}\cos 2\theta_1 \right.$$
$$\left. + \sqrt{3}\frac{R}{L}\sin 2\theta_1 \right]$$

Three-Cylinder Engine

Fig. 20.4a. Vertical type "Universal Unaflow" steam engine built by the Skinner Engine Co. *Left.* Three-cylinder crankshaft, showing counterbalances, chain sprocket and flywheel mounting. *Right.* Opposite view of flywheel, showing coupling arrangement for extended shaft. (Courtesy Skinner Engine Co.)

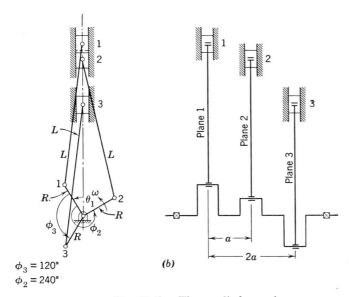

$\phi_3 = 120°$
$\phi_2 = 240°$

Fig. 20.4b. Three-cylinder engine.

354   Balancing Masses Reciprocating in Several Planes

By use of the relation $n \cos x + m \sin x = \sqrt{m^2 + n^2} \sin (x + \psi)$, where $\tan \psi = n/m$, the equation above can be given by

$$C = \frac{(W_p + W_{p'})}{2g} R\omega^2 a \left[-2\sqrt{3} \sin (\theta_1 + 60°) - 2\sqrt{3} \frac{R}{L} \sin (2\theta_1 - 60°) \right]$$

where the first quantity is the primary couple and the second quantity the secondary couple. These couples can be balanced.

Balancing the primary couple.   Figure 20.4c shows the arrangement to balance the primary couple, which is similar to the arrange-

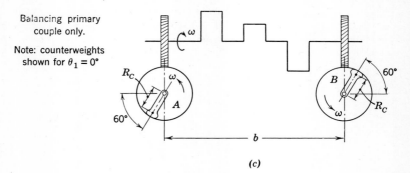

Fig. 20.4c.

ment used with the two-cylinder engine with cranks at 180 degrees, except for the positioning of the counterweight with respect to crank 1. It is left to the student to show that the necessary counterweight, W_{cb}, at a radius R_c, with the centers of the balance gears b apart, can be expressed by

$$W_{cb} \atop \text{primary} = 2\sqrt{3} (W_p + W_{p'}) \left(\frac{R}{R_c}\right)\left(\frac{a}{b}\right)$$

Balancing the secondary couple.   Figure 20.4d shows a possible arrangement to balance the secondary couple. Gears C and D rotate at twice crankshaft speed, and both gears rotate counterclockwise. The position of the counterweights are shown for the crank angle $\theta_1 = 0$. The necessary counterweight is

$$W_{cb} \atop \text{secondary} = \frac{3}{4} (W_p + W_{p'}) \frac{R^2}{R_c L} \frac{a}{b}$$

Four-Cylinder Engines

The combination of Figs. 20.4c and 20.4d may be made to effect complete balance of the three-cylinder engine.

Figure 20.4e shows the firing order that will give uniform turning effort on the crank, for both four- and two-cycle operation. Note that

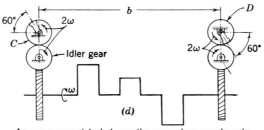

An arrangement to balance the secondary couple only
Note: counterweights shown for $\theta_1 = 0°$

Fig. 20.4d.

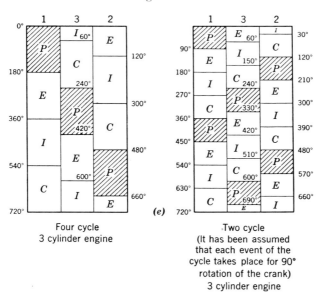

Fig. 20.4e. Firing order for four-cycle and two-cycle operation.

the diagrams are for counterclockwise rotation of the crank in Fig. 20.4b.

20.10 Four-cylinder engines

There are various arrangements of crankshafts which are possible in a four-cylinder engine, some being suitable for four-cycle operation

356 Balancing Masses Reciprocating in Several Planes

and some being suitable for two-cycle operation. Three arrangements are discussed here.

Type I. Figure 20.5 shows the schematic diagram for a four-cylinder engine suitable for four-cycle operation with a firing order of

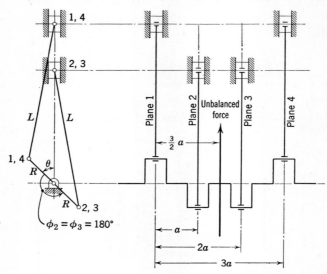

Fig. 20.5. Four-cylinder engine—type I.

1–3–4–2 or 1–2–4–3 for uniformity of power strokes. The type of unbalance is found from the following, where a is the distance between equally spaced cylinders:

ϕ	$\cos \phi$	$\sin \phi$	2ϕ	$\cos 2\phi$	$\sin 2\phi$	a	$a \cos \phi$	$a \sin \phi$	$a \cos 2\phi$	$a \sin 2\phi$
$\phi_1 = 0°$	1	0	0°	1	0	0	0	0	0	0
$\phi_2 = 180°$	-1	0	360°	1	0	a	$-a$	0	a	0
$\phi_3 = 180°$	-1	0	360°	1	0	$2a$	$-2a$	0	$2a$	0
$\phi_4 = 0°$	1	0	0°	1	0	$3a$	$3a$	0	$3a$	0
Sum	0	0	...	4	0	...	0	0	$6a$	0

The table indicates that there is a resultant force unbalance, which, from Eq. 1, Section 20.1, is

$$f = \frac{(W_p + W_p')}{g} R\omega^2 \left[\frac{R}{L} \cos 2\theta_1(4) \right]$$

Four-Cylinder Engines

The moment about plane 1 is, from Eq. 1, Section 20.3,

$$M = \frac{(W_p + W_p')}{g} R\omega^2 \left[\frac{R}{L} \cos 2\theta_1 (6a)\right]$$

The location of the resultant force is found from

$$z = \frac{M}{f} = \frac{3a}{2}$$

which indicates that the resultant force location is independent of the crank position and is located midway between cylinders 2 and 3. Refer to Figs. 20.3c and 20.3d for gearing arrangements which can be used to balance the resultant shaking force. Note that this engine can be considered as made up from two, two-cylinder engines with cranks at 180 degrees.

Type II. Figure 20.6 shows an arrangement suitable for two-cycle operation with a firing order of 1–3–4–2. The cylinders are equally spaced a apart. The unbalance is found to be a primary couple.

ϕ	$\cos\phi$	$\sin\phi$	2ϕ	$\cos 2\phi$	$\sin 2\phi$	a	$a\cos\phi$	$a\sin\phi$	$a\cos 2\phi$	$a\sin 2\phi$
$\phi_1 = 0°$	1	0	0°	1	0	0	0	0	0	0
$\phi_2 = 90°$	0	1	180°	-1	0	a	0	a	$-a$	0
$\phi_3 = 270°$	0	-1	540°	-1	0	$2a$	0	$-2a$	$-2a$	0
$\phi_4 = 180°$	-1	0	360°	1	0	$3a$	$-3a$	0	$3a$	0
Sum	0	0	...	0	0	...	$-3a$	$-a$	0	0

The unbalanced couple is found by Eq. 1, Section 20.3:

$$C = \frac{(W_p + W_p')}{g} R\omega^2 [\cos\theta_1(-3a) - \sin\theta_1(-a)]$$

$$= \sqrt{10}\, \frac{(W_p + W_p')}{g} R\omega^2 a \sin(\theta_1 - 71.57°)$$

This unbalanced couple can be balanced in the same manner that the primary couple was balanced in the three-cylinder engine and in the two-cylinder engine with cranks at 180 degrees, except for the phase relation of the counterbalancing weights with respect to the crank angle θ_1. It is left to the student to sketch the proper positioning of the counterweights to effect proper balance.

358   Balancing Masses Reciprocating in Several Planes

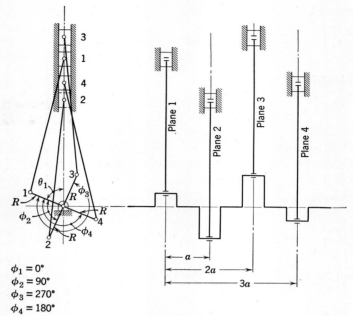

$\phi_1 = 0°$
$\phi_2 = 90°$
$\phi_3 = 270°$
$\phi_4 = 180°$

Fig. 20.6. Four-cylinder engine—type II.

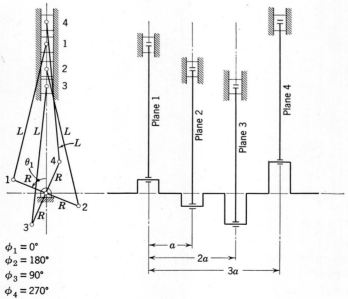

$\phi_1 = 0°$
$\phi_2 = 180°$
$\phi_3 = 90°$
$\phi_4 = 270°$

Fig. 20.7. Four-cylinder engine—type III.

Type III. Figure 20.7 shows the third arrangement. For the angles $\phi_1 = 0°$, $\phi_2 = 180°$, $\phi_3 = 90°$, and $\phi_4 = 270°$, with cranks equally spaced a apart, it is left for the student to show that the unbalance is a couple, the sum of the primary and secondary couple:

$$C = \frac{(W_p + W_p')}{g} R\omega^2 a \left[\sin(\theta_1 - 45°) - 4\frac{R}{L} \cos 2\theta_1 \right]$$

and to determine a geared arrangement whereby the couple is balanced for every position of the crank. Show, also, that the arrangement of cranks is suitable for two-cycle operation with a firing order of 1–4–2–3.

20.11 Six-cylinder engines

There are considerably more possibilities of crank arrangement in a six-cylinder engine than in a four-cylinder engine. However, space

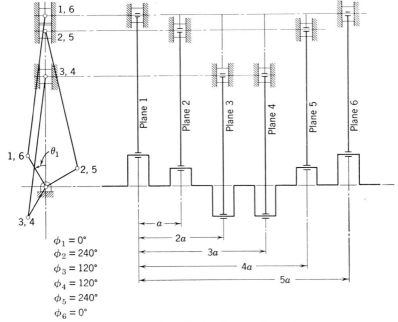

Fig. 20.8. Six-cylinder, four-cycle engine.

limitations do not allow the discussion of each arrangement. Two arrangements will be taken up, that commonly found in the four-cycle six-cylinder engine and that found in the two-cycle six-cylinder engine.

Four-cycle engine. Figure 20.8 shows the arrangement of cranks found in automobile engines, with $\phi_1 = 0°$, $\phi_2 = 240°$, $\phi_3 = 120°$, $\phi_4 = 120°$, $\phi_5 = 240°$, and $\phi_6 = 0°$. The pistons are equally spaced. Analysis shows that the engine is completely balanced.

There are several possible firing orders, as 1–5–3–6–2–4, 1–5–4–6–2–3, and 1-2-3-6-5-4, which give equally uniform power impulses to the engine. The problem of fuel distribution to each cylinder, intake and exhaust manifold design, and effect of the firing order on the "torsional windup" of the shaft have led to the adoption of 1–5–3–6–2–4 for the usual firing order.

Two-cycle engine. An engine to be discussed because of an interesting balancing arrangement is the six-cylinder engine used for

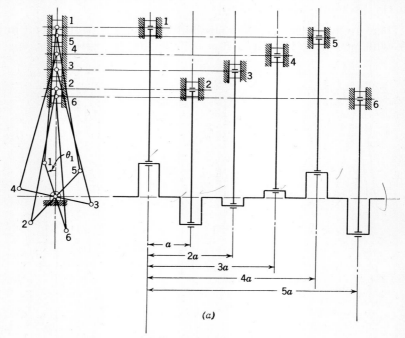

Fig. 20.9a. Six-cylinder, two-cycle engine.

two-cycle operation shown in Fig. 20.9a. For $\phi_1 = 0°$, $\phi_2 = 120°$, $\phi_3 = 240°$, $\phi_4 = 60°$, $\phi_5 = 300°$, and $\phi_6 = 180°$, the unbalance is found to be a primary couple, with $\Sigma a \cos \phi = -3a$ and $\Sigma a \sin \phi = -\sqrt{3}\, a$. The primary couple can be expressed by

$$C = 2\sqrt{3}\, \frac{(W_p + W_p')}{g} R\omega^2 a \sin(\theta_1 - 60°)$$

The actual arrangement used for balancing the primary couple by General Motors Corporation in their six-cylinder Series 71 diesel engines is shown in Figs. 20.9b and 20.9c. A train of five helical gears, mounted at the rear of the engine, is shown in Fig. 20.9b. The crankshaft gear, which is integrally mounted on the crankshaft, drives an

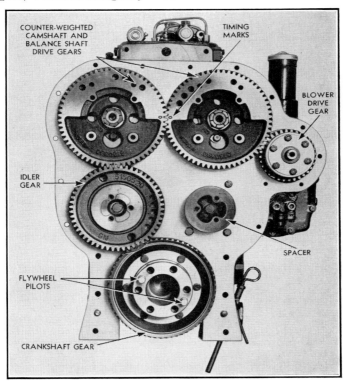

Fig. 20.9b. Rear view of the gear train of the General Motors Series 71 diesel engines. (Courtesy Detroit Diesel Engine Division, General Motors Corp.)

idler gear, which in turn drives a counterweighted camshaft gear meshing with a gear mounted on a balancer shaft. The camshaft and balancer shaft are shown in Fig. 20.9c. Balance weights are mounted integrally on the camshaft and balancer shafts on the ends opposite to where the gears are mounted. Inspection of the action of the counterweights shows that the effect is a couple which varies as the crank angle θ_1.

The arrangement described is used, with variations in balancing masses, in three-, four-, and six-cylinder two-cycle engines of the General Motors Corporation.

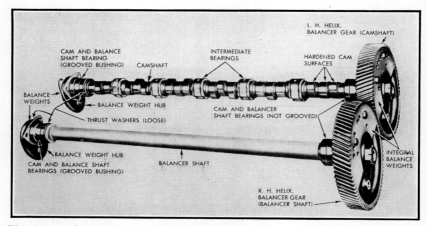

Fig. 20.9c. Camshaft and balancer shaft of the General Motors Series 71 diesel engines. (Courtesy Detroit Diesel Engine Division General Motors Corp.)

The firing order for counterclockwise rotation of the crankshaft is 1–5–3–6–2–4, with uniform power impulses.

20.12 In-line engines with more than six cylinders

The analysis of in-line engines of eight, twelve, or more cylinders with respect to unbalance is repetitive of what has already been discussed. Although there are additional problems of crankshaft design, selection of firing orders, etc., the discussion of such items is beyond the scope of this book.

20.13 Multicylinder V-type engines

There are distinct advantages of the V-type engine over the in-line type of engines, two of the advantages being a shorter crankshaft, with an increase of stiffness over a straight in-line engine with the same number of cylinders, and a shorter engine, with more power in a given length of engine. V-engines with as high as 24 cylinders have been built.

The construction of connecting rods, as discussed in Chapter 19, may be of the master rod and articulated rod type, or of the forked blade and rod type. A third possibility of construction, as used in automobile engines and some diesel engines, is the staggering of one bank of cylinders with respect to the other bank, permitting all connecting rods to be connected directly to the crankshaft. Figures 20.10a and 20.10b show the staggered arrangement used by The Caterpillar Tractor Company. All V-engines to be discussed in this chapter

Multicylinder V-type Engines

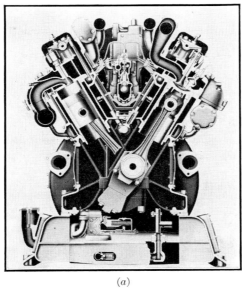

(a)

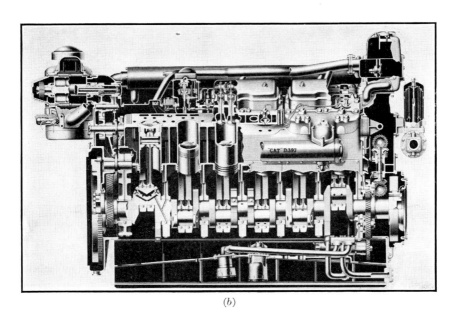

(b)

Fig. 20.10. D-397 engine. (a) Sectional view from the front end. (b) Sectional view from the right side. (Courtesy of the Caterpillar Tractor Co.)

364 Balancing Masses Reciprocating in Several Planes

will be considered to have the two pistons and the two connecting rods in a V in one plane; and it will be assumed that the two connecting rods are connected to the crankshaft at the same point.

20.14 Conditions for balance in V-engines

There are several methods of approach in determining whether a V-type engine is balanced, and, if it is unbalanced, the amount of unbalance. The first possible method is to consider the pistons in one bank of the V as forming one engine and the pistons in the other bank as forming a second engine. Each bank may be analyzed separately and the results for the two banks combined into a single result. The second method is to consider the entire engine and derive the conditions for the entire engine. Although the second method is perhaps longer initially in the derivation of the equations, the application of the equations will be simplified. Accordingly, the latter method will be used.

It was shown in Chapter 19, pages 318–324, that the equations of horizontal and vertical components of unbalance of the inertia forces of the pistons of a two-cylinder V engine were

$$f^v = 2 \frac{(W_p + W_p')}{g} R\omega^2 \left(\cos\theta \cos\psi + \frac{R}{L} \cos 2\theta \cos 2\psi \right) \cos\psi$$

$$f^h = 2 \frac{(W_p + W_p')}{g} R\omega^2 \left(\sin\theta \sin\psi + \frac{R}{L} \sin 2\theta \sin 2\psi \right) \sin\psi$$

where θ is the angle between the central plane of the engine and crank 1; ψ is one half the total angle of the V.

Figure 20.10c shows a schematic arrangement of a multicylinder V-type engine. For simplification at this time, the numbers designating a crank will refer to the cranks in one bank only, the right bank. However, the above expressions for the vertical and horizontal unbalanced forces take into account the inertia forces of the two pistons (and connecting rods) in the plane of a V.

Representing the vertical and horizontal components of the inertia forces of the two pistons in plane 1 by f_1^v and f_1^h, and θ by $(\theta_1 + \phi_1)$, where ϕ_1 is zero and θ_1 now specifically refers to the angle between the central plane and crank 1, we may write the following:

$$f_1^v = 2 \frac{(W_p + W_p')}{g} R\omega^2$$

$$\left[\cos(\theta_1 + \phi_1) \cos\psi + \frac{R}{L} \cos 2(\theta_1 + \phi_1) \cos 2\psi \right] \cos\psi$$

$$f_1{}^h = 2\frac{(W_p + W_p')}{g}R\omega^2$$
$$\left[\sin(\theta_1 + \phi_1)\sin\psi + \frac{R}{L}\sin 2(\theta_1 + \phi_1)\sin 2\psi\right]\sin\psi$$

The reason for inserting ϕ_1 is to make the expression comparable to the expressions for the forces in the other planes. $\phi_1 = 0$.

In the following expressions for the horizontal and vertical unbalanced inertia forces of the pistons in plane 2 we recognize that the

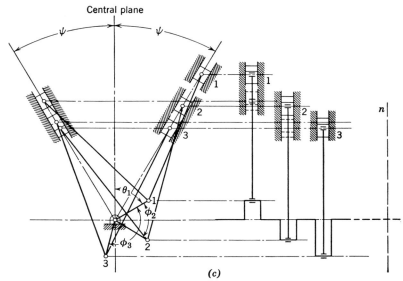

Fig. 20.10c. General arrangement of a multicylinder V-engine with n cylinders

angle between the central plane and crank 2 is $(\theta_1 + \phi_2)$ and that the expressions above may be used if the proper angle is used:

$$f_2{}^v = 2\frac{(W_p + W_p')}{g}R\omega^2$$
$$\left[\cos(\theta_1 + \phi_2)\cos\psi + \frac{R}{L}\cos 2(\theta_1 + \phi_2)\cos 2\psi\right]\cos\psi$$

$$f_2{}^h = 2\frac{(W_p + W_p')}{g}R\omega^2$$
$$\left[\sin(\theta_1 + \phi_2)\sin\psi + \frac{R}{L}\sin 2(\theta_1 + \phi_2)\sin 2\psi\right]\sin\psi$$

where ϕ_2 is the angle between crank 1 and crank 2.

366   Balancing Masses Reciprocating in Several Planes

Similarly, for crank 3,

$$f_3{}^v = 2\frac{(W_p + W_p')}{g} R\omega^2$$
$$\left[\cos(\theta_1 + \phi_3)\cos\psi + \frac{R}{L}\cos 2(\theta_1 + \phi_3)\cos 2\psi\right]\cos\psi$$

$$f_3{}^h = 2\frac{(W_p + W_p')}{g} R\omega^2$$
$$\left[\sin(\theta_1 + \phi_3)\sin\psi + \frac{R}{L}\sin 2(\theta_1 + \phi_3)\sin 2\psi\right]\sin\psi$$

And so on, for all the planes.

By expanding the trigonometric expressions, collecting terms in a manner comparable to that used in multicylinder in-line engines, we may obtain the following expressions for the amount of the components of unbalance:

$$f^v{}_{\text{resultant}} = 2\frac{(W_p + W_p')}{g}R\omega^2\left[\cos\theta_1\cos^2\psi\sum\cos\phi\right.$$
$$-\sin\theta_1\cos^2\psi\sum\sin\phi + \frac{R}{L}\cos 2\theta_1\cos\psi\cos 2\psi\sum\cos 2\phi$$
$$\left.- \frac{R}{L}\sin 2\theta_1\cos\psi\cos 2\psi\sum\sin 2\phi\right] \quad (1)$$

$$f^h{}_{\text{resultant}} = 2\frac{(W_p + W_p')}{g}R\omega^2\left[\sin\theta_1\sin^2\psi\sum\cos\phi\right.$$
$$+\cos\theta_1\sin^2\psi\sum\sin\phi + \frac{R}{L}\sin 2\theta_1\sin\psi\sin 2\psi\sum\cos 2\phi$$
$$\left.+ \frac{R}{L}\cos 2\theta_1\sin\psi\sin 2\psi\sum\sin 2\phi\right] \quad (2)$$

If the forces are to be completely balanced for every position of crank 1, regardless of the V angle, the same conditions as obtained in the in-line type of engine are obtained:

$$\Sigma\cos\phi = 0 \quad \Sigma\cos 2\phi = 0$$
$$\Sigma\sin\phi = 0 \quad \Sigma\sin 2\phi = 0$$

The expressions for the moments of the inertia force components about plane 1 may be obtained quickly by noting that, in the force equations, each component of force is multiplied by the corresponding a distance:

Balancing a Six-Cylinder, Two-Cycle V-engine

$$M^v = 2\frac{(W_p + W_p')}{g} R\omega^2 \left[\cos\theta_1 \cos^2\psi \sum a\cos\phi \right.$$

$$- \sin\theta_1 \cos^2\psi \sum a\sin\phi + \frac{R}{L}\cos 2\theta_1 \cos\psi \cos 2\psi \sum a\cos 2\phi$$

$$\left. - \frac{R}{L}\sin 2\theta_1 \cos\psi \cos 2\psi \sum a\sin 2\phi \right] \quad (3)$$

$$M^h = 2\frac{(W_p + W_p')}{g} R\omega^2 \left[\sin\theta_1 \sin^2\psi \sum a\cos\phi \right.$$

$$+ \cos\theta_1 \sin^2\psi \sum a\sin\phi + \frac{R}{L}\sin 2\theta_1 \sin\psi \sin 2\psi \sum a\cos 2\phi$$

$$\left. + \frac{R}{L}\cos 2\theta_1 \sin\psi \sin 2\psi \sum a\sin 2\phi \right] \quad (4)$$

Again, the conditions necessary to have zero moments about plane 1, regardless of the position of crank 1 and the V angle, are:

$$\Sigma a \cos\phi = 0 \qquad \Sigma a \cos 2\phi = 0$$
$$\Sigma a \sin\phi = 0 \qquad \Sigma a \sin 2\phi = 0$$

20.15 Balancing a six-cylinder, two-cycle V-engine

Figure 20.11a shows the schematic arrangement of a six-cylinder V-engine, of the type used by General Motors Corporation in the

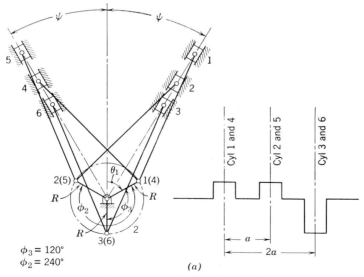

Fig. 20.11a. Six-cylinder V-engine.

368 Balancing Masses Reciprocating in Several Planes

567B series. The engine is basically two three-cylinder engines, with cranks $\phi_1 = 0°$, $\phi_2 = 240°$, $\phi_3 = 120°$. Pistons 1, 2, and 3 move in one bank; pistons 4, 5, and 6 move in the other bank. The tabular form on page 352 can be used directly, and the form shows that there is a primary couple and a secondary couple unbalance. The couples in the vertical and horizontal planes are, from Eqs. 3 and 4, Section 20.14:

$$C^v = 2 \frac{(W_p + W_p')}{g} R\omega^2 \left[\cos\theta_1 \cos^2\psi \left(-\frac{3}{2}a\right) \right.$$

$$\left. - \sin\theta_1 \cos^2\psi \left(\frac{a}{2}\sqrt{3}\right) + \frac{R}{L}\cos 2\theta_1 \cos\psi \cos 2\psi \left(-\frac{3}{2}a\right) \right.$$

$$\left. - \frac{R}{L}\sin 2\theta_1 \cos\psi \cos 2\psi \left(-\frac{a}{2}\sqrt{3}\right) \right]$$

$$C^h = 2 \frac{(W_p + W_p')}{g} R\omega^2 \left[\sin\theta_1 \sin^2\psi \left(-\frac{3}{2}a\right) \right.$$

$$\left. + \cos\theta_1 \sin^2\psi \left(\frac{a}{2}\sqrt{3}\right) + \frac{R}{L}\sin 2\theta_1 \sin\psi \sin 2\psi \left(-\frac{3}{2}a\right) \right.$$

$$\left. + \frac{R}{L}\cos 2\theta_1 \sin\psi \sin 2\psi \left(-\frac{a}{2}\sqrt{3}\right) \right]$$

Using the relation $(n \cos x + m \sin x) = \sqrt{n^2 + m^2} \sin(x + \gamma)$ and collecting terms, we can obtain

$$C^v = -2\sqrt{3} \frac{(W_p + W_p')}{g} R\omega^2 a \cos\psi$$

$$\left[\cos\psi \sin(\theta_1 + 60°) - \frac{R}{L}\cos 2\psi \sin(2\theta_1 - 60°) \right]$$

$$C^h = -2\sqrt{3} \frac{(W_p + W_p')}{g} R\omega^2 a \sin\psi$$

$$\left[\sin\psi \sin(\theta_1 - 60°) + \frac{R}{L}\sin 2\psi \sin(2\theta_1 + 60°) \right]$$

This engine may be balanced completely, regardless of the V-angle, by rotating weights that will set up couples in the vertical and horizontal planes. The basic arrangement used to balance the primary couple of the six-cylinder, two-cycle engine, as shown in Fig. 20.9c, can be used. The arrangement shown in Fig. 20.11b will permit the balancing of the primary couple in the vertical plane. Note that the horizontal components of the inertia forces of the counterweights balance each other, whereas the vertical components give a couple, which varies in magnitude so as always to balance the primary couple in the vertical plane.

Balancing a Six-Cylinder, Two-Cycle V-engine

Figure 20.11c shows an arrangement of counterweights to balance the primary couple in the horizontal plane. Analysis of the couples created, simplified in Fig. 20.11d, shows three couples: (1) a couple in the vertical plane of the two front gears; (2) a couple in the vertical plane of the two rear gears; and (3) a couple in the horizontal plane. The couples in the vertical planes balance each other, leaving only a couple in the horizontal plane.

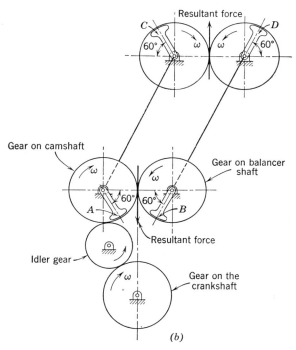

Fig. 20.11b. Arrangement to balance the primary couple in the vertical plane. (Counterweights shown for position $\theta_1 = 0°$.)

The systems shown in Figs. 20.11b and 20.11c may be combined to give just a single counterweight on each gear for complete balance of the primary couples.

No attempt is made to balance the secondary couples, although they can be completely balanced by suitable gearing arrangements.

It is interesting to note that the larger the V-angle is, the greater the unbalance in the horizontal plane, and vice versa. If the angle is small, the primary couple in the horizontal plane could be small, and therefore negligible. Thus balance of the primary couple in the vertical plane might be sufficient for smooth running of the engine.

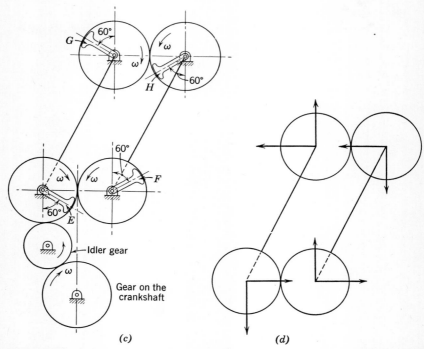

Fig. 20.11c. Arrangement to balance the primary couple in the horizontal plane. (Counterweights shown for position $\theta_1 = 0°$.)

Fig. 20.11d. Components of forces of the counterweights for balance of the primary couple in the horizontal plane.

20.16 Eight-cylinder, two-cycle engine

A crank arrangement of $\phi_1 = 0°, \phi_2 = 180°, \phi_3 = 90°,$ and $\phi_4 = 270°$ for an eight-cylinder V-engine suitable for two-cycle operation gives a system of unbalance the same as in the six-cylinder two-cycle engine of the preceding section, except for phase relation of the couples with respect to the crank angle. It is left as an exercise for the student to verify the above statements.

20.17 Eight-cylinder, four-cycle automobile V-engine

Cranks are arranged in Oldsmobile and Cadillac V-engines, shown schematically in Fig. 20.12a, with $\phi_1 = 0°$, $\phi_2 = 90°$, $\phi_3 = 270°$, $\phi_4 = 180°$. The tabular form on page 357 shows that the only unbalance is a primary couple, with $\Sigma a \cos \phi = -3a$ and $\Sigma a \sin \phi = -a$. The couples in the vertical and horizontal planes, from Eqs. 3 and 4, Section 20.14 are

Eight-Cylinder, Four-Cycle Automobile V-engine

$$C^v = 2\frac{(W_p + W_p')}{g} R\omega^2 [\cos \theta_1 \cos^2 \psi(-3a) - \sin \theta_1 \cos^2 \psi(-a)]$$

$$C^h = 2\frac{(W_p + W_p')}{g} R\omega^2 [\sin \theta_1 \sin^2 \psi(-3a) + \cos \theta_1 \sin^2 \psi(-a)]$$

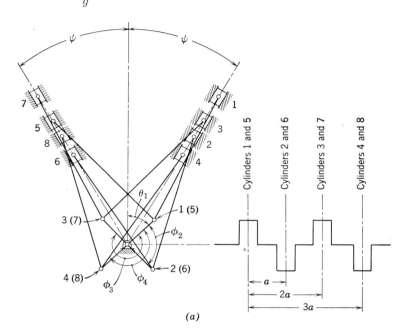

Fig. 20.12a. Eight-cylinder V-engine.

For the special case where the angle between banks is 90° ($2\psi = 90°$) the equation above may be simplified to

$$C^v = \frac{(W_p + W_p')}{g} R\omega^2 a (-3 \cos \theta_1 + \sin \theta_1)$$

$$C^h = \frac{(W_p + W_p')}{g} R\omega^2 a (-3 \sin \theta_1 - \cos \theta_1)$$

or

$$C^v = \sqrt{10}\frac{(W_p + W_p')}{g} R\omega^2 a \sin(\theta_1 - 71.57°) \quad (1)$$

$$C^h = \sqrt{10}\frac{(W_p + W_p')}{g} R\omega^2 a \cos(\theta_1 - 71.57°) \quad (2)$$

372 Balancing Masses Reciprocating in Several Planes

The couples may be combined vectorially to give the resultant

$$C = \sqrt{10}\,\frac{(W_p + W_p')}{g}\,R\omega^2 a$$

The resultant couple is of constant magnitude, and may be balanced by two weights on the crankshaft to give an opposing couple. Figure

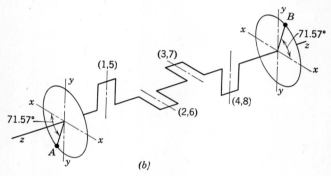

Fig. 20.12b. Counterweights on the crankshaft to balance the primary couple of constant magnitude.

Fig. 20.12c. Oldsmobile crankshaft for the "Rocket" engine. (Courtesy General Motors Corp.)

20.12b shows the arrangement of weights, in position for $\theta_1 = 0$. Examination of the couples due to the counterweights in the horizontal and vertical planes shows that the couples are in the opposite direction to that given by Eqs. 1 and 2 above. Figure 20.12c shows the counterweights on an Oldsmobile crankshaft.

20.18 Cadillac engine balance

The following information* is a description of the balancing of the two-plane crankshaft as formerly used and as now used in the Cadillac 90-degree V-engines in which a primary couple exists. Two different

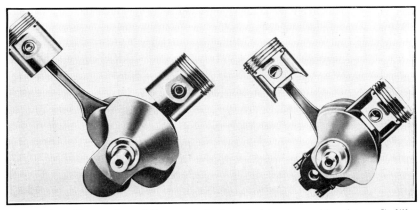

Fig. 20.13a. Front view of the crankshaft and piston assembly of two Cadillac engines, the 1948 three main bearing crankshaft on the left and the 1949 five bearing engine on the right. (Courtesy Cadillac Motor Car Division, General Motors Corp.)

Fig. 20.13b. Side view of the crankshaft and piston assembly of two Cadillac engines, the 1948 three main bearing crankshaft on the left and the 1949 five bearing crankshaft on the right. (Courtesy Cadillac Motor Car Division, General Motors Corp.)

arrangements of counterweights have been used at the Cadillac factory to balance the primary couple unbalance:

The first arrangement was handled by two large counterweights located in the proper position between the first and second crank throws and

* Given to the author by Mr. C. A. Rasmussen, General Supervisor, Engineering Laboratory, Cadillac Motor Car Division, General Motors Corp., in a letter. Quoted with permission.

between the third and fourth crank throws. This arrangement was used on the 346 cu. in. V-8 engine which was in production in 1936 through 1948. The crankshaft was a three main bearing crankshaft, which, therefore, had sufficient room in which to swing the two large counterweights.

A second method consists in using six smaller counterweights mounted opposite each throw of the shaft, which greatly reduces the bending force

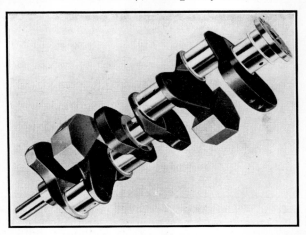

Fig. 20.13c. 1948—three main bearing crankshaft. (Courtesy Cadillac Motor Car Division, General Motors Corp.)

within the shaft. This arrangement, combined with the use of five main bearings, gives an extremely smooth operation and is the construction adopted by Cadillac in the new engine design which they brought out in 1949.

In the balancing of the 90 degree V-8 two plane crankshaft, proper ring weights are fastened around each of the crank throws to create the primary

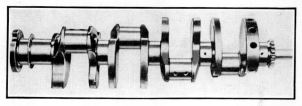

Fig. 20.13d. 1949—five main bearing crankshaft. (Courtesy Cadillac Motor Car Division, General Motors Corp.)

couple to be balanced out. By theory these weights are equal to the total of the rotating weight plus one-half of the reciprocating weight of the two connecting rod-piston assemblies normally running on each crank throw. Cadillac has learned by test experience that an additional correction of approximately 2% results in a smoother and better balance. This added correction is probably equal to the oil weight which clings to the moving parts. Cadillac production engines are, therefore, balanced slightly "heavy" as a result of this experience.

In the determination of the total rotating weight, the rotating weight of the connecting rod plus the weight of the bearing shells is used. The reciprocating weight of the connecting rod plus the weights of the piston,

Fig. 20.13e. Three main bearing crankshaft in position in a balancing machine with balancing rings weights assembled on the crank throws. (Courtesy Cadillac Motor Car Division, General Motors Corp.)

piston pin, piston pin locking devices, and piston rings make up the total reciprocating weight used.

In production, the rotating and reciprocating weights of each connecting rod are controlled by a weighing-machining operation on each end of the rod. The weights of the piston and piston pin are also controlled by weighing-machining operations.

Figures 20.13a and 20.13b show the front and side views of the piston-crankshaft assemblies of these two types of crankshaft arrangements.

376 Balancing Masses Reciprocating in Several Planes

Figures 20.13c and 20.13d show the two types of individual shafts.

Figure 20.13e and 20.13f show the crankshaft with ring weights attached as set up in a balancing machine. Figure 20.13e shows the 1948 three main bearing crankshaft, and Fig. 20.13f shows the five main bearing crankshaft as used in 1949.

Fig. 20.13f. Five main bearing crankshaft in position in a balancing machine with balancing ring weights assembled on the crank throws. (Courtesy Cadillac Motor Car Division, General Motors Corp.)

Figure 20.13g shows the five main bearing crankshaft and flywheel assembly as set up in the balancing machine for the final check. You will note here also the ring weights are shown.

Both of the types of crankshafts discussed are the two plane crankshafts used in a 90 degree V-8 type of engine. In these two engines, the cylinder number arrangement from front to rear is as follows:

Other Engines 377

Left cylinder bank as viewed from the driver's position in the car—cylinders 1,3,5,7. Right bank, 2,4,6,8.

You will note this numbering follows the position of the connecting rod in the crankshaft as counted from the front to the rear.

Fig. 20.13g. Five main bearing crankshaft and flywheel assembly in a balancing machine with balancing ring weights assembled on the crank throws. (Courtesy Cadillac Motor Car Division, General Motors Corp.)

With this cylinder arrangement, two firing orders are permissible. In the engine through 1948, the firing order used was 1,8,7,3,6,5,4,2. The firing order used with the 1949 engine is 1,8,4,3,6,5,7,2.

20.19 Other engines

There are a considerable number of other arrangements of cylinders and crankshafts in multicylinder engines, not only in the types discussed, but also in such types as the W, X, radial, etc., with more

378   Balancing Masses Reciprocating in Several Planes

than one bank of cylinders. Discussion of each type is beyond the scope of this book, and the reader is referred to the literature for further investigation. In general, however, the method of attack as regards balancing can be carried out in the same manner as outlined in this chapter and in Chapter 19.

PROBLEMS

20.1. Three two-cylinder engines are to be compared for magnitude of the shaking force. Draw the total shaking force curve, in intervals of 15 degrees of crank 1, as a function of θ_1, for the three engines:

(a) A two-cylinder engine with cranks at 0 degrees.
(b) A two-cylinder engine with cranks at 90 degrees.
(c) A two-cylinder engine with cranks at 180 degrees.

Assume that the engines are identical, except for the crank arrangement. Assume that the crankshaft is balanced in each case. Assume that the crankshafts have a speed of 1800 rpm, that each crank throw is $1\frac{1}{2}$ in., that each connecting rod length is 5 in., and that the equivalent weight of each piston is 2 lb.

20.2. (a) Consider the primary forces only of a two-cylinder engine with cranks at zero degrees. Is the resultant unbalance a force or a couple?

(b) Consider only the secondary forces of a two-cylinder engine with cranks at 90 degrees. Is the resultant of the secondary forces a single resultant force or a couple?

(c) Consider only the primary forces of a two-cylinder engine with cranks at 180 degrees. Is the resultant of the primary forces a single resultant force or a couple?

20.3. (a) If the resultant shaking force of an engine is zero for a particular angle of crank 1, the reference crank, is it true that the shaking forces are balanced? Explain.

(b) The crank radius is $1\frac{1}{2}$ in. for a two-cylinder engine with cranks at 180 degrees. The equivalent weight of each piston is 2 lb. The crank rotates at 1800 rpm. The connecting rod length is 5 in. Determine the magnitude of the shaking force for two positions: (1) $\theta_1 = 45°$ and (2) $\theta_1 = 90°$.

The distance between cylinders is 4 in. Consider the crank as balanced.

(c) For the specifications given in part b, determine the magnitude of the maximum resultant shaking force, and specify the location of the resultant shaking force with respect to the plane of cylinder 1.

20.4 (Fig. 20.4). (a) Derive a general equation for the shaking force of a two-cylinder engine with cranks at 90 degrees where the equivalent weight of each piston is *not* the same. Assume that the two crank throws are the same for the two cylinders, R. Assume that the two connecting rods are the same for the two cylinders, L. Assume that the equivalent weight of piston 1 is W, and the equivalent weight of piston 2 is $2W$.

Assume that the crank is balanced.

(b) Determine an expression for the location of the resultant force if the distance between cylinders is a inches.

20.5. A three-cylinder engine has the cranks arranged as follows, ϕ being measured counterclockwise:

$\phi_1 = 0°$, $\phi_2 = 90°$, $\phi_3 = 225°$.

Considering the plane of cylinder 1 as the reference plane, the distances to plane 1 are:

$a_1 = 0$, $a_2 = 5$ in., $a_3 = 10$ in.

The crank throw for each cylinder is the same, $R = 3$ in.
The connecting rod for each cylinder is the same, $L = 15$ in.
The crankshaft rotates at 1800 rpm.

(a) Do the primary forces cause a shaking force or a couple?
(b) Do the secondary forces cause a shaking force or a couple?
(c) If the analyses in a and b show that the resultant of the primary and secondary forces is a single force, derive an expression as a function of the crank angle of cylinder 1, θ_1, to show where the resultant force is located.

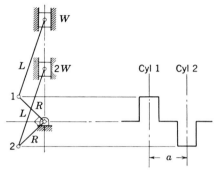

Fig. P–20.4.

20.6. An in-line engine has the cranks arranged as shown in Fig. 20.4b of the text. If $R = 1\frac{1}{2}$ in., $L = 5$ in., $a = 6$ in., crank speed = 1800 rpm, determine the magnitude of the unbalanced effect for one position of the crank: $\theta_1 = 90°$. State whether the unbalance is a force or a couple.

20.7 (Fig. 20.7). A three-cylinder, in-line, three-stage air compressor is driven at 1200 rpm. The crank relations are shown, as well as the distances between the

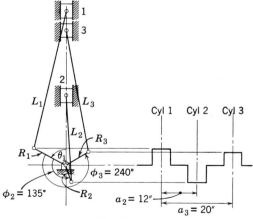

Fig. P–20.7.

380 Balancing Masses Reciprocating in Several Planes

planes. The equivalent weight of each piston is 8 lb for cylinder 1, 4 lb for cylinder 2, and 6 lb for cylinder 3. The crank radius for each piston is 4 in. for cylinder 1, 2 in. for cylinder 2, and 3 in. for cylinder 3.

The connecting rod lengths are 16 in. for cylinder 1, 10 in. for cylinder 2, and 14 in. for cylinder 3.

Determine the magnitude of the unbalanced inertia forces for one position: $\theta_1 = 30°$.

20.8. A four-cylinder in-line engine has cranks as shown in Fig. 20.6 of the text. If each crank radius is 3 in., each connecting rod is $10\frac{1}{2}$ in., the equivalent weight of each piston is 6 lb, and the crank rotates at 1800 rpm, determine a gearing arrangement to balance the engine. Show on a sketch the angular position of the weights, with respect to the crank of cylinder 1, for complete balance.

Determine, also, the magnitude of the necessary counterweights if each counterweight has a radius of 5 in. to its center of gravity. The distance between cylinders is 8 in. The distance between centers of the counterweights may be taken as 35 in.

20.9. A four-cylinder in-line engine has cranks arranged such that $\phi_1 = 0°$, $\phi_2 = 120°$, $\phi_3 = 180°$, and $\phi_4 = 300°$, the angles being measured counterclockwise. The distances from the plane of cylinder 1 are given by: $a_1 = 0$, $a_2 = 6$ in., $a_3 = 12$ in., $a_4 = 18$ in.

(a) Are the shaking forces balanced?

(b) Determine the magnitude of any resultant force or couple which may exist for the one position when $\theta_1 = 330°$. The crank radii are 3 in., the connecting rod lengths are $10\frac{1}{2}$ in. The equivalent unbalanced weight of each piston is 6 lb. The crankshaft speed is 1800 rpm.

(c) Determine the magnitude of the maximum unbalanced effect.

(d) Show how the unbalanced primary effect may be balanced by a suitable gearing arrangement.

20.10. Refer to Fig. 20.7 of the text. If $R = 6$ in., $L = 24$ in., $a = 12$ in., $2a = 24$ in., $3a = 36$ in., the equivalent weight of each piston is 9 lb, and the crank speed is 1800 rpm, determine the magnitude of the maximum unbalance. State the kind of unbalance. Show how the unbalanced effect may be balanced. Specify the magnitude of each balance weight used if the weights can be set with a distance of 6 in. to the center of gravity from the axis of rotation of each weight.

20.11 (Fig. 20.11). A five-throw crankshaft is not usually used in internal-combustion engines. However, Figs. 20.11a and b show a five-throw crankshaft

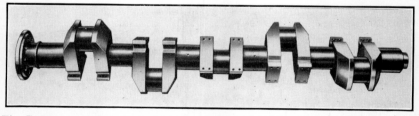

Fig. P–20.11a. Thirty-inch stroke crankshaft of a five-cylinder, 3500-hp engine. Length, 21 ft, $3\frac{1}{2}$ in. Weight, 28,600 lb. (Courtesy Skinner Engine Co., Erie, Pa.)

for a Unaflow steam engine, of no small size, as furnished by the Skinner Engine Co., Erie, Pennsylvania. Assuming that $\phi_1 = 0°$, $\phi_2 = 72°$, $\phi_3 = 144°$, $\phi_4 = 216°$,

$\phi_5 = 288°$, $a_1 = 0$, $a_2 = 2\frac{1}{2}$ ft., $a_3 = 5$ ft., $a_4 = 12\frac{1}{2}$ ft., $a_5 = 15$ ft., determine the kind of unbalance in the engine, if any. Assume that the crankshaft is balanced.

Fig. P–20.11b. Base and shaft of the marine "Unaflow," 30-in. stroke steam engine, showing counterbalances. (Courtesy Skinner Engine Co., Erie, Pa.)

20.12. A five-cylinder in-line engine uses pistons that have an equivalent weight of $4\frac{1}{2}$ lb each. Determine the magnitude of the resultant shaking force or couple when crank 1 is 90 degrees past head-end dead center. The crank rotates counterclockwise. $R = 2\frac{1}{2}$ in., $L = 10$ in. Crank speed = 1200 rpm.

$\phi_1 = 0°$, $\phi_2 = 315°$, $\phi_3 = 180°$, $\phi_4 = 225°$, $\phi_5 = 90°$, measured counterclockwise.

$a_1 = 0$, $a_2 = 4$ in., $a_3 = 8$ in., $a_4 = 12$ in., $a_5 = 16$ in.

20.13. If the equivalent weight of each piston in Fig. 20.9a of the text is 8 lb, the crank radius is 4 in., and the connecting rod length is 14 in., determine the magnitude of each counterbalance weight at a 6-in. radius to balance the engine. The distance $a = 5$ in. The distance between counterweights may be taken as 40 in.

20.14. Determine the conditions of unbalance for a six-cylinder in-line engine with the following:

$\phi_1 = 0°$, $\phi_2 = 60°$, $\phi_3 = 120°$, $\phi_4 = 180°$, $\phi_5 = 240°$, $\phi_6 = 300°$.

$a_1 = 0$, $a_2 = 4$ in., $a_3 = 8$ in., $a_4 = 12$ in., $a_5 = 16$ in., $a_6 = 20$ in.

20.15. Determine the conditions of unbalance for the six-cylinder in-line engine with the following:

$\phi_1 = 0°$, $\phi_2 = 90°$, $\phi_3 = 180°$, $\phi_4 = 270°$, $\phi_5 = 0°$, $\phi_6 = 180°$.

$a_1 = 0$, $a_2 = a$, $a_3 = 2a$, $a_4 = 3a$, $a_5 = 4a$, $a_6 = 5a$.

20.16. A six-cylinder engine runs at 3600 rpm, and each piston has a stroke of 4 in. Each piston weighs 3 lb, and the connecting rods are 8 in. long. If the total reciprocating weight per cylinder is 4 lb, determine the magnitude of the resultant force or couple when crank 1 is 30 degrees past head-end dead center, measured counterclockwise.

$\phi_1 = 0°$, $\phi_2 = 90°$, $\phi_3 = 180°$, $\phi_4 = 270°$, $\phi_5 = 45°$, $\phi_6 = 225°$.

$a_1 = 0$, $a_2 = 4$ in., $a_3 = 10$ in., $a_4 = 14$ in., $a_5 = 20$ in., $a_6 = 24$ in.

382 Balancing Masses Reciprocating in Several Planes

20.17. A seven-cylinder in-line engine has the following:

$\phi_1 = 0°$, $\phi_2 = 90°$, $\phi_3 = 180°$, $\phi_4 = 270°$, $\phi_5 = 0°$, $\phi_6 = 120°$, $\phi_7 = 240°$.

$a_1 = 0$, $a_2 = a$, $a_3 = 2a$, $a_4 = 3a$, $a_5 = 4a$, $a_6 = 5a$, $a_7 = 6a$. What type of unbalance exists?

20.18. An eight-cylinder, in-line engine has the following specifications:

$\phi_1 = 0°$, $\phi_2 = 180°$, $\phi_3 = 180°$, $\phi_4 = 0°$, $\phi_5 = 270°$, $\phi_6 = 90°$, $\phi_7 = 90°$, $\phi_8 = 270°$.

$a_1 = 0$, $a_2 = 4$ in., $a_3 = 8$ in., $a_4 = 12$ in., $a_5 = 16$ in., $a_6 = 20$ in., $a_7 = 24$ in., $a_8 = 28$ in.

$R = 3$ in., $L = 12$ in. Equivalent weight of each piston = 6 lb. The crank rotates at 1800 rpm.

Show how this engine may be balanced by a system of rotating weights. Determine the magnitude of each counterbalance weight if the distance to the center of gravity of each weight is 5 in. Assume a distance between counterweights.

20.19. Refer to the six-cylinder V-engine of Fig. 20.11a of the text. Each crank throw is 3 in., each connecting rod is $10\frac{1}{2}$ in., the equivalent weight of each piston is 6 lb, the crank speed is 1800 rpm counterclockwise, $a = 5$ in., $\psi = 30°$.

(a) Determine the type of unbalance in the engine.

(b) Determine a gearing arrangement to balance the engine.

(c) What magnitude of each counterbalance weight is necessary at a radius of 4 in. to the center of gravity of each weight? Assume a distance between counterweights.

20.20. Refer to Fig. 20.12a of the text. Each crank throw is 3 in., each connecting rod length is $10\frac{1}{2}$ in., the equivalent weight of each piston is 6 lb, the crank speed is 1800 rpm counterclockwise, $a = 5$ in., $\psi = 45°$.

(a) Determine the maximum magnitude of the unbalance of the couple in the vertical plane.

(b) Determine the magnitude of the counterweights on the crankshaft required for balance of the unbalanced couple. The distance to the center of gravity of each counterweight is 5 in. The distance between counterweights is 25 in.

CHAPTER 21

Vibrations in Shafts

A phenomenon that occurs with rotating shafts at certain speeds is an excessive amount of vibration, even though the shaft may run very smoothly at other speeds. At such speeds where the vibration becomes excessive, failure of the shaft or bearings may occur. Or the vibration may cause failure by preventing the proper functioning of parts, as might occur in a steam turbine where the clearance between the rotor and the casing is small. Such vibration may be due to what is called whirling of the shaft, or it may be due to a torsional oscillation in the shaft, or to a combination of both. Although the two phenomena are different, it will be shown that each can be handled in similar fashions by considering natural frequencies of oscillations. Inasmuch as shafts are basically elastic, and exhibit spring characteristics, analysis of a simplified mass and spring system will be used to illustrate the approach and to bring out the concepts of the basic terms used.

21.1 Mass vibrating in a horizontal plane

Figure 21.1a shows a mass weighing W pounds resting on a frictionless surface and fastened to the stationary structure through a spring.

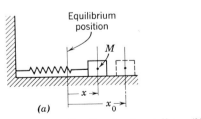
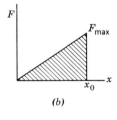

Fig. 21.1(a). Mass vibrating horizontally. (b) Work done on the spring is the area under the force-deflection curve.

The mass of the spring will be neglected in the analysis. The mass is displaced x_0 from the equilibrium position, and released. The type of motion is to be determined. Two analyses will be used, one by applying Newton's equations directly, and the other by considering energy relations.

Method I. Applying Newton's equations. For a general position, consider the instant when the center of gravity of the mass is at a distance x from the equilibrium position, Fig. 21.1a. The force exerted on the body is the spring force, kx, where k represents the spring rate, pounds per inch of deflection of the spring. The spring rate is assumed to be a constant quantity. From $F = MA$, we may set

$$-kx = \frac{W}{g}\frac{d^2x}{dt^2}$$

The negative sign is used because the spring force gives an acceleration in the negative x-direction. Or

$$\frac{d^2x}{dt^2} = -\frac{kg}{W}x$$

The general solution to the equation above is

$$x = A\cos(kg/W)^{1/2}t + B\sin(kg/W)^{1/2}t$$

which can be verified by differentiation. For the particular boundary conditions that at $t = 0$, $x = x_0$, and $\frac{dx}{dt} = 0$, the constants of integration are found to be $A = x_0$ and $B = 0$. Thus the solution to the differential equation is

$$x = x_0\cos(kg/W)^{1/2}t$$

The time for one complete cycle of motion is found by setting $(kg/W)^{1/2}t = 2\pi$, or $t = 2\pi(W/kg)^{1/2}$.

The cycles per second is given by $f = \dfrac{1}{t} = \dfrac{1}{2\pi}(kg/W)^{1/2}$

The cycles per minute is given by $N = \dfrac{60}{2\pi}(kg/W)^{1/2}$

The frequency of oscillation can be expressed differently if we determine the spring rate for a specific condition: the deflection of the spring under the action of the weight W, as though the weight were suspended from the spring in a vertical position, where the specific deflection is called the static deflection x_{st}. Thus, $k = W/x_{st}$. Substitution of the quantity in the above gives $N = \dfrac{60}{2\pi}(g/x_{st})^{1/2}$. If x_{st} is expressed in inches, and g is given by 386 in./sec^2, the frequency can be given by $N = 187.7(1/x_{st})^{1/2}$ cycles per minute. Note that the stiffer the spring, that is, the smaller the deflection for a given weight, the higher the frequency of oscillation.

Method II. Using energy relations. The use of an energy relation relates the transfer of maximum potential energy to maximum kinetic energy. The maximum potential energy stored in the spring occurs when the mass is displaced the maximum distance x_0. The potential energy in the spring can be found from the area under the force-deflection curve, Fig. 21.1b, and is $\frac{1}{2}F_{max}x_0$. Since the maximum force is kx_0, the maximum potential energy can be given by $\frac{1}{2}kx_0^2$.

It is necessary that the equation for the displacement be known to find the maximum velocity, which in turn will permit the finding of the maximum kinetic energy. The type of motion which can be assumed is a harmonic one, expressed by $x = x_0 \cos \omega t$. The velocity is found by differentiation:

$$\frac{dx}{dt} = V = -x_0 \omega \sin \omega t$$

The maximum velocity occurs when $\omega t = \frac{\pi}{2}, \frac{3\pi}{2}, \frac{5\pi}{2}$, etc., and is equal to $\pm x_0\omega$. Thus, the maximum kinetic energy is $\frac{1}{2}\frac{W}{g}V^2 = \frac{1}{2}\frac{W}{g}(x_0\omega)^2$.

Setting the maximum potential energy at one point of the cycle equal to the maximum kinetic energy at another point of the cycle, we obtain

$$\tfrac{1}{2} kx_0^2 = \frac{1}{2}\frac{W}{g}(x_0\omega)^2$$

or
$$\omega = (kg/W)^{1/2} \text{ rad/sec}$$

Recognizing that the spring rate may be defined for a particular deflection and load, the static deflection for the weight W, we may substitute $k = W/x_{st}$ into the equation above and obtain

$$\omega = (g/x_{st})^{1/2} \text{ rad/sec}$$

The cycles per minute are $N = 187.7(1/x_{st})^{1/2}$.

The result is the same as obtained by Method I. The conclusion that may be drawn is that the assumed motion is satisfactory. The same type of motion will be assumed later in the analysis of vibration of shafts.

Note that external friction, as friction due to the motion in air, and internal friction in the spring wire have been neglected. The effect of such friction would bring the mass to rest eventually. Also, the mass of the spring has been neglected, which effect would be to lower the natural frequency of oscillation.*, †

21.2 Mass vibrating in a vertical plane

Figure 21.2a shows a system comparable to that of the preceding section, except that the mass is suspended from a vertical spring. The

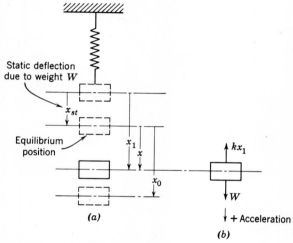

Fig. 21.2. Mass vibrating vertically has the same natural frequency as a mass vibrating horizontally, with the oscillation taking place about the equilibrium position.

weight causes the spring to deflect x_{st}. Picture the mass pulled down a distance x_0 from the equilibrium position and then released. The motion is to be found, as well as the effect of gravity.

Let x_1 represent the displacement of the mass from the position where the spring has no load applied to it, for any time t. The forces applied to the mass are the spring force and the weight of the mass. From $F = MA$,

$$W - kx_1 = \frac{W}{g} \frac{d^2 x_1}{dt^2}$$

* See *Mechanical Vibrations*, by W. T. Thompson, Prentice-Hall, first edition, pages 32–33, for an analysis taking into account the effect of a spring in vibration.
† See *Advanced Dynamics*, by Timoshenko and Young, McGraw-Hill, pages 170–171.

Let x represent the distance from the general position of the mass to the equilibrium position, or $x + x_{st} = x_1$.

Substitute x_1 in the preceding expression:

$$W - k(x + x_{st}) = \frac{W}{g}\frac{d^2 x_1}{dt^2}$$

But $W = kx_{st}$, and $\dfrac{d^2 x_1}{dt^2} = \dfrac{d^2 x}{dt^2}$, which may be seen by differentiating twice with respect to time the expression $x + x_{st} = x_1$, with x_{st} a constant for a given system.

Therefore,
$$-kx = \frac{W}{g}\frac{d^2 x}{dt^2}$$

The solution to the differential equation, as seen in the previous section, for the boundary conditions when $t = 0$, $V = 0$, and $x = x_0$, is

$$x = x_0 \cos (kg/W)^{1/2} t$$

with the cycles per minute of the oscillation given by

$$N = 187.7(1/x_{st})^{1/2}$$

Thus the same frequency of oscillation is obtained for the mass with the spring vertical or horizontal.

21.3 Whirling of shafts

The phenomenon of "whirling of shafts" will be discussed to illustrate why shafts exhibit large deflections at a particular speed of operation, although the shaft may run smoothly at reduced or increased speeds. Figure 21.3a shows a shaft L inches long supported in self-aligning bearings at the ends. A disk which is considered to be a concentrated mass and weighs W pounds is located a inches from the top bearing. The mass of the shaft will be considered negligible, gyroscopic action of the mass will be neglected, and it will be further assumed that the shaft is driven through a flexible coupling which exerts no restraint in the deflection of the shaft. The shaft is considered vertical so that gravity can be ignored, although the results to be obtained are the same whether the shaft is vertical or horizontal.

If the center of gravity of the concentrated mass is on the axis of rotation, there will not be any unbalance of any kind to cause the shaft to rotate about any other axis than the axis of the shaft. Practically, however, such a condition cannot be achieved, and the center of gravity of the disk may be assumed to be a small distance, e, from the geometric center of the disk. With the center of gravity off the axis of rotation, or the axis of the bearings, an inertia force is present to cause the shaft to deflect, the deflection of the center of the shaft being indicated by r, in Fig. 21.3b. The geometric center of the disk, O, is the same as the center of the shaft at the disk. As the shaft

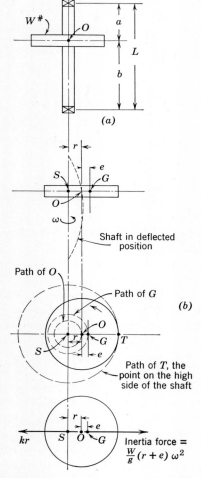

G: Center of gravity of the disk.
O: Geometric center of the disk (or the center of the shaft at the disk for the deflected position.)
S: Center of rotation of the disk.

Fig. 21.3a, b. Motion and forces for a disk on a shaft rotating about a fixed axis.

rotates, the high spot T will rotate about the axis, S, of the bearings. The inertia force of the disk is balanced by what may be called the spring force of the shaft as the shaft rotates. The inertia force, for a mass rotating about a fixed center, is

Whirling of Shafts

$$\frac{W}{g}(r+e)\omega^2$$

The spring force of the shaft may be represented by kr, where k is the shaft spring rate, pounds required per inch of deflection of the

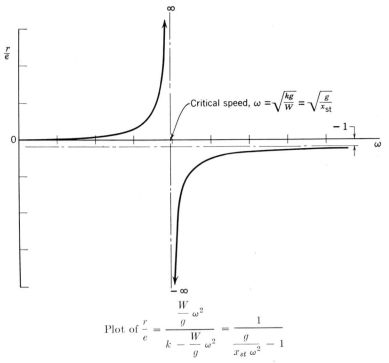

Fig. 21.4. Critical or whirling speed occurs when a shaft has a theoretical infinite radius of deflection, with friction disregarded.

shaft *at the disk*. Summing up the forces in Fig. 21.3b equal to zero, with the inertia force included, we obtain

$$\frac{W}{g}(r+e)\omega^2 - kr = 0$$

Rearranging terms,

$$\frac{r}{e} = \frac{\dfrac{W}{g}\omega^2}{k - \dfrac{W}{g}\omega^2}$$

A plot of r/e versus ω is shown in Fig. 21.4.

Vibrations in Shafts

The dangerous speed of operation of the particular shaft is indicated by the critical speed or whirling speed, the speed at which the ratio of r/e is infinite. Operation at a speed close to the critical speed is also undesirable because of the large displacement of the center of the disk from the axis of rotation. The critical speed can be obtained for the condition that the denominator of the above expression be zero:

$$k - \frac{W}{g}\omega^2 = 0 \quad \text{or} \quad \omega = (kg/W)^{1/2}$$

The constant k may be expressed in various ways, such as that obtained from the equation for the deflection of a simply supported shaft under the action of a load P:

$$r = \frac{Pab}{6LEI}(L^2 - a^2 - b^2)*$$

The ratio of P/r defines the spring rate, k. Thus,

$$k = \frac{P}{r} = \frac{6LEI}{ab(L^2 - a^2 - b^2)}$$

For the particular shaft under discussion, the critical speed can be expressed by

$$\omega = \sqrt{\frac{6LEI}{ab(L^2 - a^2 - b^2)}\frac{g}{W}} \text{ rad/sec}$$

An alternate method of expression is to specify the spring rate k in terms of a specific load and specific deflection, the load being equal to the weight of the disk, that is, $P = W$. The resultant deflection would be the static deflection of the shaft, as though the shaft were horizontal, under the action of the weight of the disk. The static deflection is called x_{st}.

Thus,
$$k = \frac{P}{r} = \frac{W}{x_{st}}$$

or
$$\omega = (kg/W)^{1/2} = \left[\frac{W}{x_{st}}\frac{g}{W}\right]^{1/2} = (g/x_{st})^{1/2} \text{ rad/sec}$$

For $g = 386$ in./sec^2, with x_{st} given in inches, the critical speed in rpm is

$$N = 187.7(1/x_{st})^{1/2}$$

---

* See any strength of materials textbook. A method of determining deflections of shafts under various loads is discussed later in this chapter.

21.4 The effect of friction on the critical speed

Although the theoretical equation developed in the preceding section indicates an infinite radius of rotation at the critical speed, such a condition is not possible practically. The shaft running at the critical speed would, of course, break or be distorted according to the results obtained, to prevent operation. However, we know that shafts running at the critical speed do not necessarily break, and may run very roughly but without even permanent distortion. The reason for the discrepancy between the analytical results and experience is that friction has been neglected, which, strangely enough, will affect r/e considerably but does not affect appreciably the value of the dangerous operating speed. Let us analyze the shaft of the preceding section with friction taken into account.*

As the shaft rotates, friction will oppose the motion. If the system is taken as in Fig. 21.5a, with points S, O, and G collinear, and if a friction force is shown as opposing the direction of motion, a system of equilibrium cannot be obtained. Note that the friction force is assumed to act through the geometric center of the disk. Thus, as long as S, O, and G are assumed to lie on a straight line, friction cannot be included. If friction is to be considered, there must be a shift of the relative position of the center of rotation of the disk, geometric center of the disk, and center of gravity of the disk.

Figure 21.5b shows an arrangement of points S, O, and G which will be assumed as a possible system wherein equilibrium can be obtained. The spring force, inertia force, and friction force give a couple effect which is balanced by a couple applied to the shaft. Note that the "heavy" and "light" sides have been shifted as compared to the case where friction has been neglected.

The friction force is assumed to be proportional to the linear velocity of the geometric center, or $F = Cr\omega$.

Summing up forces in the x- and y-directions, we obtain:

$$\frac{W}{g} p\omega^2 \cos \alpha = kr \tag{1}$$

$$\frac{W}{g} p\omega^2 \sin \alpha = Cr\omega \tag{2}$$

Introduce the angle ϕ, Fig. 21.5b, and express the geometrical relations

$$r = p \cos \alpha + e \cos \phi \tag{3}$$

$$p \sin \alpha = e \sin \phi \tag{4}$$

* See the article by R. P. Kroon, "Balancing of Rotating Apparatus," *Design Data, Applied Mechanics*, Book 2, published by the ASME.

Vibrations in Shafts

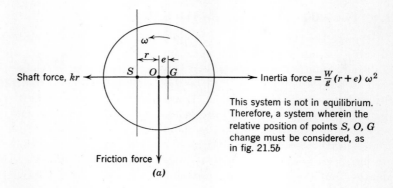

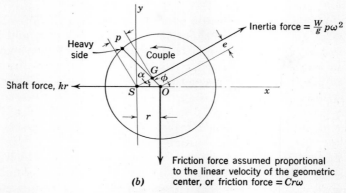

Fig. 21.5a, b. Consideration of friction requires a shift of the relative position of the center of rotation of the disk, geometric center of the disk, and center of gravity of the disk.

Substitute the values of $(p \cos \alpha)$ and $(p \sin \alpha)$ from the above into Eqs. 1 and 2:

$$\frac{W}{g}(r - e \cos \phi)\omega^2 = kr \tag{5}$$

$$\frac{W}{g}(e \sin \phi)\omega^2 = Cr\omega \tag{6}$$

From the above, $\tan \phi$ can be found:

$$\tan \phi = \frac{Cr\omega}{\dfrac{Wr}{g}\omega^2 - kr} \tag{7}$$

The Effect of Friction on the Critical Speed

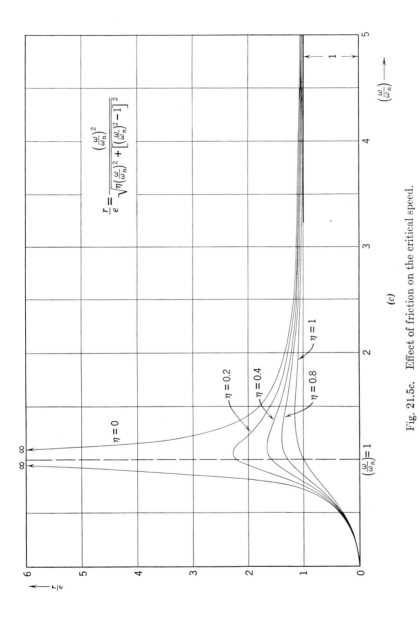

Fig. 21.5c. Effect of friction on the critical speed.

As the speed of the shaft changes, the angle ϕ changes, or the heavy side will not be the high spot as the shaft rotates. This must be recognized in the balancing of shafts, especially in trial and error balancing. If the speed of rotation is relatively low, as is used in balancing machines, the high spot and heavy side will correspond closely.

Solve for $\sin \phi$ and $\cos \phi$, Eqs. 5 and 6, substitute into $\sin^2 \phi + \cos^2 \phi = 1$, and obtain r/e:

$$\frac{r}{e} = \frac{\dfrac{W}{g}\omega^2}{\left[(C\omega)^2 + \left(\dfrac{W}{g}\omega^2 - k\right)^2\right]^{1/2}}$$

The equation above can be simplified further by substituting $\omega_n^2 = \dfrac{k}{W/g}$ and introducing $\left(\dfrac{C}{\omega_n \dfrac{W}{g}}\right)^2 = \eta$, a dimensionless number:

$$\frac{r}{e} = \frac{\left(\dfrac{\omega}{\omega_n}\right)^2}{\sqrt{\eta\left(\dfrac{\omega}{\omega_n}\right)^2 + \left[\left(\dfrac{\omega}{\omega_n}\right)^2 - 1\right]^2}}$$

The plot of r/e is shown in Fig. 21.5c. Note that the maximum ratio of r/e is not infinite when friction is taken into account. However, there is a region where the shaft whirls at a large radius, and the maximum r/e value occurs at a speed not far from that calculated with no friction. Also, the r/e value at speeds well away from the whirling speed are not too much different with or without friction. Practically, friction is usually disregarded and the whirling speed calculated for no friction, with very little error.

21.5 Whirling speed same as the natural frequency of oscillation

It will be shown in this section that the whirling speed as found previously, can be found also by consideration of the natural frequency of oscillation, which can be considered that frequency of oscillation found by deflecting the shaft with the attached mass and permitting the shaft to vibrate freely. Friction will be disregarded. Examination of Fig. 21.6, the system shown in Fig. 21.3b, where the disk whirls about the axis of rotation, shows that the geometric center rotates

Whirling Speed; Natural Frequency of Oscillation

about the axis of the bearings, with the projection of the motion of point O being simple harmonic. Therefore, let us consider the system idealized as a concentrated mass at the geometric center of the disk, and consider the projected motion only. The distance from the center of gravity to the geometric center of the disk, e, is considered zero.

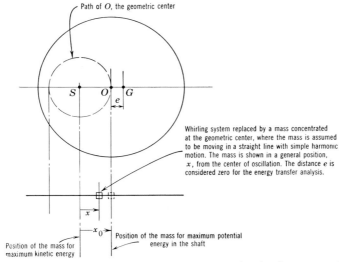

Fig. 21.6. Analysis of linear vibration gives same value for frequency of oscillation as the critical or whirling speed.

Picture the mass displaced a distance x_0 and released, as shown in Fig. 21.6, and picture the mass vibrating in a linear path. The harmonic motion may be expressed by

$$x = x_0 \cos \omega t$$

The maximum velocity of the mass is found by differentiation:

$$\frac{dx}{dt} = V = -x_0 \omega \sin \omega t$$

The maximum kinetic energy is $\frac{1}{2}\frac{W}{g} V^2 = \frac{1}{2}\frac{W}{g}(x_0\omega)^2$, which occurs when $\sin \omega t = 1$. The maximum potential energy is $\frac{1}{2}kx_0^2$, where k is the spring rate of the shaft. Relate the maximum potential energy and the maximum kinetic energy

$$\tfrac{1}{2}kx_0^2 = \frac{1}{2}\frac{W}{g}(x_0\omega)^2$$

or

$$\omega = (kg/W)^{1/2} \text{ rad/sec}$$

If k is expressed for a particular value of load and deflection, the load W and the deflection x_{st} due to the load W, as though the shaft were horizontal,

$$k = W/x_{st}$$

the above expression may be given as

$$\omega = (kg/W)^{1/2} = \left[\frac{W}{x_{st}}\frac{g}{W}\right]^{1/2} = (g/x_{st})^{1/2} \text{ rad/sec}$$

Or the natural frequency of oscillation can be expressed by

$$N = 187.7(1/x_{st})^{1/2} \text{ rpm}$$

Thus the whirling speed and natural frequency of oscillation are identical for the case of a single mass on a simply supported shaft. It is for this reason that whirling of a shaft is oftentimes called a vibration, which it is not, strictly speaking. However, we may explain the phenomenon of whirling by realizing that a shaft rotating at a speed which corresponds to the natural frequency of oscillation applies, as a result of a small displacement of the center of gravity from the axis of rotation, a periodic force in phase with the natural frequency of oscillation to cause a critical condition. This is the reason for calling the whirling speed the critical speed.

21.6 Critical speed of a multimass system

Figure 21.7a shows a shaft with any number of concentrated masses. The shaft may be considered horizontal or vertical, inasmuch as gravity does not affect the whirling or critical speed. The shape of the shaft may assume various forms of deflection, as indicated in Fig. 21.7b. The critical speed for the shape in Fig. 21.7a is found to be the lowest in magnitude, and the value found is called the primary or first-order critical speed. In this book, only the primary critical speed will be considered.

The critical speed will be found by determining the natural frequency of oscillation of the system by the application of energy considerations, as was done in the previous section. The shaft will be assumed to be deflected by the action of forces equal to the weight of each mass, and the transfer of the maximum potential energy to the maximum kinetic energy will be related, assuming harmonic motion of the shaft in a plane.

Consider static loads W_1, W_2, W_3, etc., as shown in Fig. 21.7c, applied simultaneously to the shaft. Since the forces are applied gradually, the work done by the forces on the shaft is $\frac{1}{2}W_1y_1 + \frac{1}{2}W_2y_2 + \frac{1}{2}W_3y_3 + \cdots + \frac{1}{2}W_ny_n$, which can be written $\frac{1}{2}\Sigma W_n y_n$. The deflections y_1, y_2, etc., are considered the static deflection at the cor-

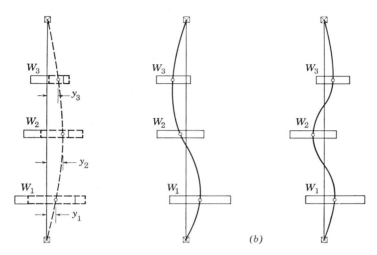

(a) First order critical speed   Higher order critical speeds

Fig. 21.7a, b. Deflection curves for various orders of critical speed.

responding load. Thus the deflection curve which has been assumed is the same as the deflection curve which results when the shaft is in a horizontal position under the action of the weight of each mass. It was pointed out in the analyses given previously that the same frequency of oscillation resulted for a mass supported by a spring for

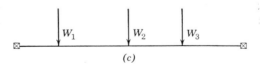

(c)

Fig. 21.7c. Analysis for critical speed of a multimass system is made as though the shaft were loaded with weights equal to the weight of each mass.

any initial displacement of the mass. For convenience, the initial displacement taken for the multimass system is that due to the weight of the disks.

For an assumed harmonic motion, the displacement of each mass can be expressed by

Vibrations in Shafts

$$y_I = y_1 \cos \omega t$$
$$y_{II} = y_2 \cos \omega t$$
$$y_{III} = y_3 \cos \omega t$$
$$\cdot \qquad \cdot$$
$$\cdot \qquad \cdot$$
$$\cdot \qquad \cdot$$

The velocity of each mass is obtained by differentiation with respect to time:
$$V_I = -y_1\omega \sin \omega t$$
$$V_{II} = -y_2\omega \sin \omega t$$
$$V_{III} = -y_3\omega \sin \omega t$$
$$\cdot \qquad \cdot$$
$$\cdot \qquad \cdot$$
$$\cdot \qquad \cdot$$

The maximum velocity of each mass will occur at the same time, when $\sin \omega t = 1$. Thus,
$$V_{1\max} = -y_1\omega$$
$$V_{2\max} = -y_2\omega$$
$$V_{3\max} = -y_3\omega$$
$$\cdot \qquad \cdot$$
$$\cdot \qquad \cdot$$

The maximum kinetic energy of each mass is $\dfrac{1}{2}\dfrac{W}{g}V^2$. Equate the total potential energy to the total kinetic energy:

$$\tfrac{1}{2}W_1 y_1 + \tfrac{1}{2}W_2 y_2 + \tfrac{1}{2}W_3 y_3 + \cdots = \frac{1}{2}\frac{W_1}{g}(-y_1\omega)^2$$
$$+ \frac{1}{2}\frac{W_2}{g}(-y_2\omega)^2 + \frac{1}{2}\frac{W_3}{g}(-y_3\omega)^2 + \cdots$$

or
$$\sum W_n y_n = \frac{\omega^2}{g}\sum W_n y_n^2$$

Solve for ω:
$$\omega = \left[\frac{g(W_1 y_1 + W_2 y_2 + W_3 y_3 + \cdots)}{(W_1 y_1^2 + W_2 y_2^2 + W_3 y_3^2 + \cdots)}\right]^{1/2} = \left[\frac{g(\Sigma W_n y_n)}{\Sigma W_n y_n^2}\right]^{1/2} \text{ rad/sec}$$

Shear Forces and Bending Moments in Beams 399

The natural frequency of oscillation, or the whirling speed, is

$$N = 187.7 \left(\frac{\Sigma W_n y_n}{\Sigma W_n y_n^2} \right)^{\!\!1/2} \text{rpm}$$

The analysis carried out for the first-order critical speed, although an approximation, inasmuch as each inertia force is assumed proportional to the mass times the static deflection at the mass, nevertheless gives good working results.

21.7 Methods of finding deflections in shafts

The problem of finding the first-order critical speed reduces to the problem of finding the deflection at each mass on a shaft, as though the shaft were statically loaded in a horizontal position, with forces equal to the weight of each mass. The deflections thus found are used, in a sense, to obtain the flexibility of the system.

Various methods are available for finding deflections, such as (1) mathematical integration, (2) area-moment method, (3) Castigliano's theorem, (4) conjugate beam or elastic loading method, (5) graphical integration. All methods, except 3, are based on the basic expression for the deflection curve: $\dfrac{d^2y}{dx^2} = \dfrac{M}{EI}$, where y is the deflection at any point along the axis of the beam, M is the bending moment at any section of the beam, E is the modulus of elasticity in tension for the material of the beam, and I is the moment of inertia about the axis of bending at the section where the bending moment is determined. Essentially, $\dfrac{d^2y}{dx^2}$ is the reciprocal of the radius of curvature of the axis of the beam at a given section of the beam, or

$$\frac{d^2y}{dx^2} = \frac{M}{EI} = \frac{1}{\rho}$$

One method will be discussed in this book, the conjugate beam method.

21.8 Shear forces and bending moments in beams

We shall digress a bit and consider the subject of shear forces and bending moments in beams preliminary to the determination of deflections in beams by the conjugate beam method, since the methods are comparable.

In any beam, regardless of the loading, or type of beam, the shear force at any section of the beam is defined as the force transverse to the axis of the beam necessary to balance the transverse forces of

400   Vibrations in Shafts

the isolated portion of the beam, and the bending moment is defined as the couple necessary to balance the couples acting on the isolated portion of the beam.* The plot of the shear forces tranverse to the axis of the beam is called the shear force diagram, and the plot of

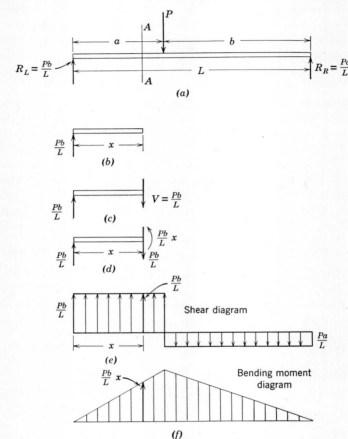

Fig. 21.8. Shear and bending moment diagrams.

the couples required for equilibrium is called the bending moment diagram.

Consider the simple system of Fig. 21.8a, where a single concentrated load P is applied to the simply supported beam. The reactions are Pb/L and Pa/L. Consider the beam cut at any section A–A. Either the left or right section can be considered, with identical results.

* Bending moment, as defined, is sometimes called the resisting bending moment.

Figure 21.8b shows the left portion isolated because there are fewer loads applied. The system is not in equilibrium, and a force $Pb/L = V$, a shear force, must be applied for balance of forces, as shown in Fig. 21.8c. The two equal and opposite forces set up a couple, which can be balanced only by another couple, Pbx/L, as shown in Fig. 21.8d. The plot of the forces and couples required for equilibrium at every section of the beam gives the shear and bending moment diagrams in Fig. 21.8e and Fig. 21.8f.

21.9 Relation of loading, shear, and bending moment

The shear and moment diagrams can be obtained for any loaded beam by the method of the preceding section by a section-to-section analysis. There are methods of obtaining the shape of the curves, with simplification of calculations. Consider a beam loaded in any fashion (Fig. 21.9a), where the loads are shown acting upwards in what is defined as the positive direction. Positive and negative shear and bending moments are as defined in Fig. 21.9b.

Isolate a small section of differential length of beam as shown in Fig. 21.9c. Positive loading, shear, and bending moment are assumed. A shear force V is acting on the left side, and a shear force $V + dV$ is acting on the right; a bending moment M is acting on the left side, and a bending moment $M + dM$ is acting on the right. A distributed loading of intensity w pounds per inch is shown. For equilibrium of forces

$$+V + w\,dx - (V + dV) = 0$$

or

$$w = \frac{dV}{dx}$$

In addition, for complete equilibrium, the moments about any point must be zero. Taking moments about any point on the right side:

$$+M + (w\,dx)\frac{dx}{2} + V\,dx - (M + dM) = 0$$

Simplifying, and recognizing that $(w/2)(dx)^2$ is a differential of higher order and therefore infinitesimal,

$$V = \frac{dM}{dx}$$

Differentiate the equation above with respect to x to obtain $\dfrac{dV}{dx} = \dfrac{d^2M}{dx^2}$.

or

$$w = \frac{dV}{dx} = \frac{d^2M}{dx^2}, \text{ with } V = \frac{dM}{dx}$$

402 Vibrations in Shafts

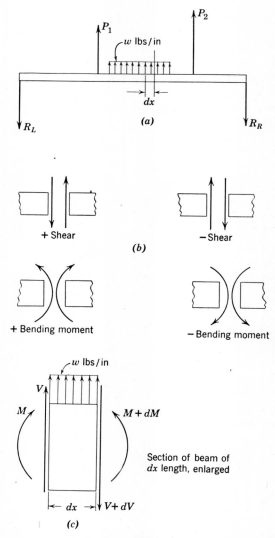

Fig. 21.9. Shear forces and bending moments at a section of a beam.

In drawing shear and moment diagrams, there are two methods of attack, using the equations above. One method is to derive from $w = \dfrac{dV}{dx}$ the expression $\int w\,dx = \int dV$ and to derive from $V = \dfrac{dM}{dx}$ the expression $\int V\,dx = \int dM$, with the interpretation that the area under the loading-displacement curve is equal to the *change of shear*

Relation of Loading, Shear, Bending Moment

and that the area under the shear-displacement curve is equal to the *change of bending moment*. Care must be exercised in the use of the expressions where discontinuities occur.

The second method is the basic interpretation of $w = \dfrac{dV}{dx}$ and $V = \dfrac{dM}{dx}$, that is, the slope of the shear diagram is equal to the intensity of loading and the slope of the moment diagram is equal to the shear force. From these basic relations, we may determine the shape of the shear and moment diagrams without any calculation. Families of curves can be found to satisfy the given loading condition, with the particular correct solution depending upon boundary conditions. Let us illustrate by reference to a simply supported beam with a single concentrated load (Fig. 21.10a). In the section to the left of the load P, the intensity of loading is zero, and therefore the slope of the shear diagram is zero. Figure 21.10b shows a possible family of curves in the left portion that have zero slopes. In a similar fashion, the intensity of loading on the right portion of the beam is zero at every point between the load and support, which means that the slope of the shear curve at every section of the beam is zero. The family of curves in the right portion is shown in Fig. 21.10c. Now the question is: Which curve of the family of curves is the correct solution? That depends entirely on what may be called the boundary conditions, the reactions at the supports in this case. Calculation of the reactions determines the shear values at two points, thus permitting the particular curves to be selected (Fig. 21.10d). The same procedure can be used to determine the moment diagram. In the portion of the beam to the left of the concentrated load, the intensity of shear is constant and positive at every point. Therefore, the slope of the moment diagram is constant and positive. The family of curves to satisfy the specifications is shown in Fig. 21.10e. Similarly, the intensity of shear at every section of the beam to the right of the load P is negative and constant. The family of curves to satisfy the conditions is shown in Fig. 21.10f. Knowing that the boundary conditions are such that the moment is zero at each reaction, we may draw Fig. 21.10g.

Thus the solution of $w = \dfrac{d^2M}{dx^2}$ for the moment diagram is obtained by determining the moment diagram of the loaded beam. Whatever the application, whether $\dfrac{M}{EI} = \dfrac{d^2y}{dx^2}$ or $A = \dfrac{d^2x}{dt^2}$, the method of integra-

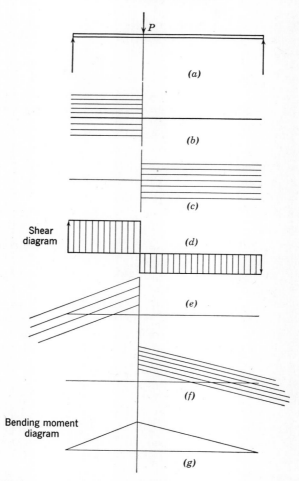

Fig. 21.10. The slope of the shear diagram at any section is equal to the intensity of loading, and the slope of the moment diagram at any section is equal to the shear force.

tion can be likened to the determination of finding a moment diagram for the proper loading. Such is the basis of the conjugate beam method for finding deflections in beams.

21.10 Deflection in beams by the conjugate beam method

Consider first the similarity of equations used in finding bending moment diagrams and deflection curves:

Conjugate Beam Applied to a Cantilever Beam

Bending moment diagrams: $w = \dfrac{dV}{dx} = \dfrac{d^2M}{dx^2}$ where $V = \dfrac{dM}{dx}$

Deflection in beams: $\dfrac{M}{EI} = \dfrac{d\phi}{dx} = \dfrac{d^2y}{dx^2}$ where $\phi = \dfrac{dy}{dx}$

The new variable, ϕ, is the slope of the deflection curve. If M/EI is considered as a loading on a fictitious beam, the deflection can be obtained by determining the proper "bending moment diagram" of the M/EI loading. Examination of the above functions shows that the intensity of the fictitious loading is equal to the slope of the slope-displacement diagram; also, the ordinate of the slope diagram is equal to the slope of the actual deflection curve.

21.11 Deflection of a cantilever beam by the conjugate beam method

Figure 21.11a shows a cantilever beam of length L and of constant cross section. The shear and bending moment diagrams are given without discussion, as well as the M/EI diagram. The M/EI diagram is considered as a loading on a fictitious beam; this type of beam is to be determined from the conditions of slope and deflection of the actual beam. Figure 21.11b shows the general deflection characteristics, where the slope and deflection are zero at the left.

Picture starting the "shear" diagram at the left of the M/EI loading. The "shear" would be zero, and since the "shear" corresponds to the slope of the actual beam, which is zero, the boundary condition is already satisfied. Picture, now, starting the "shear" diagram at the right end of the M/EI diagram. The "shear" would be zero, but a zero "shear" corresponds to a zero slope of the actual beam, which is not true. The boundary condition at the right end of the beam can be satisfied by the introduction of a "force" at the right end of the conjugate beam, the "force" being R_r'.

A second set of conditions to be satisfied is the deflection characteristic at each end. Imagine starting a "moment" diagram at the left end of the "beam." The "moment" of the M/EI loading would be zero and since the "moment" of the conjugate beam corresponds to the actual deflection in the original beam, the boundary condition is satisfied if the "beam" is left as it is on the left. If "moments" are taken, beginning from the right of the M/EI loading, the "moment" is zero, which would correspond to zero deflection. But the deflection at the right of the beam is not zero. Therefore, introduction of a "moment" M' in Fig. 21.11c will take care of the boundary condition for deflection at the right end of the beam.

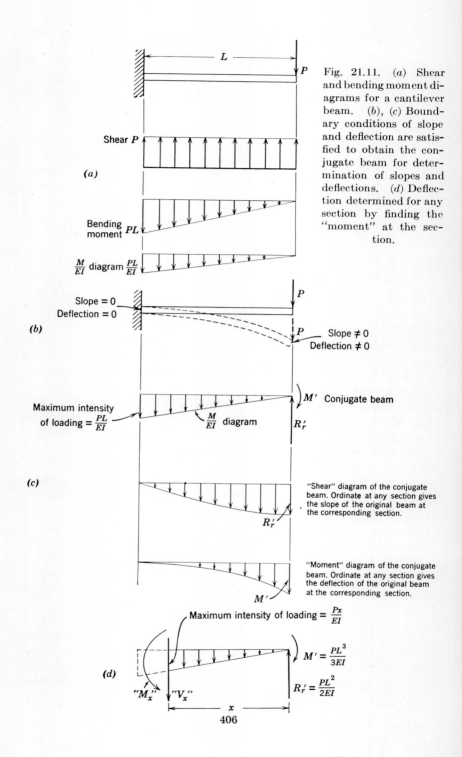

Fig. 21.11. (a) Shear and bending moment diagrams for a cantilever beam. (b), (c) Boundary conditions of slope and deflection are satisfied to obtain the conjugate beam for determination of slopes and deflections. (d) Deflection determined for any section by finding the "moment" at the section.

Conjugate Beam Applied to a Simply Supported Beam 407

Analysis of the system in Fig. 21.11c, the conjugate beam, will give quickly the slope and deflection at the right end, or at any section. The system is handled as a beam in equilibrium, with application of the equations of equilibrium. There are two unknowns, R_r' and M'. Sum up "forces" vertically equal to zero:

$$-\frac{1}{2}\left(\frac{PL}{EI}\right)L + R_r' = 0$$

or

$$R_r' = \frac{1}{2}\frac{PL^2}{EI}$$

Take "moments" about the right end for convenience:

$$-\frac{1}{2}(PL)(L)\left(\frac{2}{3}\right)L + M' = 0$$

or

$$M' = \frac{PL^3}{3EI}$$

The values of R_r' and M' are, respectively, the slope and deflection at the load, as may be verified by mathematical integration. The "shear" and "moment" diagrams for the conjugate beam are shown in Fig. 21.11c; they correspond to the slope and deflection diagrams.

The deflection at any section along the beam can be determined easily. Consider the deflection at a distance x inches from the load. The fictitious beam is cut at the section to be analyzed, and the necessary quantities for equilibrium are shown on the isolated right portion of Fig. 21.11d. By application of the equation of equilibrium, the fictitious "shear" force "V_x" and the fictitious "moment" "M_x," which correspond to the actual slope and deflection of the original beam, respectively, are

$$\text{``}V_x\text{''} = \frac{PL^2}{2EI} - \frac{Px^2}{2EI}$$

$$\text{``}M_x\text{''} = \frac{PL^3}{3EI} - \frac{PL^2x}{2EI} + \frac{Px^3}{6EI}$$

21.12 Deflection at the load of a simply supported beam with a single concentrated load by conjugate beam analysis

Figure 21.12a shows a beam of constant cross section simply supported and loaded with a single concentrated load P. Figure 21.12b shows the M/EI loading. The problem is to determine the type of reactions necessary for the determination of the deflection at the load. The boundary conditions are (1) the slope at the left support is not zero, (2) the slope at the right support is not zero, (3) the deflection at the left support is zero, and (4) the deflection at the right support is zero. If fictitious reactions, R_l' and R_r', are pictured applied at the ends, the conditions of non-zero slopes at the ends of the original beam are satisfied. Since the deflection at the ends are zero, there can be no "moment" at these points in the conjugate beam. If the beam is left as it is in Fig. 21.12c, the conditions of zero deflections will be automatically satisfied. Thus, for the case of a simply supported beam, the conjugate beam can be considered as the same, that is, one with only vertical "force" reactions.

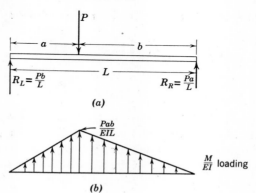

Fig. 21.12a, b. Simply supported beam with the M/EI loading.

The reactions R_l' and R_r' are found by application of the equations of equilibrium:

$$R_l' = \frac{Pab}{6EI}(a + 2b) \quad \text{and} \quad R_r' = \frac{Pab}{6EI}(b + 2a)$$

These reactions of the conjugate beam are the actual slopes at the ends.

The "moment" at any section of the conjugate beam is the actual deflection of the original beam at that section. The deflection at the load can be found by isolation of the section of the conjugate beam to the left or to the right of the load, either section giving the same result. Consider the left portion, as shown in Fig. 21.12d, where V_a and M_a are the "shear" force and "bending moment" necessary for equilibrium. To eliminate the need for finding V_a, we can sum moments about any point on the right edge equal to zero to find M_a:

$$\text{Deflection at the load} = M_a = \frac{Pa^2b}{6EI}(a + 2b) - \frac{Pa^3b}{3EIL}$$

Conjugate Beam Applied to a Simply Supported Beam

If the analysis shows that M_a is clockwise, which corresponds to a negative moment, the deflection is negative, or downwards, in the original beam.

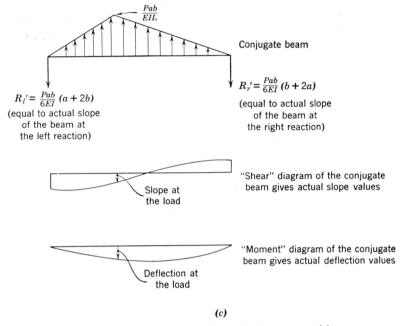

Fig. 21.12c. Conjugate beam for a simply supported beam.

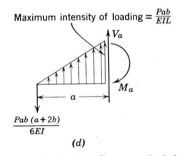

Fig. 21.12d. Deflection at the load is the "moment" of the conjugate beam at the corresponding section.

A general expression for the deflection at any section of the original beam can be written by consideration of the conjugate beam. If desired, the maximum deflection in the beam can be found by determining the section of the beam where the "shear" force of the conjugate beam is zero, or the slope of the original beam is zero, and by calculating the "moment" of the conjugate beam at the corresponding section.

21.13 Conjugate beam for an overhung beam

The overhung beam of Fig. 21.13a is selected for further illustration of setting up a conjugate beam. The beam is assumed to be of

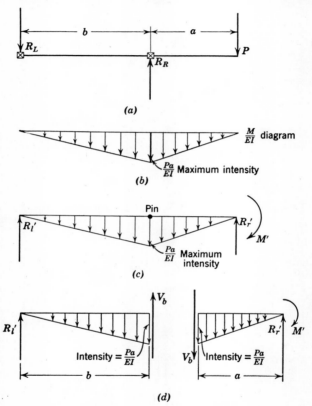

Fig. 21.13. Conjugate beam for an overhung beam.

constant cross section, with the M/EI diagram shown in Fig. 21.13b. At the left support, there is a slope of the beam, but no deflection. A single reaction R_l' (Fig. 21.13c) satisfies the boundary condition. At the load, the slope and deflection are not zero. A reaction R_r' and a bending moment M' take care of those conditions. A further stipulation is that the slope is not zero, but the deflection is zero at the right support. These conditions are taken care of if we think of a pin as connecting the two parts of the conjugate beam. Practically,

Deflection in Beams with Variable Cross-Sections

a pin can transmit a force but cannot transmit a moment, which is just what is desired in this case.

It is not possible to determine the deflection at the load immediately, for the isolated parts in Fig. 21.13d, because there are three unknowns in the right part. Determine V_b, the "shear" at the cut section, by consideration of the left portion first. A moment equation applied about the left end gives

$$V_b = Pab/3EI$$

The deflection at the load, M', can be found by taking moments about the right reaction of the right portion, setting the sum equal to zero:

$$M' = \frac{Pa^2b}{3EI} - \frac{Pa^3}{3EI}$$

which is the actual deflection at the load.

21.14 Conjugate beams of other types

Figure 21.14 shows several other beam types with the conjugate beams. They are given without further discussion. The beams are assumed to be of constant cross section.

21.15 Deflection in beams with variable cross-sections

Although the conjugate beam method for finding deflections may or may not have an advantage over other methods for simple loadings on simple beams, there is a distinct advantage to the method in beams of variable cross-sections. A numerical example is offered to illustrate the procedures. Figure 21.15a shows a steel beam made up of three different sections: 1-in. diameter for 4 in., 2-in. diameter for 8 in., and 1-in. diameter for 4 in. Concentrated masses are located at the change of sections, the masses weighing 200 lb and 400 lb, as shown. The reactions on the beam are calculated to be 250 and 350 lb. The moment diagram is shown in Fig. 21.15b.

The next step is the determination of the M/EI diagram and the conjugate beam. They are shown in Fig. 21.15c. I_1, the moment of inertia of the 1-in. diameter section, and I_2, the moment of inertia of the 2-in. diameter section, are carried through the analysis, substitutions of numerical values being made at the end.

For simplicity of calculations, the middle section, which is trapezoidal, is broken up into a triangular and a rectangular section.

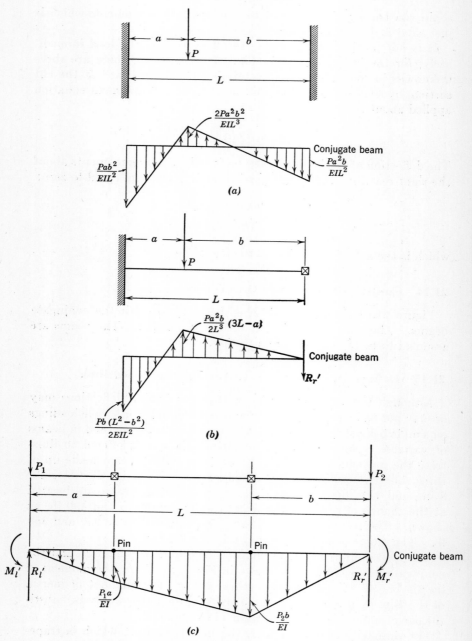

Fig. 21.14. Other beam types with the conjugate beams.

Deflection in Beams with Variable Cross-Sections

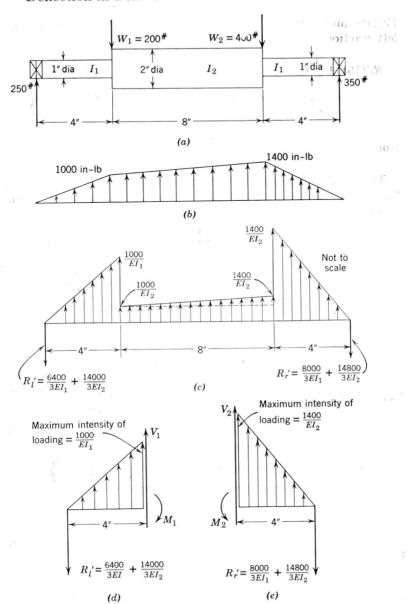

Fig. 21.15. Illustrative example for determination of deflections in a simply supported beam. (a) Loading diagram. (b) Bending moment diagram. (c) M/EI loading on the conjugate beam. (d, e) Isolated portions of the conjugate beam, the sections corresponding to the location of the loads of the original beam.

Equate moments about the right reaction equal to zero to find the left reaction:

$$-R_l'(16) + \frac{1}{2}(4)\left(\frac{1000}{EI_1}\right)\left[12 + \frac{1}{3}(4)\right] + 8\left(\frac{1000}{EI_2}\right)\left[4 + \frac{1}{2}(8)\right]$$

$$+ \frac{1}{2}(8)\left(\frac{1400}{EI_2} - \frac{1000}{EI_2}\right)\left[4 + \frac{1}{3}(8)\right] + \frac{1}{2}(4)\left(\frac{1400}{EI_1}\right)\left(\frac{2}{3}\right)(4) = 0$$

from which
$$R_l' = \frac{6400}{3EI_1} + \frac{14{,}000}{3EI_2}$$

The right "reaction" R_r' can be found in either of two ways, a "force" equation or a "moment" equation. A "moment" equation will be used, and the results checked by a "force" equation:

$$+R_r'(16) - \frac{1}{2}(4)\left(\frac{1000}{EI_1}\right)\left[\frac{2}{3}(4)\right] - 8\left(\frac{1000}{EI_2}\right)\left[4 + \frac{1}{2}(8)\right]$$

$$- \frac{1}{2}(8)\left(\frac{1400}{EI_2} - \frac{1000}{EI_2}\right)\left[4 + \frac{2}{3}(8)\right] - \frac{1}{2}(4)\left(\frac{1400}{EI_1}\right)\left[2 + \frac{1}{3}(4)\right]$$

$$= 0$$

from which
$$R_r' = \frac{8000}{3EI_1} + \frac{14{,}800}{3EI_2}$$

A check of the "forces" shows a balance:

$$\text{"Forces" down} = R_l' + R_r' = \frac{6400}{3EI_1} + \frac{14{,}000}{3EI_2} + \frac{8000}{3EI_1} + \frac{14{,}800}{3EI_2}$$

$$= \frac{4800}{EI_1} + \frac{9600}{EI_2}$$

$$\text{"Forces" up} = \frac{1}{2}(4)\left(\frac{1000}{EI_1}\right) + \frac{1}{2}\left(\frac{400}{EI_2}\right)(8) + 8\left(\frac{1000}{EI_2}\right)$$

$$+ \frac{1}{2}\left(\frac{1400}{EI_1}\right)(4) = \frac{4800}{EI_1} + \frac{9600}{EI_2}$$

Note that R_l' and R_r' are dimensionless, the slope of the beam at the reactions of the original beam.

The actual deflection at the loads can be determined now. To find the deflection at W_1, we can isolate the section shown in Fig. 21.15d. The moment M_1 can be found by taking moments about any point on the right, which eliminates the need of finding V_1:

$$-\left(\frac{6400}{3EI_1} + \frac{14{,}000}{3EI_2}\right)(4) + \frac{1}{2}\left(\frac{1000}{EI_1}\right)(4)\left[\frac{1}{3}(4)\right] + M_1 = 0$$

Conjugate Beam Applied to Kinematics

from which
$$M_1 = \frac{17{,}600}{3EI_1} + \frac{56{,}000}{3EI_2}$$

For $E = 30(10^6)$ psi, $I_1 = \dfrac{\pi D^4}{64} = \dfrac{\pi(1)^4}{64}$, the rectangular moment of inertia, and $I_2 = \dfrac{\pi(2)^4}{64}$, $M_1 = 0.00478$ in., the deflection at W_1.

To obtain the deflection at W_2, consider the isolated portion shown in Fig. 21.15e. Take moments about the left edge:

$$-M_2 - \frac{1}{2}\left(\frac{1400}{EI_1}\right)(4)\left[\frac{1}{3}(4)\right] + \left(\frac{8000}{3EI_1} + \frac{14{,}800}{3EI_2}\right)(4) = 0$$

from which $M_2 = 0.00555$ in., the actual deflection at W_2.

Critical speed calculation. The first-order critical speed for the given problem of, neglecting the mass of the shaft, is found by

$$N = 187.7\sqrt{\frac{\Sigma W_n y_n}{\Sigma W_n y_n^2}}$$

$$= 187.7\sqrt{\frac{(200)(0.00478) + (400)(0.00555)}{(200)(0.00478)^2 + (400)(0.00555)^2}}$$

$$= 2570 \text{ rpm}$$

21.16 Conjugate beam method applied to acceleration, velocity, and displacement

Since the relation of acceleration, velocity, and displacement are related in the same manner as loading, shear, and bending moment, the same techniques can be applied to the determination of displacement for a given acceleration function. The relation $A = \dfrac{dV}{dt} = \dfrac{dx^2}{dt^2}$, where $V = \dfrac{dx}{dt}$, is one where the acceleration function can be considered a fictitious loading and the displacement can be considered a fictitious moment. An exercise for the student is to set up by the conjugate beam method a system where the acceleration is constant, A; and, for the boundary conditions that at $t = 0$, $V = V_0$, and $x = x_0$ and at $t = T$, $V = V_f$, and $x = X$, show that $X = x_0 + V_0 T + \dfrac{AT^2}{2}$.

21.17 Deflections determined graphically

Inasmuch as the conjugate beam method for determining deflections is comparable to bending moment analysis, let us consider the graphical solution for moment diagrams, which method can be used to determine deflections graphically.

Bending moment diagrams determined graphically. Figure 21.16a shows a simply supported beam with three concentrated loads. Bow's notation is used to represent the forces, where each "space" is designated by a letter. The forces are drawn to scale in Fig. 21.16b. The steps to find the reactions $A-E$ and $E-D$ are as follows:

(1) Select a pole O, Fig. 21.16b, at any convenient location.

(2) Draw the rays from O to the beginning and end of each force vector.

(3) At any point on the line of action of force $A-B$ (Fig. 21.16a) break up the force $A-B$ into two components, $a-o$ and $o-b$.

(4) Where the force $o-b$ intersects $B-C$, break up the force $B-C$ into components $b-o$ and $o-c$. At this point, the forces $A-B$ and $B-C$ can be replaced by the two components, $a-o$ and $o-c$.

(5) Break up force $C-D$ into two components $c-o$ and $o-d$, where $c-o$ intersects $C-D$.

(6) Extend $o-d$ until it intersects $E-D$ at point M. Extend $a-o$ until it intersects $A-E$ at N.

(7) The force components of $A-E$ are $o-e$ and $o-a$. The force components of $E-D$ are $o-d$ and $o-e$. Thus the force component $o-e$ must pass through points M and N.

(8) Determine point E in Fig. 21.16b by drawing a line parallel to the ray $o-e$ through the pole O.

Note that the forces are in equilibrium since they add up to zero, and each force component is balanced by an equal, opposite, and collinear force component.

The string polygon, or so-called funicular polygon, obtained in Fig. 21.16a has another characteristic: an intercept, as **RS**, or any other intercept, is proportional to the bending moment at the section where the intercept is measured. The proof will be given for the particular intercept **RS**, the proof being the same for any other intercept. The moment M at the load $B-C$ is found by considering the isolated section shown in Fig. 21.16c and is

$$M = AE(m) - AB(n)$$

where m and n are the distances shown.

Bending Moment Diagrams Determined Graphically

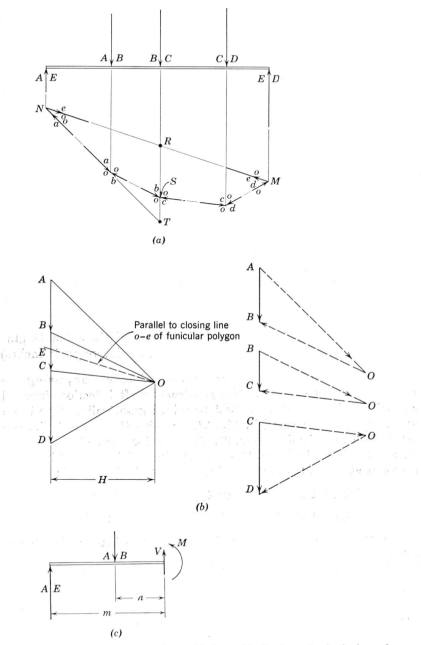

Fig. 21.16. Bending moments determined graphically from the funicular polygon.

From similar triangles,

$$\frac{AE}{H} = \frac{RT}{m} \quad \text{or} \quad AE = \frac{RT(H)}{m}$$

and

$$\frac{AB}{H} = \frac{ST}{n} \quad \text{or} \quad AB = \frac{ST(H)}{n}$$

In the equation above, H is considered a force.

The moment M can be expressed by

$$M = \left[\frac{RT(H)}{m}\right]m - \left[\frac{ST(H)}{n}\right]n$$
$$= H(RT - ST)$$
$$= H(RS)$$

Which shows that the bending moment is proportional to the intercept. The actual bending moment is determined by introducing the proper scale factors. Let s = the space scale to which the beam is drawn (1 in. = s inches) and let p = the force scale to which the force polygon is drawn (1 in. = p lb). The actual moment, inch-pounds, is

$M = (p)$(the actual pole distance H, in inches) (s)(the actual length of the intercept, in inches)

Example of graphical determination of deflection, using the conjugate beam method. The problem of shaft deflection discussed analytically in Section 21.15 will be solved graphically. The beam and loading are repeated in Fig. 21.17a, with the moment diagram determined graphically in Fig. 21.17b. The force scale is 1 in. = 100 lb, and the beam is drawn to a scale of 1 in. = 3 in., H = 4 in. The moments at W_1 and W_2 are found from $(p)(H)(s)$(intercept):

Moment at $W_1 = (100)(4)(3)(0.83) = 1000$ in-lb

Moment at $W_2 = (100)(4)(3)(1.167) = 1400$ in-lb

The M/EI diagram is determined by calculations at the change points, with the result shown in Fig. 21.17c.

A distributed load can be considered made up of a series of concentrated loads. If the loading is imagined as an infinite number of concentrated loads, the true deflection curve may be obtained correctly. Practically, a finite number of concentrated loads must be considered, in which case the deflection curve becomes an approximation. Let us determine an approximate curve by considering the

Deflections Determined Graphically

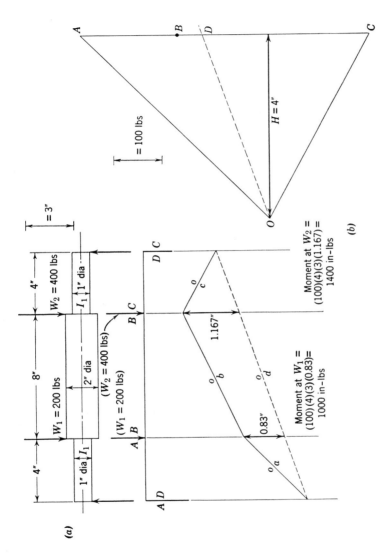

Fig. 21.17a, b. Illustrative example. Bending moment diagram determined graphically.

420 Vibrations in Shafts

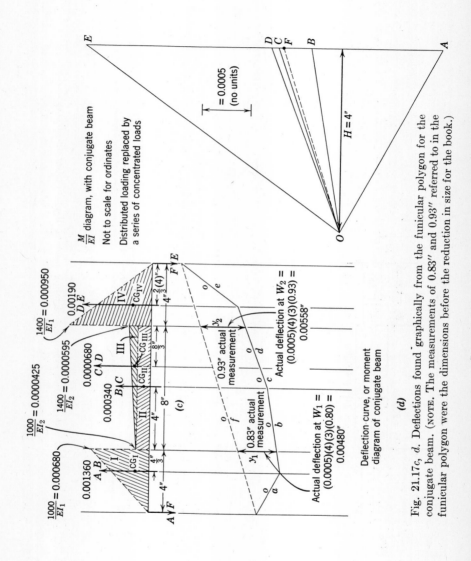

Fig. 21.17c, d. Deflections found graphically from the funicular polygon for the conjugate beam. (NOTE. The measurements of 0.83" and 0.93" referred to in the funicular polygon were the dimensions before the reduction in size for the book.)

Effects of the Weight of the Shaft

distributed loadings replaced by concentrated forces at the center of gravity of each section, the convenient geometric sections I, II, III, and IV in Fig. 21.17c. The "load" for each section is

I: Resultant load $= \frac{1}{2}(4)(0.000680) = 0.001360$

II: Resultant load $= (8)(0.0000425) = 0.000340$

III: Resultant load $= \frac{1}{2}(8)(0.0000595 - 0.0000425) = 0.0000680$

IV: Resultant load $= \frac{1}{2}(4)(0.000950) = 0.00190$

with each load being dimensionless.

The funicular polygon is shown in Fig. 21.17d, drawn as though a bending moment diagram were being found. The funicular polygon gives the actual deflection diagram, as a close approximation, however, since the loading was approximated by concentrated loads.

The magnitude of the deflections at the concentrated loads W_1 and W_2 of the original beam, y_1 and y_2, are found by applying the proper scale relationships. By analogy with the graphical determination of bending moments, the following will give the true deflections:

True deflection $= (p')$(the actual pole distance, in inches)(s)
(the actual length of the intercept, inches)

where p' represents the scale to which the "force," or M/EI, diagram has been drawn and s is the space scale to which the beam has been drawn. For this illustrative problem, the M/EI diagram has been drawn to a scale of 1 in. = 0.0005 (with no units), or $p' = 0.0005$; and the beam has been drawn to a scale of 1 in. = 3 in., or $s = 3$. The pole distance $H = 4$ in. Thus,

$y_1 = (0.0005)(4)(3)(0.80) = 0.00480$ in.

$y_2 = (0.0005)(4)(3)(0.93) = 0.00558$ in.

In spite of the fact that concentrated loads were used, the deflections correspond very closely to those calculated: 0.00478 in. and 0.00555 in. The deflection curve of Fig. 21.17d, although not a smooth one, can be made one by the use of more concentrated loads.

21.18 Effect of the weight of the shaft

The weight of the shaft has been neglected in the problem of the previous sections. If the weight of the shaft is to be considered, a method that might be used is to consider the shaft broken up into a series of concentrated loads contributing to the deflection in the shaft. In this manner, the procedures of the preceding sections would be exactly the same, except for the number of loads considered.

21.19 Critical speed of a simply supported shaft

Let us determine the critical speed, using the energy method, for a simply supported shaft of constant cross section. The deflection curve due to the weight of the shaft alone is given by:

$$y = \frac{wx}{24EI}(L^3 - 2Lx^2 + x^3)$$

where x represents the distance to any point along the beam, as shown in Fig. 21.18, E is the modulus of elasticity in tension, and I is the

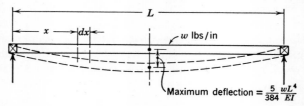

Fig. 21.18. Simply supported shaft of constant cross-section and w pounds per inch.

rectangular moment of inertia about an axis of bending. $I = \dfrac{\pi d^4}{64}$, where d is the diameter of the shaft. w is the weight of the shaft in pounds per inch. Substitute the equation of deflection into the critical speed equation, consider the shaft as made up of an infinite number of concentrated weights, each weight being equal to $(w)(dx)$, and replace the "sum" sign by an integral sign since differential lengths are considered:

$$N = 187.7\sqrt{\frac{\Sigma Wy}{\Sigma Wy^2}}$$

$$= 187.7\sqrt{\frac{\int_0^L w(dx)\left[\dfrac{wx}{24EI}(L^3 - 2Lx^2 + x^3)\right]}{\int_0^L w(dx)\left[\dfrac{wx}{24EI}(L^3 - 2Lx^2 + x^3)\right]^2}}$$

Expanding and collecting terms, we obtain finally

$$N = 1855(EI/wL^4)^{1/2} \text{ rpm}$$

Higher-Order Critical Speeds

The expression may be given as a function of the maximum deflection of the shaft:

$$y_{\max} = \frac{5}{384} \frac{wL^4}{EI}$$

$$N = 211.8(1/y_{\max})^{1/2} \text{ rpm}$$

21.20 Dunkerley's equation

In some cases it might be desirable to break up the analysis into parts to simplify the work, or to investigate the effect of changes. An approximate equation due to Dunkerley* is

$$\frac{1}{N^2} = \frac{1}{N_1^2} + \frac{1}{N_2^2} + \frac{1}{N_3^2} + \frac{1}{N_4^2} + \cdots$$

where N is the critical speed of an entire system, N_1 is the critical speed with mass 1 only on the shaft, N_2 is the critical speed with mass 2 only on the shaft, etc. The reader is referred to the reference for the analysis in the development of the equation and for the assumptions made.

21.21 Critical speed calculations are approximations

The procedure used in obtaining the critical speed, where energy transfer is considered in the lateral vibration of a shaft, a method due to Rayleigh, is an approximation inasmuch as the deflection curve is assumed to be the same as the static deflection curve. It can be shown that the critical speed calculated is always higher than the actual critical speed if the deflection curve assumed is incorrect, although for most applications the percentage difference is very small and therefore negligible.

21.22 Higher-order critical speeds

It has been mentioned previously that it is possible to obtain various shapes of curves in rotating shafts at different critical speeds. Even a shaft of constant cross-section, and simply supported, exhibits various critical speeds, although the critical speed increases rapidly as the shaft assumes a greater number of reverse curves. In other words, a shaft rotating with the shape shown in Fig. 21.19b has a higher critical speed than the shaft shown in Fig. 21.19a; and the shaft in Fig. 21.19c has a higher critical speed than the one shown in Fig.

---

* See *The Theory of Machines*, by Thomas Bevan, published by Longmans, Green and Co.

424   Vibrations in Shafts

21.19b. Actually, it has been shown mathematically that the ratio of critical speeds are as 1:4:9, etc., for the shaft of constant diameter simply supported.

The procedure of determining higher-order critical speeds is essentially a trial and error process, and the reader is referred to *Steam and Gas Turbines*, by Stodola and Loewenstein, for a discussion of a procedure of attack.

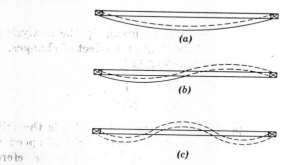

Fig. 21.19. Higher-order critical speeds.

Practically, the usual design is based on operation below the first critical speed because of the difficulty encountered in passing through a critical speed, although operation above the first-order critical speed is desirable from the standpoint of stability.

21.23 Torsional vibrations

Another type of phenomenon that may occur simultaneously with or separately from critical or whirling speeds is a torsional vibration. Torsional vibrations are a definite type of vibration or oscillation, as contrasted with critical speeds, and are due directly to the spring action of materials. Torsional vibration has been encountered already in this book in an elementary form in the determination of the moment of inertia of a body by experimental means, pages 199–204.

A mass, suspended as shown in Fig. 21.20 and displaced an angle θ, will execute a vibration about its axis. The motion will be harmonic, if friction is neglected, giving a natural frequency of oscillation. If impulses are applied to the mass which are in phase with the natural frequency of oscillation, large amplitudes of motion will result, with the consequent high stresses in the supporting shaft. If the system were part of a rotating system with a speed of operation that corresponded to the natural frequency of oscillation, a critical condition would result. It is the purpose of this and the following sections to determine the natural frequency of oscillation for various arrangements. Let us compute the natural frequency of oscillation of the arrangement in Fig. 21.20, assuming that the shaft is of constant diameter, D inches, and is L inches long. The material is assumed to be steel, with a modulus of elasticity in torsion of G. The mass of the shaft is assumed negligible.

Picture the mass displaced θ_0 radians and then released. The potential energy stored in the shaft will be transferred to kinetic energy as the shaft returns to its equilibrium position, with the mass overshooting the equilibrium position until the kinetic energy is transferred back to potential energy again. The process repeats itself, giving a true vibration.

From strength of materials, $\theta = TL/JG$, where θ is the angle of twist of the shaft in radians, T is the couple applied to the shaft in

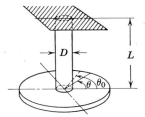

Fig. 21.20. Torsional vibration of a single mass displaced from its quilibrium position.

pound-inches, L is the length of shaft in inches, J is the polar moment of inertia of the shaft in inches^4, and G is the modulus of elasticity in torsion in pounds per square inch. Thus, $T = \dfrac{JG}{L} \theta$.

Applying $T = I\alpha$, we obtain

$$-\frac{JG}{L} \theta = I \frac{d^2\theta}{dt^2}$$

where the minus sign is used because the couple from the shaft is in the opposite direction to the positive angle θ, and where I, the mass moment of inertia of the mass, has units of pound-inch-second^2. For the boundary conditions that at $\theta = \theta_0$, $t = 0$, and $\omega = 0$, the solution to the above is

$$\theta = \theta_0 \cos (JG/LI)^{\frac{1}{2}} t$$

which shows that the motion is a harmonic one. The frequency of oscillation is given by $\dfrac{1}{2\pi} (JG/LI)^{\frac{1}{2}}$ cycles per second. Note that the frequency of oscillation is independent of the initial angular displacement.

21.24 Torsional oscillation of a two-mass system

A two-mass system is shown in Fig. 21.21a. The shaft is supported in bearings which have no effect on the torsional vibration, except for friction which is disregarded. The natural frequency of oscillation is to be found.

The mass moment of inertia for mass A about its axis of rotation is given as I_a, and that for mass B is given by I_b. The distance between the masses is L, and the shaft diameter, which is assumed

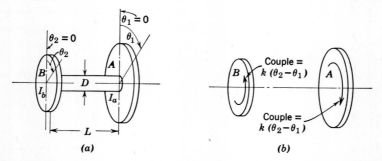

Fig. 21.21. Two-mass system with the couples applied to each mass as a result of a relative angular displacement.

constant, is given by D. From $\theta = TL/JG$, we may obtain the spring rate in torsion, $\dfrac{T}{\theta} = \dfrac{JG}{L}$ inch-pounds per radian, called k.

At an instant of time, picture the masses as shown in Fig. 21.21a, where mass A is at an angle θ_1 from an arbitrary vertical axis, and mass B is at an angle θ_2 from an arbitrary vertical axis. The twist in the shaft is $(\theta_2 - \theta_1)$. The couple in the shaft is a function of $(\theta_2 - \theta_1)$, and is $k(\theta_2 - \theta_1)$. If θ_2 is assumed larger than θ_1, the couple applied to mass A is clockwise as shown in Fig. 21.21b, and, by action and reaction, the couple applied to mass B is counterclockwise. From $T = I\alpha$, we may write for the two masses the following:

$$\text{Mass } A: \quad k(\theta_2 - \theta_1) = I_a \frac{d^2\theta_1}{dt^2} \qquad (1)$$

$$\text{Mass } B: \quad -k(\theta_2 - \theta_1) = I_b \frac{d^2\theta_2}{dt^2} \qquad (2)$$

Torsional Oscillations of a Two-Mass System 427

where the minus sign is used in the second expression because the couple acts in the opposite direction to the positive θ_2.

Before proceeding with the solution to the equations above, add the two:

$$0 = I_a \frac{d^2\theta_1}{dt^2} + I_b \frac{d^2\theta_2}{dt^2}$$

which may be interpreted as prescribing two conditions: (1) the externally applied couple is zero, which is the initial assumption in that the masses are in motion as the result of the elasticity of the shaft, and (2) the sum of the inertia couples must always be equal and opposite. The latter may be used to show that the frequency of oscillation of the two masses is the same since the two masses must always move in opposite directions to each other. When one mass has a zero acceleration, the other one must have a zero acceleration; when the acceleration of one mass is maximum, the acceleration of the other mass must be a maximum. If it were possible for one mass to "catch up" and "pass" the other mass, it would follow that the accelerations of the two masses would be in the same direction at some time, which is impossible, however, from the equation above. Although the frequency of oscillation for the two masses is the same and although the angular speeds of the two masses are always in opposite directions, the instantaneous magnitudes of the angular speeds are not necessarily the same. Therefore, let us assume that the motions are harmonic with the same frequency of oscillation and can be expressed by

$$\theta_1 = \theta_a \cos \omega t$$

$$\theta_2 = \theta_b \cos \omega t$$

where θ_a and θ_b are the maximum angular displacements for masses A and B, respectively. Differentiation twice of the equations above gives

$$\frac{d^2\theta_1}{dt^2} = -\theta_a \omega^2 \cos \omega t$$

$$\frac{d^2\theta_2}{dt^2} = -\theta_b \omega^2 \cos \omega t$$

Substitution of the equations above into Eqs. 1 and 2 gives

$$k(\theta_b \cos \omega t - \theta_a \cos \omega t) = -I_a \theta_a \omega^2 \cos \omega t$$

$$-k(\theta_b \cos \omega t - \theta_a \cos \omega t) = -I_b \theta_b \omega^2 \cos \omega t$$

Collect terms:
$$[\theta_b k - \theta_a(k - I_a\omega^2)] \cos \omega t = 0$$
$$[-\theta_b(k - I_b\omega^2) + \theta_a k] \cos \omega t = 0$$

Since the functions must be zero for every instant of time,
$$\theta_b k - \theta_a(k - I_a\omega^2) = 0$$
$$-\theta_b(k - I_b\omega^2) + \theta_a k = 0$$

Solve for θ_b/θ_a from each expression, equate the two, and obtain
$$\omega^2[I_a I_b \omega^2 - k(I_a + I_b)] = 0$$

The solutions for ω are:
$$\omega = 0$$
$$\omega = \left[\frac{k(I_a + I_b)}{I_a I_b}\right]^{1/2}$$

The first solution indicates no relative motion of the masses; for the second value the frequency of oscillation of each mass is
$$N = \frac{60}{2\pi}\left[\frac{k(I_a + I_b)}{I_a I_b}\right]^{1/2} \text{ cycles per minute}$$

The oscillatory motion could occur with the shaft stationary or the motion could occur with the shaft rotating, the oscillatory motion then being superimposed upon the steady-state motion.

21.25 Torsional oscillation of a three-mass system

Figure 21.22a shows a three-mass system, with the mass moment of inertia of masses A, B, and C given by I_a, I_b, and I_c, respectively, and with the masses shown at angles θ_1, θ_2, and θ_3 from the vertical. The spring rate for the right shaft is k_1, and the spring rate for the left shaft is k_2. Assume that θ_3 is larger than θ_2, and θ_2 larger than θ_1 for purposes of setting up the equations.

$$\text{Mass } A: k_1(\theta_2 - \theta_1) = I_a \frac{d^2\theta_1}{dt^2} \quad (1)$$

$$\text{Mass } B: k_2(\theta_3 - \theta_2) - k_1(\theta_2 - \theta_1) = I_b \frac{d^2\theta_2}{dt^2} \quad (2)$$

$$\text{Mass } C: -k_2(\theta_3 - \theta_2) = I_c \frac{d^2\theta_3}{dt^2} \quad (3)$$

Assume that the motion of each mass is

$$\theta_1 = \theta_a \cos \omega t \quad (4)$$

$$\theta_2 = \theta_b \cos \omega t \quad (5)$$

$$\theta_3 = \theta_c \cos \omega t \quad (6)$$

where θ_a, θ_b, θ_c are, respectively, the maximum amplitude of motion of masses A, B, and C.

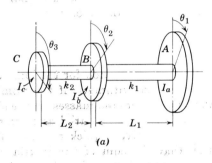

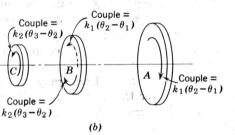

Fig. 21.22. Three-mass system with the couples applied to each mass as a result of relative angular displacements.

Substitution of Eqs. 4, 5, and 6, and also the second derivative of each expression with respect to time, into Eqs. 1, 2, and 3 gives

$$\frac{\theta_a}{\theta_b} = \frac{k_1}{k_1 - I_a \omega^2} \quad (7)$$

$$k_2 \frac{\theta_c}{\theta_b} + k_1 \frac{\theta_a}{\theta_b} = k_2 + k_1 - I_b \omega^2 \quad (8)$$

$$\frac{\theta_c}{\theta_b} = \frac{k_2}{k_2 - I_c \omega^2} \quad (9)$$

Substitute Eqs. 7 and 9 into 8:

$$\omega^6 - \omega^4 \left[\frac{k_1 I_c (I_a + I_b) + k_2 I_a (I_b + I_c)}{I_a I_b I_c} \right] + \omega^2 \left[\frac{(k_1 k_2)(I_a + I_b + I_c)}{I_a I_b I_c} \right] = 0$$

If ω^2 is factored out, the quadratic equation may be applied to solve for two other values of ω^2. Thus, there will be three solutions for ω, one of which will be zero. The answers for a particular problem would best be determined by the substitution of numerical values in the solution of the equation above.

21.26 Holzer's method

It should be apparent that the solution for torsional critical speeds becomes exceedingly complex in arithmetic if the number of masses is large. If, for instance, there were ten masses on the shaft, an equation with ω^{20}, ω^{18}, etc., would result. A trial and error method presented by Holzer gives comparatively quick results. The procedure is based on the relation that the sum of the inertia couples of the masses must be zero. By expressing the inertia couple as a function of the angular speed, we may assume an angular speed and the angular displacement of the first mass and proceed from one mass to the next until the last mass is considered. If the couple on the last mass is not zero, the assumed angular speed is incorrect. Several trials are usually sufficient to give the needed accuracy. It would be well to point out first that whatever the amplitude of the masses, provided the ratio of amplitudes is correct, the natural frequency of torsional oscillation will be the same.

Before proceeding to the method of solution using Holzer's method, let us examine an alternate method of expression for

$$I_a \frac{d^2 \theta_1}{dt^2} + I_b \frac{d^2 \theta_2}{dt^2} + I_c \frac{d^2 \theta_3}{dt^2} + \cdots = 0 \tag{1}$$

Assuming that the motion of the masses may be expressed by

$$\theta_1 = \theta_a \cos \omega t$$
$$\theta_2 = \theta_b \cos \omega t$$
$$\theta_3 = \theta_c \cos \omega t$$
$$\vdots \qquad \vdots$$

Holzer's Method

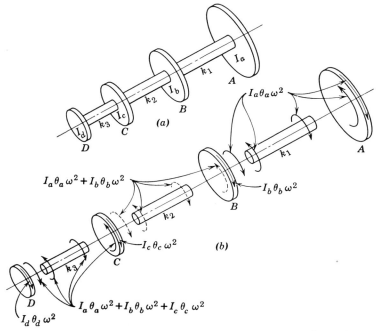

Fig. 21.23a, b. Four-mass system with the inertia couples and couples applied by the shafts shown on isolated parts.

we may differentiate each expression twice and obtain

$$\frac{d^2\theta_1}{dt^2} = -\theta_a\omega^2 \cos \omega t$$

$$\frac{d^2\theta_2}{dt^2} = -\theta_b\omega^2 \cos \omega t$$

$$\frac{d^2\theta_3}{dt^2} = -\theta_c\omega^2 \cos \omega t$$

. . .
. . .
. . .

Substitution of the equations above into Eq. 1 gives:

$$(I_a\theta_a\omega^2 + I_b\theta_b\omega^2 + I_c\theta_c\omega^2 + \cdots)(-\cos \omega t) = 0$$

If the expression is to hold true for any time,

$$I_a\theta_a\omega^2 + I_b\theta_b\omega^2 + I_c\theta_c\omega^2 + \cdots = 0$$

where each term can be considered to be the maximum inertia couple of the corresponding mass, at the time of maximum angular displacement.

The method of attack will be illustrated by reference to a four-mass system, as shown in Fig. 21.23a, where the mass moment of inertias are I_a, I_b, I_c, and I_d, with the shaft sections having spring rates expressed by k_1, k_2, and k_3.

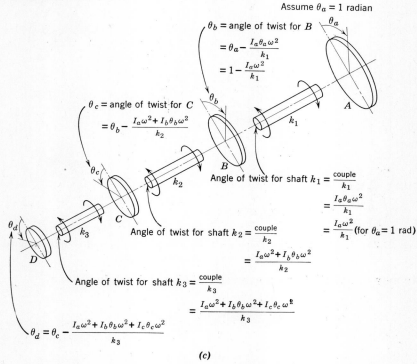

Fig. 21.23c. Relation of angular displacements for the various masses.

Figure 21.23b shows the system broken up into a series of free bodies. The inertia couple of mass A is given by $I_a \theta_a \omega^2$, with the shaft couple that is applied being equal and opposite. The same couple is transmitted through the shaft of k_1, and applied to mass B. The inertia couple of mass B, $I_b \theta_b \omega^2$, together with the couple applied by shaft k_1, requires a couple of $(I_a \theta_a \omega^2 + I_b \theta_b \omega^2)$ to be applied by the shaft of k_2. The procedure is carried out to the last mass, with the values as shown. Note that all the inertia couples are shown clockwise just for simplification in setting up the system. The total couple on the last disk must be zero, with no externally applied couple.

Figure 21.23c shows the same system with the relation of angular

Numerical Example of Holzer's Method

displacement for the various masses. Mass A is assumed to have an angular displacement of $\theta_a = 1$ radian. The angular displacement of the shaft of k_1 is $\dfrac{\text{couple}}{k_1} = \dfrac{I_a \theta_a \omega^2}{k_1} = \dfrac{I_a \omega^2}{k_1}$. Thus, $\theta_b = \theta_a - \dfrac{I_a \omega^2}{k_1}$.
The angle θ_b can be calculated numerically at this point and also the inertia couple of mass B. Knowing the inertia couple of mass B, we can find the total couple transmitted through the shaft of k_2, which in turn will permit the numerical determination of the angle of twist in the shaft of k_2:

$$\frac{\text{couple}}{k_2} = \frac{I_a \omega^2 + I_b \theta_b \omega^2}{k_2}. \quad \text{Thus, } \theta_c = \theta_b - \left[\frac{I_a \omega^2 + I_b \theta_b \omega^2}{k_2}\right]$$

Knowing θ_c, we can find the inertia couple of the disk C. In a similar manner, the inertia couple of mass D can be found. The sum of the inertia couples of all masses should be zero if the correct angular speed is assumed.

21.27 Numerical example of Holzer's method

An application of the method to a numerical example will serve to further illustrate the method of attack. A four-mass system is shown in Fig. 21.24a, where it is assumed that

$I_a = 200$ in.-lb-sec^2 $k_1 = 4(10)^6$ in.-lb per radian

$I_b = 125$ " $k_2 = 2(10)^6$ "

$I_c = 75$ " $k_3 = (10)^6$ "

$I_d = 50$ "

Since Holzer's method is a trial and error one, it will be necessary to assume an initial value for ω. To save time, we may, in general, obtain an approximate starting point by considering the system reduced to a two-mass system, with a single shaft connecting the masses. Let us consider the original four masses replaced by two, with the inertia of one 200 and the inertia of the other $(125 + 75 + 50)$, or 250. Using the expression for a two-mass system, page 428, and considering k as $3(10)^6$, we find

$$\omega \approx \left[k \frac{(I_a' + I_b')}{I_a' I_b'}\right]^{1/2}$$

$$\omega \approx \left[\frac{3(10)^6 (200 + 250)}{200(250)}\right]^{1/2} \approx 100 \text{ rad/sec}$$

Let us try as a first approximation $\omega = 100$ rad/sec. It will be convenient to use the tabular form for the calculations. It is sug-

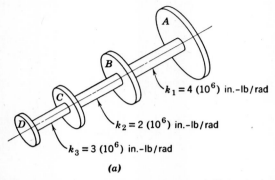

Fig. 21.24a. Data for illustrative example of Holzer's method.

$$I_A = 200 \text{ lb-in.-sec}^2$$
$$I_B = 125 \text{ lb-in.-sec}^2$$
$$I_C = 75 \text{ lb-in.-sec}^2$$
$$I_D = 50 \text{ lb-in.-sec}^2$$

gested that the reader follow the procedure of Figs. 21.23b and 21.23c in the step-by-step process, the summary of which is given below:

TRIAL I
($\omega = 100$ rad/sec $\omega^2 = 10{,}000$)

Mass	I	$I\omega^2$	θ	Inertia Couple $I\theta\omega^2$	Sum of Inertia Couples $\Sigma I\theta\omega^2$	k	Shaft Twist $\dfrac{1}{k}\sum I\theta\omega^2$
A	200	$2(10)^6$	1	$2(10)^6$	$2(10)^6$	$4(10)^6$	$\dfrac{2(10)^6}{4(10)^6} = 0.5$
B	125	$1.25(10)^6$	$[1-0.5]$ 0.5	$0.625(10)^6$	$[2(10)^6 + 0.625(10)^6]$ $2.625(10)^6$	$2(10)^6$	$\dfrac{2.625(10)^6}{2(10)^6} = 1.313$
C	75	$0.75(10)^6$	$[0.5-1.313]$ -0.813	$-0.610(10)^6$	$[2.625(10)^6 - 0.610(10)^6]$ $2.015(10)^6$	$3(10)^6$	$\dfrac{2.015(10)^6}{3(10)^6} = 0.672$
D	50	$0.50(10)^6$	$[-0.813-0.672]$ -1.485	$-0.743(10)^6$	$[2.015(10)^6 - 0.743(10)^6]$ $1.272(10)^6$		

The values bracketed are given as intermediate steps to aid in following the procedure.

The final value of the sum of the inertia couples, $1.272(10)^6$ in.-lb, shows that the initial assumption of frequency is incorrect. A plot of relative angular displacement for the various masses (Fig. 21.24b) shows that we are either above or below the first natural frequency of oscillation, since there is only one node in the shaft. Whether the natural frequency is greater or smaller than the initially assumed value of $\omega = 100$ rad/sec can be found by reference to Fig. 21.24c, which is a plot of the sum of the inertia couples versus frequency.

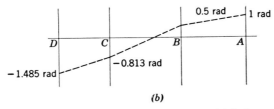

(b)

Fig. 21.24b. Plot of angle of twist of each mass for trial I shows only one node (between B and C), or the assumed value of ω is near the first natural frequency of oscillation.

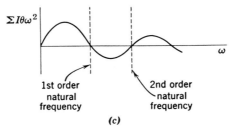

(c)

Fig. 21.24c. General curve of the sum of the inertia couples versus frequency to determine whether the assumed frequency is above or below a given order of natural frequency.

Since the sum of the inertia couples, $1.272(10)^6$ is positive, the assumed value is less than the first-order natural frequency.

For a second try, let us assume $\omega = 120$ rad/sec.

TRIAL II
($\omega = 120$ rad/sec $\omega^2 = 14{,}400$)

Mass	I	$I\omega^2$	θ	$I\theta\omega^2$	$\Sigma I\theta\omega^2$	k	$\dfrac{1}{k}\sum I\theta\omega^2$
A	200	$2.88(10)^6$	1	$2.88(10)^6$	$2.88(10)^6$	$4(10)^6$	0.72
B	125	$1.80(10)^6$	0.28	$0.504(10)^6$	$3.384(10)^6$	$2(10)^6$	1.692
C	75	$1.08(10)^6$	-1.412	$-1.525(10)^6$	$1.859(10)^6$	$3(10)^6$	0.616
D	50	$0.72(10)^6$	-2.028	$-1.460(10)^6$	$0.399(10)^6$		

The sum, $0.399(10)^6$, indicates that the angular frequency is still low.

436   Vibrations in Shafts

Let us try, for the third calculation, $\omega = 130$ rad/sec.

Trial III
($\omega = 130$ rad/sec $\omega^2 = 16,900$)

Mass	I	$I\omega^2$	θ	$I\theta\omega^2$	$\Sigma I\theta\omega^2$	k	$\frac{1}{k}\sum I\theta\omega^2$
A	200	$3.389(10)^6$	1	$3.389(10)^6$	$3.389(10)^6$	$4(10)^6$	0.845
B	125	$2.113(10)^6$	0.155	$0.328(10)^6$	$3.717(10)^6$	$2(10)^6$	1.859
C	75	$1.268(10)^6$	-1.704	$2.161(10)^6$	$1.556(10)^6$	$3(10)^6$	0.519
D	50	$0.845(10)^6$	-2.223	$-1.878(10)^6$	$-0.322(10)^6$		

Trial IV
($\omega = 126$ rad/sec $\omega^2 = 15,876$)

Mass	I	$I\omega^2$	θ	$I\theta\omega^2$	$\Sigma I\theta\omega^2$	k	$\frac{1}{k}\sum I\theta\omega^2$
A	200	$3.175(10)^6$	1	$3.175(10)^6$	$3.175(10)^6$	$4(10)^6$	0.794
B	125	$1.985(10)^6$	0.206	$0.409(10)^6$	$3.584(10)^6$	$2(10)^6$	1.792
C	75	$1.191(10)^6$	-1.586	$-1.889(10)^6$	$1.695(10)^6$	$3(10)^6$	0.565
D	50	$0.794(10)^6$	-2.151	$-1.708(10)^6$	$-0.013(10)^6$		

The negative sum of the inertia couples, $-0.322(10)^6$, indicates that the frequency is high. A closer approximation to the frequency may be obtained by assuming that the variation of $\Sigma I\theta\omega^2$ is linear for the small change in frequency from 120 rad/sec to 130 rad/sec. From similar triangles, Fig. 21.24d,

$$\frac{0.39(10)^6}{0.322(10)^6} = \frac{x}{10-x}$$

from which x is approximately 6. Thus, for a fourth trial, $\omega = 126$ rad/sec will be assumed.

The final couple, $-0.013(10)^6$, indicates that the frequency is a little less than 126 rad/sec. However, any more refinement is unnecessary, and the final answer for the first-order natural frequency of

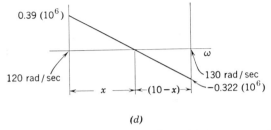

Fig. 21.24d. Straight-line variation assumed to obtain a new value of angular frequency for checking.

oscillation may be given as 126 rad/sec, or $\dfrac{(126)(60)}{2\pi} = 1200$ cycles per second.

21.28 Second-order natural frequency of oscillation

The second-order natural frequency of oscillation will be computed now. A value of $\omega = 250$ rad/sec will be assumed for the first trial.

Trial I
($\omega = 250$ rad/sec $\omega^2 = 62{,}500$)

Mass	I	$I\omega^2$	θ	$I\theta\omega^2$	$\Sigma I\theta\omega^2$	k	$\dfrac{1}{k}\Sigma I\theta\omega^2$
A	200	$12.50(10)^6$	1	$12.50(10)^6$	$12.50(10)^6$	$4(10)^6$	3.13
B	125	$7.813(10)^6$	-2.13	$-16.64(10)^6$	$-4.14(10)^6$	$2(10)^6$	-2.07
C	75	$4.688(10)^6$	-0.06	$-0.281(10)^6$	$-4.32(10)^6$	$3(10)^6$	-1.44
D	50	$3.125(10)^6$	1.38	$4.31(10)^6$	$0.01(10)^6$		

The final value, $0.01(10)^6$, is sufficiently small to require no further calculations. It is to be noted that the second-order natural frequency of oscillation of 250 rad/sec is by coincidence, in this case, about twice the first-order natural frequency.

To be sure that the second-order natural frequency is the one that has been found, we can make a plot as shown in Fig. 21.24e, where θ is shown for the various masses. For the first-order frequency, there is one node, with masses A and B moving opposite to masses C and D, while for the second-order there are two nodes, with masses A and D moving opposite to masses B and C.

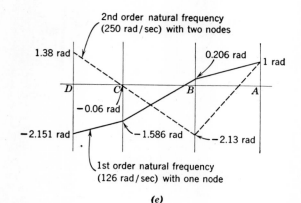

Fig. 21.24e. Angular displacement of masses for first- and second- order natural frequencies of oscillation.

21.29 Higher-order natural frequencies of oscillation

The same method can be applied to find the higher-order natural frequencies of oscillation, which will not be found for the illustrative problem.

21.30 Dangerous operating speeds

It is well to point out that the natural frequencies of vibration for shafts are not the only dangerous speed of operation in a given system. The actual dangerous speeds of operation are determined by those impulses which may be applied to the shaft. If impulses are applied in phase with the natural frequency of vibration, trouble will be encountered. The problem of analyzing an internal-combustion engine, for instance, and breaking up the crankshaft torque curve into component harmonic functions to determine the dangerous speeds of operation is a lengthy procedure. The reader is referred to textbooks on vibrations for detailed analysis in the continuation of vibrations, as well as discussions on the vibration of coupled systems, as is found with gear trains.

PROBLEMS

Note. In the following problems on critical speed, neglect the weight of the shaft, unless otherwise specified. Use $E = 30{,}000{,}000$ psi for steel.

21.1. A weight of 15 lb is suspended from a spring which has a spring rate of 300 lb/in. What is the frequency of oscillation in cycles per minute if the weight is displaced 1 in. from the equilibrium position and released? What is the frequency of oscillation if the weight is displaced 2 in. from the equilibrium position and released?

Problems

21.2. An I-beam, 8 ft long, supports a motor at its midpoint. The beam deflects 0.12 in. at the motor under the action of the weight of the motor. What is the natural frequency of oscillation of the beam?

21.3. The following data are taken from the static deflection curve for a shaft supported on two bearings. Distance X along the shaft is measured from the center of the left bearing. The distance between bearing centers is 16 in.

X, inches	Deflection, inches
2	0.00105
4	0.00185
6	0.00204
8	0.00182
10	0.00148
12	0.00106
14	0.00052

The shaft carries a 50-lb rotor 3 in. to the right of the left bearing and a 70-lb rotor 12 in. to the right of the left bearing. The shaft is made of steel. What is the critical speed?

21.4. A shaft of 2 in. diameter is simply supported on bearings which are 10 in. apart. A rotor, which weighs 150 lb, is located 4 in. from the left bearing. What is the critical speed?

21.5. The 18-in. midsection of a steel shaft has a diameter of $2\frac{1}{2}$ in. On each side of the midsection the shaft has a diameter of 2 in. for a distance of 8 in. The shaft is 34 in. long. If concentrated weights of 600 lb are applied at each change of section, determine the critical speed in rpm.

21.6. A solid shaft 3 in. in diameter has keyed to it a flywheel weighing 700 lb. The critical speed has been found to be 1350 rpm. If the weight of the flywheel were increased to 900 lb and the shaft diameter were made $3\frac{1}{2}$ in., what would be the critical speed?

21.7. A 26-in. long shaft simply supported on two bearings has a diameter of 2.00 in. for a length of 10 in. to the right of the left-hand bearing. The remainder of the shaft has a diameter of 2.50 in. A 1200-lb rotor is mounted on the shaft 14 in. to the right of the left-hand bearing. What is the critical speed?

21.8. A uniform shaft is 40 in. long. It has a weight of 500 lb located 10 in. from the left end and an 800 lb weight located 15 in. from the right end. If the critical speed is 1890 rpm, determine the diameter of the shaft. The shaft is simply supported in the bearings.

21.9. A steel shaft simply supported in bearings 24 in. apart carries a concentrated weight of 200 lb 8 in. from the right bearing and a weight of 200 lb 8 in. from the left bearing. If the shaft diameter is to be constant along the shaft, find the diameter of the shaft if the shaft is connected by a flexible coupling to a motor running at 1800 rpm. The shaft is to run at 50% of the critical speed. (Note that the shaft is designed on the basis of a condition other than strength, although the strength of the shaft would have to be checked before a final answer should be given. No strength check is required for this problem, however.)

21.10. A steel shaft, simply supported in bearings at its ends, is to be run 25% above its critical speed. The shaft is to have a constant diameter of $1\frac{1}{2}$ in. for 8-in. length, the next 12 in. having a constant diameter of 2 in., the next 8 in. of shaft has a diameter of $1\frac{1}{2}$ in. Thus the shaft is 28 in. between bearing centers,

A single concentrated weight is located at the midpoint of the shaft. What is the critical speed?

21.11. A shaft is made up of two sections. The first section is 10 in. long and has a diameter of d_1; the second section is 14 in. long and has a diameter of $2d_1$. If a weight of 1000 lb is located on the larger section, 4 in. from the change of section, and if the critical speed is 1400 rpm, determine the diameter of each section of the shaft. Consider the shaft as simply supported at the ends.

21.12. The steel shaft of an engine is 24 in. long between bearings. For a distance of 8 in. from the left bearing, the shaft is 2 in. in diameter. The remainder of the shaft is 3 in. in diameter. A weight of 1000 lb is concentrated at a distance of 10 in. from the left bearing, and a weight of 2000 lb is concentrated at a distance of 15 in. from the left bearing. If the shaft is to run at least 40% below the critical speed, what is the maximum permissible speed of operation of the shaft? Consider the shaft as simply supported in the bearings at ends of the shaft.

21.13. A 3 in. diameter shaft 4 ft long is simply supported on bearings at each end and carries a 5000-lb disk at a distance of 18 in. from the left end. The left half of the shaft is steel ($E = 30[10^6]$ psi) and the right half of the shaft is bronze ($E = 12[10^6]$ psi) rigidly fastened together at the midpoint.

Determine the maximum speed at which the shaft should run without exceeding 50% of the critical speed.

21.14. A steel shaft simply supported in bearings at the ends is made up of a 2 in. diameter section 10 in. long and a $1\frac{1}{2}$ in. diameter section 10 in. long. Assuming that a concentrated weight of 500 lb is located at the change of section, determine the critical speed. If the weight is reduced to 250 lb, what is the critical speed?

21.15. A 4 in. diameter shaft is simply supported in bearings 48 in. apart. At the center of the shaft is a gear weighing 3000 lb. A 2 in. diameter hole in the center of the cross section extends 18 in. towards the gear from the left bearing. At this point the hole is decreased to $1\frac{1}{2}$ in. diameter, and the hole is continued through the remaining length of the shaft. What is the critical speed of the shaft?

21.16. A 2 in. diameter shaft is simply supported on bearings 36 in. apart. A 1000-lb pulley is supported 18 in. to the right of the left bearing. Under these conditions the operating speed coincides with the critical speed. A 25% change in the critical speed is required, and it has been proposed that the pulley be relocated 9 in. from the left bearing. Will the new arrangement produce the desired effect? Determine the percent change in critical speed (based on the original value), and state whether the critical speed will be increased or decreased.

20.17. A steel shaft supported on two bearings has an overhung section. The distance between bearings is 20 in. A rotor weighing 900 lb is located on the overhang 15 in. from the nearer bearing. The shaft is of constant diameter, 3 in.

(a) Assuming that the overhung portion is considered a cantilever beam, determine the critical speed of the shaft.

(b) Considering the entire beam, with the proper bearing action, determine the critical speed.

(c) Compare the results of a and b.

21.18. A 4 in. diameter shaft 40 in. long simply supported on bearings 32 in. apart carries a 500-lb flywheel 8 in. to the right of the right bearing.

(a) What is the critical speed of the shaft?

(b) What would the critical speed be if an additional flywheel weighing 500 lb were located 8 in. from the left bearing?

(*Note.* For part *b* the weights should be considered as acting in opposite directions, with the consequent deflection at each load considered positive in the critical speed equation.)

21.19 (Fig. 21.19). Determine the natural frequency of oscillation of the two mass system shown.

Fig. P-21.19.

$I_A = 300$ lb-in.-sec^2
$I_B = 500$ lb-in.-sec^2
$k_1 = 2(10^6)$ in.-lb/rad

21.20 (Fig. 21.20). Determine the two natural frequencies of oscillation of the system shown.

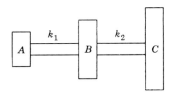

Fig. P-21.20.

$I_A = 300$ lb-in.-sec^2 $\quad k_1 = 2(10^6)$ in.-lb/rad
$I_B = 500$ lb-in.-sec^2 $\quad k_2 = 6(10^6)$ in.-lb/rad
$I_C = 1200$ lb-in.-sec^2

21.21 (Fig. 21.21). Determine the three natural frequencies of oscillation of the system shown.

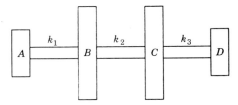

Fig. P-21.21.

$I_A = 400$ lb-in.-sec^2 $\quad k_1 = 2(10^6)$ in.-lb/rad
$I_B = 800$ lb-in.-sec^2 $\quad k_2 = 3(10^6)$ in.-lb/rad
$I_C = 800$ lb-in.-sec^2 $\quad k_3 = 2(10^6)$ in.-lb/rad
$I_D = 400$ lb-in.-sec^2

CHAPTER 22

Gyroscopes

Although the principles of the gyroscope have been known for a long time and practical applications were seen in gyroscopic compasses and gyroscopes for stabilizing large ships, the requirements of World War II were such that considerable new applications were found for the gyroscopes: bombsights, stabilization of tank turrets, control of airplanes and guided missiles, to mention a few.* In comparison, to the use of gyroscopes for control and stabilization, we find, on the other hand, undesirable effects from gyroscopic action: increase of bearing reactions in crankshafts of automobiles as the automobile travels around a curve, increase of bearing reactions in ship propellers as the ship pitches and rolls in a heavy sea, increase of bearing reactions in the engine shaft of a jet airplane as the airplane changes direction of motion, to mention a few.

Space limitations prevent the discussion of the practical applications of gyroscopes, which in itself is a complete topic for analysis. The point of view to be taken in this chapter is the machine designer's point of view in determining the effects of gyroscopic action in machinery. Let us examine the motion which takes place under the effect of gyroscopic action, together with the terms which are used, before we proceed to an analysis.

22.1 A toy gyroscope

Figure 22.1 represents a toy gyroscope: a rotating mass, supported in gimbal rings (only one being shown in the sketch), is rotating with an angular velocity, called spin, ω_s rad/sec, as shown. The projection of the gimbal is supported on what is assumed as a frictionless surface. At first glance, it would be assumed that the torque created by the weight of the unit would cause the mass to fall vertically downwards. However, as determined from an analysis and confirmed by experi-

---

* *Gyroscopes and Their Applications*, by K. A. Oplinger, *Westinghouse Engineer*, May, 1948, pages 75–79.

A Toy Gyroscope

mental investigation, the mass executes a motion quite contrary to a first reaction: the axis of the rotating shaft rotates about the vertical z-axis with an angular velocity, called precession, ω_p. If friction is considered zero, the axis of rotation of the rotating mass would rotate in a horizontal plane of the x- and y-axes. If friction is considered, energy would be dissipated in friction, which would come from a loss of potential energy of the rotating mass. Thus, the mass would gradually drop to provide the energy used in friction.

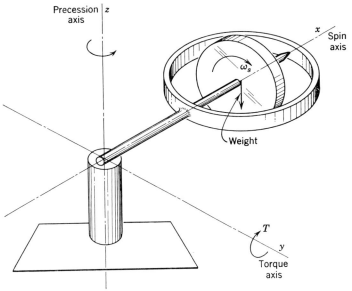

Fig. 22.1. Gyroscopic action causes the axis of the rotating weight to rotate in the x-y plane.

As seen from the simple toy gyroscope illustration, there are three axes to be considered, each axis mutually perpendicular: (1) the spin axis, (2) the torque axis, and (3) the precession axis.

The same action occurs in propeller shafts of ships or crankshafts of automobiles, except that the precession is what is called "forced," which in turn sets up couples acting on the shaft. As an illustration of forced precession, think of an automobile moving around a curve at constant speed, while the crankshaft is rotating about its own axis. The motions are prescribed, which require a torque about a third axis to cause the prescribed motion.

Let us examine the relation of angular velocity about the spin axis, ω_s, angular velocity of precession, ω_p, and torque, T. The analysis

444 Gyroscopes

will be given, using Newton's basic relations. A second analysis to be given will use angular momentum relations.

22.2 Gyroscopic relations, using Newton's laws directly

Figure 22.2a shows a mass rotating about its axis, the z-axis, with an angular velocity of ω_s radians per second. At the same time, the

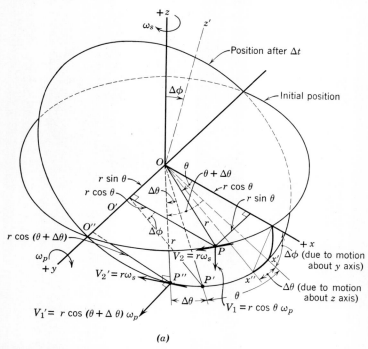

Fig. 22.2a. The velocity vectors are shown for point P initially due to rotation about the z-axis and y-axis with ω_s and ω_p and finally due to rotation about the z'-axis and y-axis with ω_s and ω_p. Point P moves to P' as a result of rotation about the y-axis; motion from P' to P'' results from rotation about the z'-axis.

mass is rotating about an axis in the body, the y-axis, with an angular velocity of ω_p radians per second. The angular velocity about the axis of the mass, ω_s, is called the spin velocity, whereas the angular velocity about an axis perpendicular to the axis of the mass, ω_p, is called the precessional velocity.

The problem is to determine what has to be applied to cause the prescribed motions.

Before we consider the motion of a point on the body, let us look

Gyroscopic Relations, Using Newton's Laws Directly

at the motion of the x-axis in a time interval of Δt seconds. The simultaneous motion about the z-axis and the y-axis may be considered as broken up into two separate motions: (1) the displacement of the x-axis due to rotation about the y-axis, represented by an angular change of $\Delta\phi$ due to ω_p bringing the axis to the position marked as $O-x'$; (2) the displacement of $O-x'$ to $O-x''$, represented by $\Delta\theta$, and due to the spin velocity, ω_s, of the body.

Consider a point P, taken for convenience in making the drawing as a point on the circumference. The point P is located on the body at the beginning of the time interval being considered. The body may be considered as horizontal at the beginning of the time interval.

The velocity of point P is made up of two components:
(1) The velocity due to rotation about the y-axis, expressed as

$$V_1 = r \cos \theta \omega_p$$

where θ is the angle made by the radial line from O to P with the x-axis, and r is the length of the radius from O to P.

(2) The velocity due to rotation about the z-axis, expressed as

$$V_2 = r\omega_s$$

V_1 is perpendicular to the plane of the mass at P, whereas V_2 is in the plane of the mass perpendicular to the radius.

Now consider the two separate motions, ω_s and ω_p, as affecting the position of P. Owing to rotation about the y-axis through an angle $\Delta\phi$, the body moves so that P goes to P'. Or, $O'P$ rotates through an angle $\Delta\phi$ to $O'P'$. As a result of the angular velocity, ω_s, about the axis of the mass, the mass rotates through an angle $\Delta\theta$, causing point P' to move to P''.

The velocity of P'' is made up of two components:
(1) The velocity of rotation about the y-axis, expressed as

$$V_1' = r \cos (\theta + \Delta\theta)\omega_p$$

where the distance from O'' to P'' for the mass in the new position is given by $r \cos (\theta + \Delta\theta)$. This velocity component is perpendicular to the line $O''P''$: the line from P'' perpendicular to the y-axis.

(2) The velocity due to rotation about the axis perpendicular to the mass (in this case axis z'), the velocity being expressed by

$$V_2' = r\omega_s$$

This velocity is the same as V_2 in magnitude, but the direction is different as a result of the rotation of the disk and tipping of the disk.

The problem now is to determine the changes of velocity, in both magnitude and direction, so that the accelerations may be determined, which will permit finding the force that has to be applied to the particle to cause the prescribed motion. Let us consider the change in velocity due to ω_p first, or the change from V_1 to V_1'.

Change of velocity from V_1 to V_1'. First, it must be recognized that the planes determined by $O'-P-V_1$ and $O''-P''-V_1'$ are parallel. Thus the projections of V_1 and V_1' as vectors on a plane perpendicular to the y-axis will give the true lengths of the vectors, as shown in Fig. 22.2b. The changes of velocity in the vertical (z-direction) and

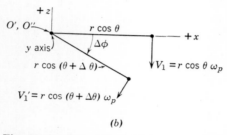

(b)

Fig. 22.2b. Velocities initially and finally due to rotation about the y-axis with ω_p.

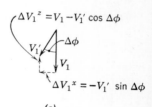

(c)

Fig. 22.2c. Changes of velocity in the x- and z-directions due to rotation about the y-axis.

horizontal (x-direction), as can be seen from Fig. 22.2c, are, respectively, from final velocity minus initial velocity:

$$\Delta V_1{}^z = -V_1' \cos \Delta\phi - (-V_1) = V_1 - V_1' \cos \Delta\phi$$
$$\Delta V_1{}^x = -V_1' \sin \Delta\phi - 0$$

It can be recognized at this point that $\Delta V_1{}^z$ is positive, or in the upward direction. The minus sign is used with $\Delta V_1{}^x$ to indicate a change of velocity in the negative direction of the x-axis. Substitution of $V_1 = r \cos \theta \, \omega_p$ and $V_1' = r \cos(\theta + \Delta\theta)\omega_p$ into the above gives

$$\Delta V_1{}^z = r \cos \theta \omega_p - [r \cos(\theta + \Delta\theta)\omega_p] \cos \Delta\phi$$
$$\Delta V_1{}^x = -[r \cos(\theta + \Delta\theta)\omega_p] \sin \Delta\phi$$

Expansion of $\cos(\theta + \Delta\theta) = \cos\theta \cos\Delta\theta - \sin\theta \sin\Delta\theta$, substitution into the above, dividing through by Δt, and taking the limit as Δt approaches zero, we obtain:

Gyroscopic Relations, Using Newton's Laws Directly

$$\lim_{\Delta t \to 0} \frac{\Delta V_1{}^z}{\Delta t} = \lim_{\Delta t \to 0} \frac{r \cos\theta \omega_p}{\Delta t} - \lim_{\Delta t \to 0} \frac{r\omega_p \cos\Delta\phi \cos\theta \cos\Delta\theta}{\Delta t}$$
$$+ \lim_{\Delta t \to 0} \frac{r\omega_p \cos\Delta\phi \sin\theta \sin\Delta\theta}{\Delta t}$$

$$\lim_{\Delta t \to 0} \frac{\Delta V_1{}^x}{\Delta t} = -\lim_{\Delta t \to 0} \frac{r\omega_p \sin\Delta\phi \cos\theta \cos\Delta\theta}{\Delta t}$$
$$+ \lim_{\Delta t \to 0} \frac{r\omega_p \sin\Delta\phi \sin\theta \sin\Delta\theta}{\Delta t}$$

Recognizing that the two expressions are, respectively, the accelerations in the z- and x-directions due to the change in direction and magnitude of V_1, recognizing also that in the limit $\sin\Delta\phi = d\phi$, $\cos\Delta\theta = 1$, and $\sin\Delta\theta = d\theta$, we may write, disregarding differentials of higher order:

$$A_1{}^z = r\omega_p \sin\theta \frac{d\theta}{dt}$$

$$A_1{}^x = -r\omega_p \frac{d\phi}{dt} \cos\theta$$

Or, since $\dfrac{d\theta}{dt} = \omega_s$ and $\dfrac{d\phi}{dt} = \omega_p$, we may write:

$$A_1{}^z = r\omega_p \omega_s \sin\theta$$

$$A_1{}^x = -r\omega_p{}^2 \cos\theta$$

Before we interpret the two equations above, let us determine the accelerations due to the change of direction of V_2 to V_2'.

Change of velocity from V_2 to V_2'. Since $V_2 = r\omega_s$ and $V_2' = r\omega_s$, or the velocities are the same in magnitude, any accelerations present come from a change of direction of velocity. Consider obtaining the components of the velocities in the original x-, y-, and z-directions for a quick evaluation of the changes of velocities taking place.

Consider, first, the components of V_2 in the x- and y-directions. As seen in Fig. 22.2d, we may write for the components:

$$V_2{}^x = -r\omega_s \sin\theta$$

$$V_2{}^y = +r\omega_s \cos\theta$$

where the minus sign is used for $V_2{}^x$ because the velocity component is in the negative direction.

Next, consider the components of $V_2' = r\omega_s$ in the x-, y-, and

z-directions. Picture first the components of the velocity of V_2' in the plane of the mass, or in the x'- and y-directions, as shown in Fig. 22.2e. The components are

$$V_2'^{x'} = -r\omega_s \sin(\theta + \Delta\theta)$$

$$V_2'^{y} = +r\omega_s \cos(\theta + \Delta\theta)$$

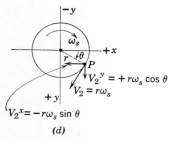
(d)

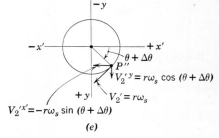
(e)

Fig. 22.2d. Velocity components in the x- and y-directions of point P initially due to ω_s.

Fig. 22.2e. Velocity components of the point P'' due to ω_s in the x'-y plane of the mass after rotation of $\Delta\phi$ about the y-axis, and $\Delta\theta$ about the z'-axis.

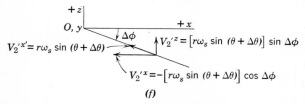
(f)

Fig. 22.2f. Velocity vector $V_2'^{x'}$ in the x'-direction replaced by two components in the original x- and z-directions.

Next, consider obtaining the components of $V_2'^{x'}$ in the original x- and z-directions (Fig. 22.2f):

$$V_2'^{x} = -[r\omega_s \sin(\theta + \Delta\theta)] \cos \Delta\phi$$

$$V_2'^{z} = [r\omega_s \sin(\theta + \Delta\theta)] \sin \Delta\phi$$

The changes of velocity in the x-, y-, and z-directions are

$$\Delta V_2^x = V_2'^x - V_2^x = -[r\omega_s \sin(\theta + \Delta\theta)] \cos \Delta\phi - [-r\omega_s \sin \theta]$$

$$\Delta V_2^y = V_2'^y - V_2^y = +r\omega_s \cos(\theta + \Delta\theta) - [r\omega_s \cos \theta]$$

$$\Delta V_2^z = V_2'^z - V_2^z = [r\omega_s \sin(\theta + \Delta\theta)] \sin \Delta\phi - 0$$

By expanding the trigonometric functions, dividing through by Δt, and taking the limit as Δt approaches zero, we may obtain the final

Gyroscopic Relations, Using Newton's Laws Directly

equations for the acceleration of point P due to the change of position of the spin velocity $r\omega_s$:

$$A_2{}^x = -r\omega_s{}^2 \cos\theta$$
$$A_2{}^y = -r\omega_s{}^2 \sin\theta$$
$$A_2{}^z = r\omega_s\omega_p \sin\theta$$

Interpretation of results. Let us repeat the acceleration components that have been found:

$$A_1{}^z = r\omega_p\omega_s \sin\theta \qquad\qquad A_2{}^z = r\omega_s\omega_p \sin\theta$$
$$A_1{}^x = -r\omega_p{}^2 \cos\theta \qquad\qquad A_2{}^x = -r\omega_s{}^2 \cos\theta$$
$$\qquad\qquad\qquad\qquad\qquad A_2{}^y = -r\omega_s{}^2 \sin\theta$$

The total acceleration of point P is the vector sum of the above five components. However, let us add the components in the following order:

$$A_1{}^z + A_2{}^z = A^z = 2r\omega_p\omega_s \sin\theta$$
$$A_2{}^x + A_2{}^y = A^n = r\omega_s{}^2 \cos\theta + r\omega_s{}^2 \sin\theta$$
$$A_1{}^x = -r\omega_p{}^2 \cos\theta$$

Since $r \sin\theta = y$, $r \cos\theta = x$, and $r\omega_s{}^2 \cos\theta + r\omega_s{}^2 \sin\theta = r\omega_s{}^2$, the acceleration components may be written as:

$$A^z = 2y\omega_p\omega_s$$
$$A^n = r\omega_s{}^2$$
$$A_1{}^x = -x\omega_p{}^2$$

Couple Due to A^z Acceleration. Figure 22.2g shows the components of A^z in each quadrant, with the forces applied to each particle causing a couple. Note that the direction of the force applied to each particle is determined by the sign of the y-coordinate. The resultant force in the z-direction is zero if the y-axis is taken through the center of gravity, as may be seen from

$$F = \int dM(2y\omega_p\omega_s) = 2\omega_p\omega_s \int y\, dM$$

If the moment of each force is taken about the x-axis, the magnitude of the resultant couple can be obtained:

$$T_x = \int dM(2y\omega_p\omega_s)y = 2\omega_p\omega_s \int y^2\, dM = 2\omega_p\omega_s I_x$$

where I_x is defined as the mass moment of inertia of the mass about the x-axis. Since, for moments of inertias, $I_z = I_x + I_y$, where I_z

is the moment of inertia about the z-axis and I_y is the moment of inertia about the y-axis, and for a cylinder $I_x = I_y$, we may write $I_x + I_y = 2I_x$, or $I_z = 2I_x$. Thus,

$$T_x = I_z \omega_p \omega_s$$

Force Due to A^n Acceleration. If the body is rotating about an axis through the center of gravity, the resultant force due to the

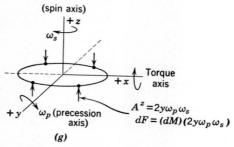

Fig. 22.2g. Acceleration of four particles located symmetrically with the x- and y-axes and the force applied to each particle as a result of its acceleration show that a couple about the x-axis results.

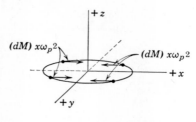

Fig. 22.2h. The forces in the x-y plane balance each other if the y-axis passes through the center of gravity.

normal acceleration, $A^n = r\omega_s^2$, of each particle is zero. (This is quickly seen by going back to the original components of acceleration, $A_2{}^x$ and $A_2{}^y$, setting up the expressions for the total forces in the x- and y-directions, and showing that the resultant forces are zero.)

Force Due to $A_1{}^x$ Acceleration. The last component of acceleration to be considered, $A_1{}^x = -x\omega_p^2$, is due to rotation about the y-axis. Figure 22.2h shows the forces applied to a particle in each quadrant. The resultant force is

$$F = \int (dM)(-x\omega_p^2)$$

which is zero if the y-axis passes through the center of gravity of the body. The moment about the z-axis is

$$M = \int (dM)(-x\omega_p{}^2)(y) = -\omega_p{}^2 \int xy(dM)$$

The integral of the foregoing expression is recognized as the product of inertia of the mass, which integral is zero if the body is rotating about an axis of symmetry. For a mass of circular cross section, the product of inertia is zero about the axis of the mass.

Conclusions. For a thin disk as used in the above analysis rotating about an axis through the center of gravity, the torque that has to be

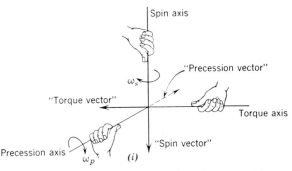

Fig. 22.2i. Right-hand rule relating the direction of torque with the directions of angular velocity of spin and angular velocity of precession: The direction of precession causes the spin vector to coincide with the torque vector.

applied is $T = I_z \omega_p \omega_s$, where the torque axis is perpendicular to the spin and precession axes. By examination of Fig. 22.2g, a convenient rule may be obtained relating the direction of the torque for given directions of spin and precession. If the right-hand rule is used to express the angular velocities of precession and spin, where the fingers represent the direction of rotation, the thumb will represent the angular velocity. Figure 22.2i shows the right-hand rule applied in expressing the angular velocities of precession and spin. The method of determining the direction of torque is as follows:

Picture the precession axis being rotated (in the direction of the precession) to cause the spin vector to rotate through 90 degrees. The spin vector will then correspond to the direction of the torque vector. Or, expressed differently, the direction of precession is such to cause the spin vector to coincide with the torque vector. It is suggested that the student always draw the spin and torque vectors away from the intersection of the axes, for consistency and simplification.

Example. As an illustration of forced precession, let us examine the effect of gyroscopic action in the following situation: the effect on the bearings of a crankshaft of an automobile in traveling around a curve. We shall idealize the arrangement of rotating crankshaft, flywheel, and oscillating connecting rod by considering the system as a rotating disk, with a moment of inertia about the axis of rotation of 5 lb-sec^2-ft. Let us assume that the crankshaft is rotating at 3400 rpm and that the car is moving at 60 miles per hour. If the car is traveling on a straight road, no gyroscopic forces will be present. If the

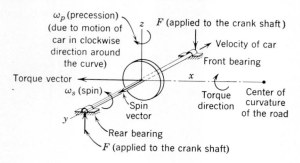

(j)

Fig. 22.2j. A schematic arrangement of a rotating crankshaft of an automobile traveling around a curve. The crankshaft is rotating at ω_s, and the forced clockwise precession of the car is ω_p due to motion about the center of curvature of the road.

car is traveling around a curve, a forced precession is caused which will set up additional forces on the bearings. Assuming that the car travels around a curve of 300 ft radius, determine the couple applied to the crankshaft. What is the direction of each force?

It will be assumed that the crankshaft is rotating counterclockwise, as seen from the rear of the engine.

The torque applied to the shaft is found from

$$T = I\omega_p\omega_s$$

$$= (5)\left(\frac{88}{300}\right)\left[\frac{(3400)(2\pi)}{60}\right]$$

$$= 522 \text{ ft-lb}$$

Alternate Analysis, Using Angular Impulse

where $\omega_p = \dfrac{V}{R}$ and $V = 88$ ft/sec (for a velocity of 60 miles per hour) and $R = 300$ ft. The units of ω_p are radians per second. The angular velocity of spin is $\omega_s = \left(\dfrac{(3400)(2\pi)}{60}\right)$ radians per second.

Figure 22.2j is a schematic arrangement of the crankshaft. Only two bearings are assumed for simplicity. The car is assumed to be traveling clockwise around the curve, as viewed from above. The crankshaft is assumed to be rotating counterclockwise, as viewed from the rear of the car. The spin vector is shown as determined from the right-hand rule. The torque vector direction is determined by

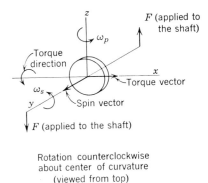

(k)

F8g. 22.2k. Same as Fig. 22.2j except that the forced precession is counterclockwise.

considering the rotation of the precession axis to cause the spin vector to correspond with the torque vector. The direction of the couple applied to the crankshaft is clockwise, as seen from the right. Since the couple determined is the couple applied *to* the shaft, the forces applied to the shaft at the bearings are as indicated by F. Consequently, the force applied *to* the rear bearing is downwards, and the force applied *to* the front bearing is upwards.

If the car is traveling around the curve in a counterclockwise direction, the forces on the bearing will be reversed in direction from the preceding case, as shown in Fig. 22.2k.

22.3 Alternate analysis, using angular impulse

An alternate method of approach to the gyroscopic relations uses angular impulse and angular momentum relations, an alternate interpretation of Newton's equation for rotation which gives a shorter analysis than that of the preceding section.

It was shown in Chapter 13 that $T = I\alpha$, where T represents the moment about the center of gravity and I represents the moment of inertia of the body about the center of gravity axis of the body; or T represents the moment about the axis of rotation and I represents the moment of inertia of the body about the axis of rotation. Expressing the equation as $T = I\dfrac{d\omega}{dt}$, multiplying both sides by dt, and integrating, we obtain: $\int T\,(dt) = \int I\,(d\omega)$.

If the moment of inertia remains constant,

$$\int T\,(dt) = I\,(\Delta\omega)$$

where $(\Delta\omega)$ represents the change of angular velocity. The expression $\int T\,(dt)$ is called the angular impulse, while $I\omega$ is called the angular momentum (also called the moment of momentum). Thus, from the equation above, the angular impulse is equal to the change of angular momentum. The discussion has been limited to a body having plane motion. It is possible to have a body moving with a constant angular velocity, and yet require a torque to cause a prescribed motion, as seen in the case where a body is rotating about an axis of the body, but the axis is itself moving. A more generalized form of the above equation is given by

$$T = \frac{d}{dt}(\vec{I\omega})$$

which is expressed as the applied torque is equal to the rate of change of the angular momentum vector.

The torque and angular momentum are vector quantities, and may be expressed by the right-hand rule. The torque is expressed as a vector by curving the fingers in the direction of the torque, with the thumb being the vector for the torque. Similarly, if the fingers are curved in the direction of the angular velocity, the thumb will represent the vector for the angular momentum.

Consider a mass rotating with an angular velocity of ω_s rad/sec (the spin velocity) and having a mass moment of inertia of I about the axis of rotation through the center of gravity. Figure 22.3a shows the angular momentum vector. Picture the axis of the mass rotating, in the plane of the paper, with an angular speed of ω_p rad/sec (the speed of precession). Figure 22.3b shows the angular momentum vector in the original position and in the position after a time interval of Δt in which the axis of rotation has moved through an angle of $\Delta\phi$.

Alternate Analysis, Using Angular Impulse

From Fig. 22.3c, the change of angular momentum is

$$2(I\omega_s) \sin \frac{\Delta\phi}{2}$$

The rate of change of angular momentum is given by

$$2 \frac{(I\omega_s)}{\Delta t} \sin \frac{\Delta\phi}{2}$$

which, in the limit as $\Delta t \to 0$, is equal to $I\omega_s \dfrac{d\phi}{dt}$. Since $\dfrac{d\phi}{dt}$ is defined as the velocity of precession, ω_p, the rate of change of the angular momentum is equal to $I\omega_s\omega_p$. But the rate of change of the angular

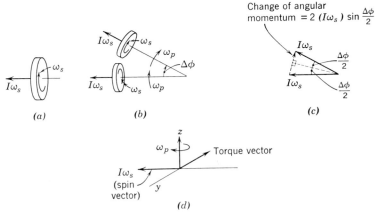

Fig. 22.3. Gyroscopic relations determined from the relation that the rate of change of the angular momentum is equal to the applied torque.

momentum is, from Newton's equation, defined as the applied torque. Therefore,

$$T = I\omega_s\omega_p$$

where the vector for the applied torque, *which corresponds to the change of angular momentum vector in direction*, is perpendicular to the spin vector.

The same rule for determining the direction of the torque vector as given in the preceding section can be deduced from the vectors shown in Fig. 22.3d. The direction of precession (ω_p) is such as to cause the spin vector to coincide with the torque vector.

PROBLEMS

22.1. A ship turbine shaft is mounted with its axis parallel to the longitudinal axis of the ship. If the shaft is rotating clockwise, as viewed from the stern of the ship, determine the direction of the forces applied to the bearings of the turbine shaft if the ship is pitching so that:

(a) The bow is moving down, and the stern is moving up.

(b) The bow is moving up as the stern is moving down.

22.2. The generator shaft of an automobile is mounted with its axis parallel to the longitudinal axis of the car. The rotor weighs 8 lb, and has a moment of inertia about the axis of rotation equal to 0.005 lb-sec^2-in. If the shaft is rotating at 1800 rpm counterclockwise, as viewed from the rear of the car, while the car goes around a clockwise curve of 500 ft at 60 mph, what are the forces, in magnitude and direction, which are applied to the two bearings of the generator shaft as a result of the gyroscopic action? Show on a sketch the direction of the forces applied.

22.3. A motorcycle rider tips the motorcycle clockwise as he is moving in a straight path. In what direction will a couple be applied to the motorcycle as a result of the gyroscopic action?

22.4 (Fig. 22.4). (a) A bevel gear of a planetary gear system is rotating about two axes simultaneously. The bevel gear is rotating at 1200 rpm about axis

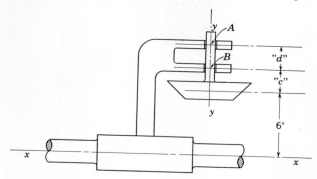

Fig. P–22.4.

y–y, the axis of the bevel gear, while the gear is rotating at 900 rpm about the x–x axis. Determine the forces on the bearings A and B, in direction and magnitude.

The bevel gear has a moment of inertia of 0.10 lb-sec^2-in. about the y–y axis The distance $c = 4$ in., and the distance $d = 4$ in.

(b) Same as (a) except that $c = 2$ in., and $d = 6$ in.

INDEX

Acceleration, angular, 60
  defined, 4
  Atkinson engine, 87
  auxiliary point, 118
  coincident point, 96
  completely graphical solution, 84
  Coriolis' component, 96, 101, 102, 105
  defined, 4
  four-link mechanism, 76
  image, 76
  jaw crusher mechanism, 81
  link rotating about a fixed center,
    analytic analysis, 58
    graphical analysis, 64
  modified shaper mechanism, 122
  normal, 61, 65, 68
  oscillating roller follower, 112
  pole, 74
  Powell engine, 79
  radial, 61, 68
  relative, 68, 98
    two points on a rigid link, analytic
      analysis, 61
      graphical analysis, 65
  rolling cylinder, 90
  shaper mechanism, 109
  slider-crank mechanism, 71
    analytic analysis, 238–241
  special position, 83
  supplementary component, 101
  tangential, 61, 65, 68
  Watt "walking beam" mechanism, 118
d'Alembert's principle, 192
Allis-Chalmers A-C Industrial Press, 82
Angular acceleration, 4, 60
Angular impulse, 453
Angular momentum, 453
Angular speed, 4

Angular velocity, 4
  relative, 174
Articulated rod, 322, 332
Atkinson engine, accelerations, 87
Automobile six-cylinder engine, 360
Auxiliary point, acceleration, 118
  acceleration analysis, modified shaper mechanism, 122
  Watt "walking beam" mechanism, 118
  velocity, 44
  velocity analysis, modified shaper mechanism, 49
  Stephenson mechanism, 51
  Wanzer needle-bar mechanism, 52
  Watt "walking beam" mechanism, 44

Balance, conditions, for multicylinder in-line engine, 343–345
  for multicylinder V-engine, 364–367
  for rotating masses, 284
Balancing, Brush arrangement, 317
  Cadillac engine, 373
  dummy pistons for, 308
  dynamic, 284
  flexible rotors, 292
  gearing arrangements, for primary couple, three-cylinder engine, 354
    two cylinder engine, cranks at 180 degrees, 349
    V-engine, six-cylinder, two-cycle, 368
  primary force, single-cylinder engine, 314
  secondary couple, three-cylinder engine, 354
  single-cylinder engine, 316

Balancing, gearing arrangements,
secondary couple, two-cylinder
engine, cranks at 180 degrees,
349
  Lanchester arrangement, 350
  machines, 282–292
    Cadillac engine shaft, 375, 376
    portable, 290
  masses reciprocating in a plane, 294
  methods for rotating masses, 285
  primary and secondary forces, single-
    cylinder engine, 318
  radial engine, 336
  reciprocating masses, counterweights
    for, 304
  rotating masses, analytic method,
    264
    dynamic balance, 261
    dynamic balancing machines, 288
    equations for equilibrium, 265
    graphical analysis, 269
    moments as vectors, 269
    multi-weight system, 261
    single rotating mass, 259
    static balance, 260, 261, 265, 266,
      284
    static balancing machines, 286,
      290
    trial and error, 285
    two planes required, 263
    two rotating masses, 260
    types of, 284
    vector equations for equilibrium,
      269
  rotors, by electrical networks, 292
  single-cylinder engine, single
    rotating weight for, 308
  six-cylinder engine, two-cycle, 361
  V-engine, eight-cylinder, four-cycle,
    370
    six-cylinder, two-cycle, 367
    two-cylinder, couterbalances used
      for, 324–328
  W-engine, 328
  X-engine, 329–332
Beam, conjugate, 401–404
  applied to acceleration, velocity,
    and displacement, 415
  for a cantilever beam, 405–407
  for an overhung beam, 410–411

Beam, conjugate, for a simply supported
  beam, 407–410
  various types of, 411–412
  deflections, in, 405–421
    found graphically, 416–421
    variable cross-sections, 411
  radius of curvature, 400
Bending moment, defined, 400
  graphically determined, 416
  positive and negative, defined, 401
  relation to shear, 401
  scale in graphical determination, 418
  slope of bending moment diagrams,
    403
Binomial theorem, 239
Brush arrangement for balancing the
  primary force, 317

Cadillac engine, balancing of, 370–373
  crankshaft, 373–374
  firing order, 377
Cam curvature, 134
Cams, equivalent mechanisms, 127–132
Camshaft, six-cylinder engine, 361
Cantilever beam, deflection by con-
  jugate beam method, 405
Caterpillar Tractor Co., 362–363
Center of percussion, 192
Coaxial type, opposed engine, 310
Cocking, 169
Coefficient, of fluctuation of speed,
  defined, 244
  of friction, variations of, 164
Coincident points, relation of accelerations, 96
  relation of velocities, 28
Compression stroke, 351
Conjugate beam, 401–404
  applied to acceleration, velocity, and
    displacement, 415
  deflections by, 401–404
  for a cantilever beam, 405–407
  for a simply supported beam, 407–
    410
  for an overhung beam, 410–411
  various types of, 411–412
Connecting rod, articulated, 322, 332
  interlocking, blade and fork con-
    struction, 319
  kinetically equivalent system for, 295

Index

Connecting rod, master, 322
  replaced by two weights, analytical analysis, 295
Constant velocity, 3
Cooper-Bessemer Corp., 321
Cooper-Bessemer engine, 321
Coriolis' component of acceleration, 96, 101, 102, 105, 122
Counterbalanced link, 198
Counterbalances (or counterweights),
  eight-cylinder V-engine, automobile-type, 372
  five-cylinder engine, 381
  radial engine, 336
  single-cylinder engine, 304–308
  six-cylinder, two-cycle, 361–362
  six-cylinder, two-cycle V-engine, 368–370
  two-cylinder V-engine, 324–328
  X-type engine, 332
Couple, defined, 136
  moment about any point, 136
Crankshaft, Cadillac engine, 373–374
  five-cylinder engine, 380–381
  Oldsmobile engine, 372
  two-plane, 373
Crank shaper mechanism, acceleration analysis, 109
  dynamic analysis, 223–224
  force analysis, 167
  inertia forces, 197
  modified, 122
  velocity analysis, 29–31
Critical speed, 390
  deflection curves for various orders of, 397
  Dunkerley's equation, 423
  effect of friction on, 391
  effect of weight of shaft, 411
  first-order, 396
  higher-order, 423
  multimass system, 396
  primary, 396
  Rayleigh, 423
  same as natural frequency, 395
  simply supported shaft, 422
  torsional, 424–438
Curvature, cam, 134
  radius of, determined, 90
Cutler Hammer, Inc., 186

Deflection, by conjugate beam, 401–404
  graphically, 416
  methods for finding, 399
  static, 384
  variable cross-section beams, 411
Diesel cycle, 257
Dynamic analysis, combined static and inertia forces, 219–224
  defined, 219
  flywheel determination, 250–254
  Powell engine, 221–222
  separate static and inertia forces, 226–229
  shaper mechanism, 223–224
  slider-crank mechanism, 219–221
Dynamic balancing, 284
Dynamic balancing machine, 288
Dynamics of machinery, defined, 1
Dummy pistons, 308
Dunkerley's equation, 423

Edwards Brothers, 315, 316
Eight-cylinder, four-cycle V-engine, 370–377
  two-cycle V-engine, 370
Energy in punching, 247
Engine, Cadillac, 370
  Cadillac, firing order, 377
  Cooper-Bessemer, 321
  eight-cylinder V-type, 370–377
  five-cylinder, 380–381
  four cylinder, 355–359
  Gobron-Brillé, 309, 312
  Junker, 309, 313
  Oldsmobile, 370
  opposed, 308, 310–314
  quarter-crank, 346
  radial, 332, 377
  six-cylinder, 359–362
  V-engine, 367–370
  three-cylinder, 352
  two-cylinder, comparison of, 350
    cranks at zero degrees, 350
    cranks at 90 degrees, 346
    cranks at 180 degrees, 347
    firing order of, 351
    V-type, 318
  Unaflow, 304, 353
  unbalance, multicylinder in-line, 343
    V-type, 366

Index

Engine, V-type, 318–328, 362–377
  W-type, 328, 377
  X-type, 329, 377
Equilibrium, defined, 2
  equations of, 135
  five or more parallel forces, 140
  four non-parallel forces, 138
  parallel forces, 140, 142
  resolution of forces, 142
  three non-parallel forces, 137
Equivalent crank weight, 300
Equivalent mechanisms, 127
  kinematically, 127
  kinetically, 204
    slider-crank mechanism, 206
Exhaust stroke, 351

Firing order, Cadillac engine, 377
  events in, 351
  three-cylinder engine, four-cycle, 355
  three-cylinder engine, two-cycle, 355
  two-cylinder engines, 351
Flexible rotors, balancing of, 292
Floating link, 31, 44
Flywheel, 230
Flywheel analysis, kinetically equivalent system used in, 252
  maximum excess or deficiency of energy, 254
  punch press, 241, 246
  purpose, 243, 244
  weight required, 245–246, 250
Force, a vector, 136
Force analysis, combined inertia and static, 219
  slider-crank mechanism, 219–221
  couples applied to two links in a mechanism, 159
  crank shaper mechanism, 167
  dynamic, Powell engine, 221–222
  with separate static and inertia forces, 226–229
  four-link mechanism, 158
    with friction circles, 175
  press mechanism, 154
    with friction circles, 176
  riveter mechanism, 156
  slider-crank mechanism, 152
    with friction, 164
    with friction circles, 173

Force analysis, slider-crank mechanism, with separate static and inertia forces, 227–229
  Zoller double-piston engine, 161
Forced precession, 443
Forces, acting in planes, 135
  gear, 150, 156
  inertia, 188, 191
    in a four-link mechanism, 196
    in a shaper quick-return mechanism, 197
    in a slider-crank mechanism, 192–196
  parallel, 140
  pin, 150
  primary, 301
  resultant, due to motion, 192–196
  reversed resultant, 191
  secondary, 301
  shaking, 198, 219, 229, 294
  shear, 399
  sliding members, 151
  static, 135
  superposition of, 227
  three non-parallel, 137
  types of, 150
Four-cycle engines, 350
  firing order, 355
Four-cylinder engines, 355–359
Four-link mechanism, accelerations in, 76
  force analysis, 158
    with friction circles, 175
  inertia forces, 196
  velocity analysis, 20–23
Frequency of oscillation, 384
  higher order in torsional vibrations, 438
  natural, 394
  same as whirling speed or critical speed, 394
  second-order in torsional vibrations, 437
  torsional vibrations, 425
Friction, 163
  angle, 166
  causing cocking, 169
  circle, 173
  direction of resultant force, 165

Friction, dry surfaces, 164
  effect of impending motion on direction, 166
  effect on critical speed, 391
  pin joint, 171
  variation of coefficient of, 164
  with lubrication, 164
Funicular polygon, 416

Gear forces, 150
Gearing arrangements, balance of the primary couple, three-cylinder engine, 354
  two-cylinder engine with cranks at 180 degrees, 349
  balance of the secondary couple, three-cylinder engine, 354
  balance of the secondary forces, three-cylinder engine with cranks at 180 degrees, 349
  balance of a single-cylinder engine, primary force, 314
  secondary force, 316
  balance of a six-cylinder, two-cycle V-engine, 368
  Lanchester arrangement, 350
General Motors Corp. 367, 372–377
  Detroit Diesel Engine Division, 361, 362
  Electro-Motive Division, 310, 367
Gobron-Brillé engine, 309, 312
Governor, 213–214, 244
Gyration, radius of, 191
Gyroscope, 442

Harmonic motion, 6
Higher order critical speeds, 423
Holzer's method, 430

Image, acceleration, 76
  velocity, 20
Impending motion, direction from velocity analysis, 168
Indicator card, 250
  diesel cycle, 257
  Otto cycle, 256
Inertia, mass moment of, 190, 199–204
  product of, 451
Inertia force, 188, 191
  four-link mechanism, 196

Inertia force, shaper quick-return mechanism, 197
  slider-crank mechanism, 192–196
  special position, 198
Intake stroke, 351

Jaw crusher mechanism, 81
Joy locomotive valve gear, 23, 25
Junker engine, 309, 313

Kinematically equivalent mechanisms, 127–132
Kinetically equivalent systems, 204
  connecting rod, 295
  flywheel analysis, 252
  slider-crank mechanism, 206

Lanchester balancer, 350
Link, counterbalanced, 198
  floating, 31, 44
Locomotive Cyclopedia, 34, 35, 52, 53

Mass moment of inertia, defined, 190
  determination, 199–204
Master rod, 322, 332
Mechanisms, equivalent, 127
Modified shaper mechanism, acceleration analysis, 122–125
  velocity analysis, 49–51
Moment, as a vector, 269
  conjugate beam, 405
  of a couple, 136
  of inertia, mass, 190, 199–204
  of momentum, 454
  right-hand rule, 270

Natural frequency of oscillation, 394
Newton's laws, 1
Node, 434
Normal acceleration, 61, 64, 68
Normal component of acceleration, graphically, 84

Offset type engine, 311
Oldsmobile engine, balance in, 370
  crankshaft, 372
Opposed engines, 308–313
Oscillating roller follower, acceleration analysis, 112
Oscillation, frequency, 384
  torsional, single-mass system, 424

462   Index

Oscillation, torsional, three-mass system, 428
    two-mass system, 426
Osculating circle, 61 (footnote)
Otto cycle, 256
Ounce-inch unbalance, defined, 282
Overhung beam, conjugate beam for, 410

Parallel forces, 140
    resultant of, 142
Pin forces, 150
Pin joint friction, 171
Pistons, dummy, 308
Plane motion, 11
Planetary gear system, Model T Ford transmission, 185
Polar diagram, 301
Polar shaking force curves, single-cylinder engine, 303–307
    two-cylinder V-engine, 325
Pole, acceleration polygon, 74
    velocity polygon, 19
Powell engine, acceleration analysis, 79–81
    dynamic analysis, 221–222
    velocity, 23–24
Power stroke, 351
Precession, 443
    axis, 443, 444
    forced, 443
Press mechanism, force analysis, 154
    with friction circles, 176
Pressure line, 150
Primary critical speed, 396
Primary force, 301
    Brush arrangement for balancing of, 317
Product of inertia, 451
Punching, energy for, 247
Punch press, flywheel, 241
    flywheel analysis, 246

Quarter-crank engine, 346
Quick-return mechanism, shaper, 29
    Whitworth, 215

Radial acceleration, 61, 68
Radial component of acceleration, graphically, 84

Radial engine, 332, 377
    balancing, 336
Radius of curvature, determined, 90
    in a beam, 399
Radius of gyration, 191
Rayleigh, 423
Relative acceleration, 68, 98
    two coincident points, 96–105
    two points on a rigid link, 61–68
Relative angular velocity, 174
Relative velocity, 8, 28
    two coincident points, 23, 25–29
    two points on a rigid link, 10–14
Resolution of forces, 142
Resultant force, defined, 142
    direction, with friction, 165
    multicylinder in-line engine, 345
    to cause a prescribed motion, 190
    V-engine, 366
Resultant of forces nearly parallel, 145
Resultant of parallel forces, 142, 144
Reversed resultant force, 191
Right-hand rule, gyroscopes, 451
    moments, 270
Riveter mechanism, force analysis, 156
Rolling cylinder, accelerations, 90
Rotating masses, balancing, 259

Scales, bending moments, graphically, 418
    deflections, graphically, 421
    normal component of acceleration, graphically, 85
Scotch yoke, 38
Secondary force, 301
Shaking force, 198, 219, 229, 294, 301, 343, 366
    defined, 229
    location of, in a multicylinder in-line engine, 344
    polar diagrams of, 303–307, 325
    slider-crank mechanism, 300
Shaper mechanism, *see* Crank shaper mechanism
Shear force, corresponding to slope in a conjugate beam, 405
    diagram, slope of, 403
    defined, 399
    positive and negative, defined, 401
    relation to loading, 401–403

Index

Shear force, relation to bending moment, 401–403
Simmons-Boardman Corp., 34, 35, 52, 53
Single-cylinder engine, balance of primary and secondary forces, 318
 balancing by means of auxiliary gearing, 314–318
 theoretical balancing of, by means of one rotating weight, 308
Six-cylinder engine, automobile, 360
 camshaft, 361
 firing order, 360
 four-cycle, 360
 two-cycle, 360
 V-type, 367
Skinner Engine Co., 305, 353, 380, 381
Slider-crank mechanism, accelerations, 71
 analytically, 238–241
 combined inertia and static force analysis, 219–221
 counterbalancing, 304–308
 effort for a prescribed motion, 224–226
 equivalent mechanism, 130
 force analysis, 152
  with friction, 164
  with friction circles, 173
  with separate static and inertia forces, 227–229
 inertia forces in, 192–196
 resultant forces due to motion, 192–196
 reversed resultant forces in, 192–196
 velocity analysis, 17–20
Special position, acceleration analysis, 83
 inertia forces, 198
 velocity analysis, 32
Speed, angular, 4
 coefficient of fluctuation, 244
 critical, 390
Spin, 443
Spin axis, 443, 444
Spring constant, 202
Spring rate, 384
Spur gear forces, 156
Static balancing, 284
Static balancing machine, electrical, 290

Static balancing machine, Taylor, 286
 Tinius Olsen, 288–290
Static deflection, 384
Static forces, 135
 in dynamic analysis, 226–229
Stephenson mechanism, 51
String polygon, 416
Suction stroke, 351
Superposition of forces, 227
Supplementary component of acceleration, 101

Tangential acceleration, 61, 65, 68
Taylor Dynamometer and Machine Co., 286, 287
Taylor static balancing machine, 286
Three-cylinder engine, 352
 firing order, 355
Tinius Olsen dynamic balancing machine, 288
Tinius Olsen static balancing machine, 290
Torsional vibrations, 424
 four-mass system, 432
 frequency of oscillation, 425, 428
 Holzer's method, 430
 second order, 437
 single mass system, 425
 three-mass system, 428
 two-mass system, 426
Turbine blade, velocity analysis, 29
Turning effort, 253
Two-cycle, eight-cylinder engine, 370
 firing order in, 352, 355
 four-cylinder engine, 357, 359
 six-cylinder engine, 360
Two-cylinder engine, comparison of, 350–352
 cranks at zero degrees, 350
 cranks at 90 degrees, 346
 cranks at 180 degrees, 347
 firing order, 350
Two-force members, 144

Unaflow engine, 305, 353
 crankshaft for a five-cylinder engine, 380, 381
Unbalance, causes of, 282
 equations for, multicylinder in-line engine, 343–345

Unbalance, equations for, multicylinder V-engine, 364–367
ounce-inch, 282

Vector, force, 136
Velocity analysis, four-link mechanism, 20–23
  modified shaper mechanism, 49–51
  shaper mechanism, 29–31
  slider-crank mechanism, 17–20
  special positions, 32
  using auxiliary points, 44
  Watt "walking beam" mechanism, 44–48
Velocity, angular, 4
  constant, 2
  defined, 3
  image, 20
  linear, 2
  pole, 19
  relative, 8, 10–14, 23, 25–29
V-engine, connecting rod construction, 362
  counterblancing, 324–327
  eight-cylinder, automobile, 370
  equations for unbalance, multicyliner, 364–367
  equations for unbalance, two-cylinder, 323–324

V-engine, interlocking, blade and fork construction, connecting rods, 319
  two-cylinder, 318
  two-cylinder, 90-degree, 327
Vibration, effect of friction, 386
  shaft, 383
  single mass system, 383–387
  torsional, 424

Walschaert valve gear, 33–35
Wanzer needle-bar mechanism, 52
Watt "walking beam" mechanism, acceleration analysis, 118
  velocity analysis, 44–48
W-engine, 328, 377
Westinghouse Electric Corp., 284
Westinghouse-Gisholt, 292
Whirling, considered as a vibration, 396
  of shafts, 387
Whirling speed, effect of friction, 394
  equation for, multimass system, 399
  same as natural frequency, 394
Whitworth quick-return mechanism, 215

X-engine, 329, 377

Zoller double-piston engine, force analysis, 161

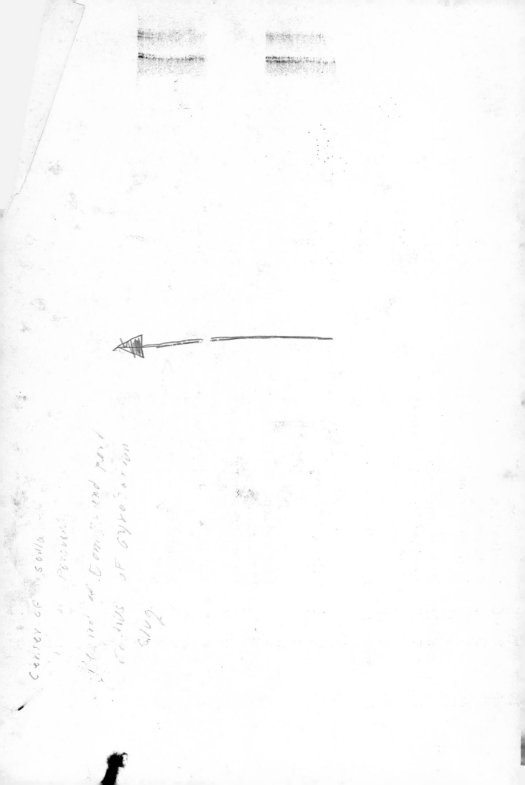

Center of gravity Somlla
in Percussion
Planes at Convulsed part
Causes of gyration
slug

Complete Graphically 304 §305

19	T	5	Examination	
20	W	6	Discussion	
21	F	8	Inertia Forces	
22	M	11		
23	T	12		Ch. 13
24	W	13	Dynamic Analysis	Ch. 14 410B
25	F	15	Accel. (Analytic Method)	Ch. 15 418
26	M	18	Flywheel Dynamics	Ch. 16 416, 417
27	T	19		~16.9 in text
28	W	20	Balancing Rot. mass	Ch. 17 16.1 text
29	F	22		503-
30	M	25		5-10
31	T	26	Balancing Recip. mass	Ch. 19 505 § 506
32	W	27		521 § 526 § 527
33	F	3-1	Examination	
34	M	4	Discussion	
35	T	5	Balancing-several planes	Ch. 20 521 § 522
36	W	6		
37	F	8	Vibrations in shafts	Ch. 21 513
38	M	11		801, 803
39	T	12	Gyroscopes	Ch. 22 807, 809 21.3
40	W	13	Review 701 § 902; 903	
41	F	15		

E 312

Dynamics of Machines

Texts: Dynamics of Machines, Holowenko
Kinematics of Machines, Hinkle

Problems: Problems in Dynamics of Machinery, Leutwiler
Problems in Kinematics of Machines, Fellinger
(section on gears)

Period	Date	Subject	Hinkle	Holowenko	Problems
1	F 1-4	Rolling Circles	Ch. 8		48 & 49
2	M 7	Gearing	Ch. 9.1-9.7		51 & 52
3	T 8		9.8-9.14		
4	W 9		9.15-9.18		
5	F 11		9.19-9.22		
6	M 14		9.23-9.26		
7	T 15	Gear Trains	Ch. 11.1-11.7		
8	W 16		11.8-11.10		
9	F 18	Examination			
10	M 21	Review Kinematics	Ch. 2.1-3.9	pp. 1-32	132-134
11	T 22		3.10-4.4	44-58	137
12	W 23		5.1-5.7	59-89	209, 214
13	F 25	Coriolis accel.	5.8	Ch. 8.1-8.2	236
14	M 28			8.3-8.4	
15	T 29	Complex mech.	5.9	Ch. 9, 10	238
16	W 30	Static Forces		Ch. 11	242
17	F 2-1				301 & 306
18	M 4				